Fossil Nonmarine Ostracoda of the United States

FURTHER TITLES IN THIS SERIES

1. A.J. Boucot
EVOLUTION AND EXTINCTION RATE CONTROLS

2. W.A. Berggren and J.A. van Couvering
THE LATE NEOGENE-BIOSTRATIGRAPHY, GEOCHRONOLOGY AND PALEOCLIMATOLOGY OF THE LAST 15 MILLION YEARS IN MARINE AND CONTINENTAL SEQUENCES

3. L.J. Salop
PRECAMBRIAN OF THE NORTHERN HEMISPHERE

4. J.L. Wray
CALCAREOUS ALGAE

5. A. Hallam (Editor)
PATTERNS OF EVOLUTION, AS ILLUSTRATED BY THE FOSSIL RECORD

6. F.M. Swain (Editor)
STRATIGRAPHIC MICROPALEONTOLOGY OF ATLANTIC BASIN AND BORDERLANDS

7. W.C. Mahaney (Editor)
QUATERNARY DATING METHODS

8. D. Janóssy
PLEISTOCENE VERTEBRATE FAUNAS OF HUNGARY

9. Ch. Pomerol and I. Premoli-Silva (Editors)
TERMINAL EOCENE EVENTS

10. J.C. Briggs
BIOGEOGRAPHY AND PLATE TECTONICS

11. T. Hanai, N. Ikeya and K. Ishizaki (Editors)
EVOLUTIONARY BIOLOGY OF OSTRACODA. ITS FUNDAMENTALS AND APPLICATIONS

12. V.A. Zubakov and I.I. Borzenkova
GLOBAL PALAEOCLIMATE OF THE LATE CENOZOIC

13. F.P. Agterberg
AUTOMATED STRATIGRAPHIC CORRELATION

14. J.C. Briggs
GLOBAL BIOGEOGRAPHY

15. A. Montanari, G.S. Odin and R. Coccioni (Editors)
MIOCENE STRATIGRAPHY: AN INTEGRATED APPROACH

Developments in Palaeontology and Stratigraphy, 16

Fossil Nonmarine Ostracoda of the United States

Frederick M. Swain

Department of Geology and Geophysics
University of Minnesota, Minneapolis, MN, USA

1999
ELSEVIER
Amsterdam – Lausanne – New York – Oxford – Shannon – Singapore – Tokyo

ELSEVIER SCIENCE B.V.
Sara Burgerhartstraat 25
P.O. Box 211, 1000 AE Amsterdam, The Netherlands

First edition 1999

Library of Congress Cataloging in Publication Data
A catalog record from the Library of Congress has been applied for.

ISBN: 0-444-50089-8

⊗ The paper used in this publication meets the requirements of ANSI/NISO Z39.48-1992 (Permanence of Paper).
Printed in The Netherlands.

To Frances, Annette, and Suzanne

Preface

The Ostracoda are small aquatic bivalved Crustacea that live in both marine and nonmarine environments. Many are characteristic of estuarine and other tidal habitats, but only a few occupy hypersaline waters. One or two kinds are found in wet soils or in leaf or flower cups in tropical rain forests, a few live in caves and others are commensal in gills of fish and other aquatic animals. In size ostracodes generally range from about 0.5 to 1.5 mm but some are only about 0.1 mm or less and others may attain several centimeters.

As presently, and somewhat doubtfully defined, Ostracoda are found sparingly in certain Cambrian shales and limestones, but became abundant in Ordovician and younger rocks and in Holocene habitats. In the oceans ostracodes are most abundant on the continental shelves, but exist down the slopes and are present in greatly reduced numbers in the abyssal realm above the carbonate compensation level where they compete for test-building calcium carbonate with the foraminifera and coccolithophorids.

The ostracode animal typically possesses seven pairs of appendages: antennules, antennae, mandibles, maxillae and three pairs of thoracic legs, the latter subject to much modification. An additional pair of copulatory appendages is found in many forms. Most of the appendages are modified to perform more than one function (see Appendix 1). The ostracode shell or carapace is composed of chitin that is impregnated with calcium carbonate to a greater or lesser extent. Certain planktonic marine ostracodes are only chitinous, and some nonmarine or tidal forms inhabiting low-calcium lakes, marshes and tidal rivers have little or no calcium carbonate in their carapaces.

Growth in Ostracoda takes place by ecdysis involving eight or nine molt stages. Sexual maturity is attained by the fourth or fifth instar. There is little information on the degree to which fossil or Holocene shells in sediments represent final or penultimate growth stages, but it seems reasonable that many, if not most, observed shells are not the final stage. Shell characters of the smaller instars lack the development of the adult shells, particularly that of hingement and marginal shell structures important in classification of genera and species.

Because of their small size and durable shells, ostracodes have long been recovered from cores and drill cuttings of bore holes. Their relatively rapid rates of evolution have made them useful in subsurface stratigraphy and their sensitivity to environmental changes has provided a means of recognizing variations in rock facies. In nonmarine aquatic rocks they are commonly the most easily recoverable microfossils, and have been widely used in petroleum exploration, notably in China, Russia, Brazil and the western United States.

Fossil nonmarine Ostracoda in the United States have been recorded taxonomically in publications for more than 100 years. T.R. Jones' and F. Chapman's early work, before 1900, was followed by the Paleozoic studies of W.C. Holland, H.W. Scott, C.H. Summerson, and of the Mesozoic by H.C. Vanderpool, R.E. Peck, Robert Roth, and F. Harper and A.H. Sutton in the 1930s to 1950s. I.G. Sohn continued his major contributions to Paleozoic nonmarine ostracode studies in the latter part of the century and expanded his work into the Mesozoic and Cenozoic during that period. K.K. Kietzke and S.G. Lucas have added to knowledge of early Mesozoic nonmarine species in the western United States.

F.L. Staplin, K.H. Lister, R.H. Benson and E.D. Gutenberg, S.P. Cameron and R.F. Lundin made important contributions to knowledge of Pleistocene nonmarine faunas in the 1960s and 1970s. The present writer began work on nonmarine assemblages in the 1940s and has continued to the present time. Otherwise, relatively few taxonomic papers have been published on fossil nonmarine ostracodes in the United States. In recent years, workers have emphasized the environmental importance of nonmarine Ostracoda and their use in the interpretation of climatic changes.

In the present compilation the writer has attempted to bring together data on the stratigraphic and geographic distribution of pre-Holocene nonmarine ostracodes in the United States, and summaries of shell characteristics, of the species. A great many forms are left in open nomenclature and await examination of additional collections.

Acknowledgments. Permission to reproduce illustrations has been granted by the organizations listed in the plate descriptions. I.G. Sohn, S.G. Lucas, and K.K. Kietzke provided copies of several of their publications. D.E. Peterson provided for study examples of the Pliocene ostracodes from the Merced Formation near Berkeley, California. Sharon Kressler and Julia Bleakney typed most of the manuscript and Gary Fouty assisted in the search for references. Michelle Elston, Kathy Ohler, Greg C. Brown, Annette Swain, Bret Eayrs, and Dirk Slawinski assisted in processing the manuscript.

F.M. Swain
Minneapolis, Minnesota
August, 1998

Contents

Preface vii
Contents ix
Introduction 1
Stratigraphic Distribution of Paleozoic Non-Marine Ostracoda 3
  Devonian 3
    Devonian Catskill Magnafacies, New York 3
  Mississippian 3
    Mississippian Bluestone Formation, West Virginia and Virginia 3
  Pennsylvanian 3
    Pennsylvanian Kanawha Series, Kentucky and Tennessee 4
    Pennsylvanian Monongahela Formation, Pennsylvania and West Virginia 4
    Pennsylvanian Atokan through Virgilian Series, Illinois 5
    Pennsylvanian Allegheny Group, Pennsylvania 5
  Permian 5
    Permian Nineveh Limestone, Pennsylvania and West Virginia 5
    Permian Wellington Formation, Kansas 6
Stratigraphic Distribution of Mesozoic Non-Marine Ostracoda 6
  Triassic 6
    Triassic Newark Supergroup, North Carolina, Virginia and Pennsylvania 6
    Triassic and Jurassic? Culpeper Group, Virginia 7
    Triassic Moenkopi Formation, New Mexico 7
    Triassic Chinle Group, Arizona, New Mexico and Texas 8
    Triassic Shublik Formation, Alaska 8
  Jurassic 8
    Jurassic Todilto Limestone, New Mexico 8
    Jurassic? Kayenta Formation, Arizona 9
    Jurassic Morrison Formation, South Dakota, Wyoming, Montana and Colorado 9
    Jurassic Shongaloo Formation, Arkansas 10
  Cretaceous 10
    Cretaceous Draney and Peterson Limestones, Wyoming and Idaho 10
    Cretaceous 'Lower Bear River Formation', Wyoming 11
    Cretaceous Kootenai Formation, Montana 11
    Cretaceous Lakota Formation and Minnewaste Limestone, South Dakota 12
    Cretaceous Cloverly Formation, Wyoming 12
    Cretaceous 'Newark Canyon Formation', Nevada 13
    Cretaceous strata, North Carolina, Arkansas, Louisiana and Texas 13
    Cretaceous subsurface 'Washita Group', Florida 14

Cretaceous Bear River Formation, Wyoming 14
Stratigraphic Distribution of Paleogene Non-Marine Ostracoda 15
Paleocene 15
Late Cretaceous?, Paleocene and Eocene? Sheep Pass Formation, Nevada 15
Paleocene Fort Union Formation, Montana 16
Eocene 16
Eocene Green River, Colton, Wasatch and Flagstaff Formations, Utah, Colorado and Wyoming 16
Eocene and Oligocene Elko Formation, Nevada 19
Eocene or Oligocene Twin Bridges and Illipah Formations, Nevada 19
Oligocene 21
Oligocene Indian Well Formation, Nevada 21
Oligocene Hells Bells Canyon Formation, Nevada 21
Oligocene? Edwards Creek Tuff, Nevada 21
Oligocene? Titus Canyon Formation, Nevada and California 22
Oligocene Pruett Formation, Texas 22
Oligocene Chadron Formation, Nebraska 23
Oligocene? of Madison County, Montana 23
Oligocene-Miocene Renova Formation, Montana 23
Stratigraphic Distribution of Neogene Pre-Holocene Non-Marine Ostracoda 24
Miocene 24
Oligocene? and Miocene Humboldt and North Creek Formations, Nevada 24
Miocene Esmeralda Formation, Nevada 28
Miocene Horse Camp Formation, Nevada 32
Miocene Horse Spring Formation, Nevada 32
Miocene? New Pass Tuff, Nevada 33
Miocene? Truckee Formation, Nevada 33
Pliocene 33
Pliocene Cache Valley Formation, Idaho and Utah 33
Pliocene Hay Ranch Formation, Nevada 35
Pliocene Glenns Ferry Formation, Idaho 36
Miocene and Pliocene Salt Lake Group, Utah and Idaho 38
Pliocene in Beartrap Canyon Quadrangle, California 40
Pliocene at localities in California, Arizona and Utah 40
Pliocene Merced Formation, California 41
Pliocene Teewinot Formation, Wyoming 42
Pliocene Laverne and Rexroad Formations, Kansas 42
Pliocene and Pleistocene? Idaho Group, Oregon 43
Pleistocene 43
Pleistocene in Illinois and Adjacent Areas 43
Pleistocene, Matayuma-Age Deposits, Illinois 51
Pleistocene, Pre-Illinoian, Indiana 52
Pleistocene (Kansan), Kansas and Nebraskan? Stage, Nebraska 52
Pleistocene Bone Bed, Missouri 53
Pleistocene Wisconsin-age Deposits, Missouri 53

Pleistocene, Wrenshall Lake Clays, and Lake Itasca marl, Minnesota 54
Pleistocene Rita Blanca Lake Deposits, Texas 54
Pleistocene Archeological Sites, Texas and New Mexico 54
Pleistocene Lake Bonneville Group, Utah 55
Pleistocene, Lake Cochise Area, Arizona 57
Pleistocene Gubik Formation, Alaska 58
Pleistocene, Lake Lahontan, Nixon Area, Nevada 59
Pleistocene, Owens Lake, California 59
Suprageneric Classification of Fossil Nonmarine Ostracoda in the USA 59
Systematic Paleontology of Fossil Non-Marine Ostracoda 61
References 251
Plates 1–34 272
Appendix A, Features of nonmarine Ostracoda animal and shell 383
Appendix B, Range Charts of Significant Species 386
Index to Species Illustrated on Plates 393
Index to Stratigraphic Units 399

Introduction

Nonmarine ostracodes of leperditiid type are reported to occur in the Upper Devonian of New York (Knox, 1996; Lundin, 1998). In China (Shi, 1964) and Russia (Zanina, 1960) more typical nonmarine carbonitid ostracodes have been recorded in the Upper Devonian, but in the United States these forms appear in Carboniferous shales (Sohn, 1985) associated with coal beds. The shales represent swampy ponds or lakes with abundant vegetation into which fine clastics were transported by slow moving streams. The climate was warm and humid and relief was low, along, or not far inland from, coastlines of inland seas. In the United States rocks of late Mississippian age contain the first definite nonmarine Ostracoda, which are of simple cypridacean types. In older strata, there are several examples of marine ostracodes migrating into, or originating in, brackish water estuarine or lagoonal habitats, but these are recognizable as diverse marine kinds of ostracodes that have adapted to the brackish environment. The Middle Silurian Bloomsburg Formation (Hoskins, 1961) red shale and sandstone is an example of a fauna of marine ostracodes that apparently developed in a lagoonal or brackish water habitat.

Non-marine Ostracoda remained obscure in the United States until the late Jurassic. The late Paleozoic, Triassic and early Jurassic were marked by smooth-shelled forms of the Carbonitidae, a few Cyprididae and primitive Darwinulidae. The late Jurassic saw the introduction of the notched Cyprideidae and of ornamented Limnocytheridae such as Theriosynoecum. Among freshwater flora, distinctive oogonia of the Characeae accompanied the Ostracoda in some habitats.

These changes were probably related to the progression from Carboniferous-Triassic swampy environments to open-water late Jurassic lacustrine conditions. The reason for the widespread distribution of Cypridea and its relatives in the Jurassic and early Cretaceous of parts of the northern and southern hemispheres lies in the appearance of the Purbeck-Wealden lithofacies. This facies of variegated lacustrine shales and limestone and associated coarser clastics is perhaps the result of the progressive development of upland plants. The floral conditions contributed to the development of upland soil profiles and the resulting erosion of such soils provided nutrients to large lakes that formed in the late Jurassic and early Cretaceous in the western United States as well as in many other parts of the world. The lakes provided habitats that favored development of the Cypridea-Theriosynoecum-Trapezoidella faunas as well as Characeae of our western interior region. Bisulcocypridea of the late Cretaceous and early Tertiary was accompanied by, and succeeded by, a small variety of Cyprididae in river flood plains and major lakes of the Green River and other lacustrine deposits of the Colorado Plateau and eastern Great Basin. Volcanic deposits began to be significant associates of these lakes, their faunas and microfloras. By the Oligocene, diatoms appeared in lacustrine beds of the Great Plains.

A major increase in variety of non-marine lacustrine ostracodes took place in the Miocene and Pliocene of the Great Basin area where, also, volcanic activity was high. Many of the lacustrine deposits of the Neogene of Nevada, Idaho and Utah, rich in ostracodes are interbedded with, or consist of, tuff and ash beds. The writer believes the volcanics provided nutrients and chemical environments that were a major factor in the evolutionary development of the Neogene Ostracoda.

The Pliocene marked a major increase in the Candonidae in the northern Great Basin, Idaho and Utah, that continued into the Pleistocene. Cypridopsidae, Cyclocyprididae, other Cyprididae and Limnocytheridae that, along with Candonidae, dominate the Pleistocene and Holocene freshwater faunas in the United States, were among the last groups to show evolutionary expansion.

Considerable progress has been made in the environmental determination on non-marine ostracodes in the recent years in the United States. Most of these studies have dealt with Holocene faunas (Smith, 1987, 1991, 1993; Schwalb et al, 1995; Forester, 1983, 1985, 1987; Smith et al, 1992; Forester and Smith, 1994), but a few have traced environmental indicators back into the Pleistocene (Miller et al, 1993, 1994; Carter, 1997) and Pliocene (Forester, 1991).

The extent of conversion of the amino acids L-isoleucine to D-alloisoleucine and the paleomagnetic signatures of Pleistocene deposits in Illinois and Indiana have been used as indications of the age and depositional environments of the deposits. Future applications of such geochemical and geophysical data should be useful in younger Pleistocene ostracode-bearing strata.

The employment of chemical studies in ostracode shells, such as the ratios of Sr and Mg to Ca, and isotopic ratios ($^{18}O/^{16}O$, $^{13}C/^{12}C$) in environmental studies has provided some useful results. Holmes (1996), however, has found variations in such data from species to species at the same horizon that raise caution in their interpretation as environmental indicators.

STRATIGRAPHIC DISTRIBUTION OF PALEOZOIC NONMARINE OSTRACODA

DEVONIAN

Devonian Catskill Magnafacies, New York

Ostracoda of Leperditicopida - type have been reported to occur in nonmarine, probably fluvial, deposits in the Upper Devonian Catskill "redbed" magnafacies of New York (Knox, 1996; Lundin and Friedman, 1998). Sollenella ? sp. was recorded in the Gilboa Formation, Upper Middle Devonian, Greene Co., N.Y. in association with a freshwater bivalve Archanodon.

Ostracodes generally considered to be nonmarine were recorded in the Devonian of southern China, i.e., Carbonita lentiforma Shi, 1964; and of the Russian Platform, i.e., Carbonita lipetzkensis Zannina, 1960.

MISSISSIPPIAN

Mississippian Bluestone Formation, West Virginia and Virginia

Nonmarine Ostracoda were recorded from the Bramwell member of the Bluestone Formation in West Virginia and Virginia by Sohn (1985). The Bramwell Member represents the uppermost Mississippian of West Virginia and is believed to be of Namurian A age. The Bramwell consists of about 30 m of fine grained shale, siltstone, and sandstone that coarsens upward, as well as changing from nonmarine to marine upwards (Repetski and Henry, 1983; Englund et al., 1981). The top of the Bramwell is believed to correlate with, or be slightly younger than, thc uppcrmost Chesterian of the Illinois Basin, and to correspond to the Namurian A-Namurian B boundary in western Europe (Pfeffercorn and Gillespie, 1981).

According to Sohn (1985) the basal Bramwell from which the ostracodes were obtained, represents a freshwater lake or a brackish water marsh. The following species were recorded: Carbonita, n. sp., Carbonita? sp., Gutschickia spp., Pruvostina? sp. 1, Whipplella? sp. 1, Geisinidae, gen indet., sp.2, Darwinula sp. 1, D. sp.2, D. sp.3.

The collections were made in dark shales in Wyoming County, West Virginia, Crumpler Quadrangle; in Mercer County, Western Virginia, Flat Top $7^1/_2$ minute Quadrangle; in Tazewell County, Virginia, Bramwell Quadrangle, by I.G. Sohn and others.

PENNSYLVANIAN

Pennsylvanian Kanawha Series, Kentucky and Tennessee

Strata of the Lower Pennsylvanian Kanawha Provincial Series approximately equivalent to Westphalian B of Europe were found to contain a nonmarine ostracode fauna in the southern Appalachian region (Scott and Summerson, 1943). Bivalves, crustaceans, insects, worm tubes, fish remains, and plants are found with the ostracodes. Shallow lakes and swamps probably represent the habitats of the fossils. According to the authors the presence of *Candona* and *Cypridopsis* in the collections suggests waters less than 10 feet (3 m) and having rich growth of aquatic vegetation.

The following formations were collected in search of ostracodes in Kentucky and Tennessee (Scott and Summerson, 1943): (in Kentucky) Hance, Breathitt, and Mingo; (in Tennessee) Briceville, Jellico, and Scott (Moore et al., 1944).

The Hance Formation contains: *Cypridopsis fabulina* (Jones and Kirkby), *Pruvostina wanlessi* Scott and Summerson, *Candona kentuckyensis* Scott and Summerson, *Jonesina? arcuata* (Bean), *Eocypridina radiata* (Jones, Kirkby, and Brady), and *E. aciculata* (Scott and Summerson).

The Breathitt Formation did not yield ostracodes. The Mingo Formation contained *Jonesina? bispinosa* Scott and Summerson.

In Tennessee, the Briceville formation which is in part equivalent to the Hance and lower Mingo of Kentucky yielded *Cypridopsis fabulina* and *Jonesina arcuata*.

The Jellico Formation which correlates with the upper Mingo of Kentucky contained *Cypridopsis fabulina* and *Candona kentuckyensis*. The Scott Formation, late Kanawha and early Allegheny (late Westphalian B, early Westphalian C) yielded *Cypridopsis fabulina*, *Hilboldtina multipunctata* Scott and Summerson, *Pruvostina wanlessi*, and *Candona kentuckyensis* as well as unidentified forms.

The occurrence of *Jonesina* in nonmarine strata is unusual and their presence in these deposits may indicate brackish or local marine conditions in the cyclothemic deposits. *Eocypridina radiata* occurs in the Westphalian in Europe, mainly in Westphalian A, rarely in B and C. Its occurrence in the Hance Formation is a basis for correlating that formation with the early Westphalian (Scott and Summerson, 1943).

Pennsylvanian Monongahela Formation, Western Pennsylvania and West Virginia

The late Pennsylvanian Monongahela Formation in western Pennsylvania and West Virginia includes shales and calcareous nonmarine beds with freshwater ostracodes (Sohn, 1977). The following preliminary identifications were recorded by Sohn from Monongahela

County, West Virginia: Gutschickia sp. and Darwinula sp. These are described and illustrated after Sohn in the systematic section.

Pennsylvanian Atokan, Des Moinesian, Missourian and Virgilian Series, Illinois

Pennsylvanian cyclothemic underclay limestones and shales of Illinois yielded a variety of freshwater ostracodes (Cooper, 1964; Scott, 1944; Sohn, 1977a). The faunas are characterized by many individuals in only a few species. The marine portions of the cyclothems contain numerous marine ostracode species.

The Atokan (Lower Pennsylvanian) Seville Limestone contains Carbonita orbiculata (Cooper). The Des Moinesian Summum cyclothem yielded Hastifaba robusta (Cooper), Candona planidorsata (Cooper), C. salterina (Jones and Kirkby), Gutschickia levis Cooper, G. ovata Cooper; the Jamestown Limestone has Hastifaba pervulgata Cooper; the Gimlet cyclothem has Carbonita agnes (Jones). The Missourian "Centralia" Limestone yielded freshwater species not named. The Virgilian (Late Pennsylvanian) Cohn and Bogota cyclothems and the Omega and Newton Limestones contain Hastifaba spinosa Cooper, Candona bairdioides (Jones and Kirkby), C. salterina (Jones and Kirkby), Whipplella cuniformis Holland, Carbonita tenuis Cooper, Gutschickia sp., G. subangulata (Jones and Kirkby) and Whipplella carbonaria Scott. The last 5 species were referred to the "Carbonita humilis group" by Bless and Pollard (1975).

The present writer questions that Hastifaba, a kloedenellid genus represents a true freshwater form, and suggests it may represent a brackish environment similar to that of the Middle Silurian kloedenellids in Appalachian red beds referred to in the introduction.

Pennsylvanian Allegheny Group, Pennsylvania

The Middle Pennsylvanian Allegheny Group of western Pennsylvania contains lacustrine limestones in a cyclothemic sequence of deposits (Weedman, 1994). The upper Freeport Limestone is reported to contain unidentified ostracodes.

PERMIAN

Permian Nineveh Limestone, Pennsylvania and West Virginia

The early Permian Dunkard Group of northwestern Pennsylvania and northwestern West Virginia contains freshwater limestones in the Greene Formation, Nineveh Limestone Member (Holland, 1934; Scott, 1944; Sohn, 1975a, b, 1977).

According to Scott (1944) the lakes and ponds in which the Nineveh ostracodes lived were very shallow, 2-3 m deep, with bottoms of calcareous mud, decayed vegetable matter and algae.

The following species have been recorded from the Nineveh Limestone: Whipplella cunieformis (Holland), W. depressa (Holland), W. carbonaria (Scott), Cypridopsis ovata (Holland), C. parvula (Holland), Candona magnitata (Holland), Gutschickia deltoidea (Holland) and Darwinula hollandi (Scott).

Sohn (1977) recorded the following additional forms from the Greene Formation, Wetzel County, West Virginia: Whipplella sp. (upper Rockport Limestone Member) and W. sp. (Windy Gap Limestone Member).

Permian Wellington Formation, Kansas

Nonmarine ostracodes were recorded from the early Middle Permian Wellington Formation in Dickinson and Marion Counties, south-central Kansas, by Tasch (1963). The Wellington consists of up to 700 feet (213 m) of shale, salt and freshwater limestone. According to Tasch the limestones formed in discontinuous fresh- to brackish-water ponds and puddles lying in belts along a coastal marine swamp. The ostracodes occur in limestone coquinas with snails and clams, 42 to 53 feet above the Annelly Gypsum.

The following ostracodes were recorded in the Wellington Formation: Gutschickia wellingtonensis Tasch, G. convexa Tasch, Cypridopsis cf. C. parvula (Holland), C. dickinsonia Tasch, and Candona paleostraca Tasch. The charophyta Iclistochara occurs with the ostracodes. An environmental similarity to the Dunkard Formation, Permian, of West Virginia and Pennsylvania is suggested by the author (Tasch, 1963).

STRATIGRAPHIC DISTRIBUTION OF MESOZOIC NONMARINE OSTRACODA

TRIASSIC

Triassic and Jurassic? Newark Supergroup, North Carolina, Virginia, and Pennsylvania

Upper Triassic nonmarine Ostracoda were recorded from two localities in North Carolina (Swain and Brown, 1972) in the Newark Supergroup: one near Gulf, Chatham County; the other near Carpenter, Wake County.

According to Froelich and Olsen (1984) rocks of the Newark Supergroup in North Carolina lie in two belts: one, the Deep River Basin, lies along the eastern edge of the Piedmont

Plateau and includes the following formations in ascending order: Pekin, Cumnock, and Sanford. The other basin areas lie in the north-west part of the Piedmont Plateau and are the Davey County and Dan River Basins.

The Newark strata are reddish brown and dark gray lacustrine and fluvial shales, sandstones and conglomerates totaling several thousand feet thick. The recorded ostracodes occur in thin lacustrine beds, typically dark gray but also in red shales.

The ostracodes recorded here came from the Deep River Basin. Near Gulf, the specimens occur in black shales of the Cumnock Formation exposed in a roadside borrow pit lying just south of Bethany Baptist Church, east of the Gulf water-treatment plant. Near Carpenter the ostracodes were obtained from red shales in undifferentiated Newark deposits exposed in mine pit of Triangle Brick Co. north of Carpenter on Highway 55. At both localities the ostracodes occur with the conchostracan Howellites berryi Bock and with plant and fish remains. Six species of Darwinula were recorded: D. subquadrata Swain and Brown, D. rogersii (Jones), and D. spp. 1-4. The collection is considered to be late Triassic. Farther north, the upper Newark is placed in the early Jurassic on the basis of palynomorphs. There is no ostracode evidence for a Jurassic age of any of the Newark deposits known to the present writer.

Jones (1862) recorded Candona rogersii Jones (=Darwinula rogersii (Jones)) from black and red shales of the Newark Supergroup near Richmond, Henrico County, Virginia and in Culpeper (red shale) and Prince Edward Counties, Virginia; in North Carolina from Deep River Chatham County; and in Pennsylvania near Phoenixville, Chester County, all associated with "Estheria" (?=Howellites). Cypris-like forms were also noted, but not described or illustrated. The occurrences in North Carolina and Virginia were mainly abundant flattened shells on the shale bedding planes. In Pennsylvania the occurrences were less numerous.

Triassic and Jurassic? Culpeper Group, Virginia

According to Gore (1994) the Lower Jurassic? Buckland and Waterfall Formations, Culpeper Group, of the Culpeper Basin, Virginia contains unidentified ostracodes. The non-marine Culpeper Group is stated to be Triassic and Jurassic in age and the lower (Triassic) part also contains unidentified ostracodes.

Triassic Moenkopi Formation, New Mexico

The early and middle Triassic (Sythian-Anisian) Moenkopi Formation in central New Mexico, near Albuquerque, occurs in an outlier in the Lucero Uplift (Kietzke, 1989). About 10 m of the Moenkopi redbeds are exposed in the Mesa Gallina section, sec. 30, T. 7 N., R. 4 W., and 7 m in the White Ridge section, sec. 10, T. 5 N., R. 4 E., Cibola County. The following ostracodes were recorded by Kietzke in these exposures: Darwinula sp., A, D. sp., B, Gerdalia sp., Darwinuloides sp., A, and D. sp. B. An early to middle Triassic age of the collection is suggested by the ostracodes, particularly the Darwinuloides and associated charophytes and Spirorbis when compared to European assemblages (Kietzke, 1989).

Triassic Chinle Group of Arizona and New Mexico and the Tecovas Formation of Texas

Nonmarine ostracodes and a charophyte were recorded from the Chinle Group in the Petrified Forest National Park by Lucas and Kietzke (1993). The fossils were found in the Painted Desert member of the Petrified Forest Formation and include Darwinula sp., aff. D. liulingchuanensis and Gerdalia cf. triassica and the charophyte Porochara abjecta. The authors suggest a resemblance of the assemblage to those of the Norian (late Triassic) of New Mexico and China. Indicated environment: highly mineralized permanent lake with water depths of about six to eight meters or less.

Other Upper Triassic ostracodes have been recorded from Texas (Lucas, 1991) in the Tecovas Formation, Crosby County and from New Mexico in the Chinle and Sloan Canyon Formations (Kietzke, 1987).

Triassic Shublik Formation, Arctic Stage, Alaska

A small collection of ostracodes recorded, but not described or illustrated, by Sohn (1964) from northern Alaska included Hungarella sp., Paracypris? sp., and Darwinula? sp. The first two are marine forms, and Darwinula is a non-marine form. It may have been derived from a non-marine deposit. The occurrence is in the Shublik Formation, Upper Triassic, on Dodo Creek, 2 to 2.3 miles above junction with Sadlerochit River in foothills of Sadlerochit Mountains, northern Alaska.

JURASSIC

Jurassic Todilto Limestone, New Mexico

The nonmarine Todilto Limestone is the basal member of the Morrison Formation in northeastern New Mexico. In its type area, McKinley County, New Mexico the Todilto rests unconformably on the Wingate Sandstone (Lower Jurassic) (Baker, Dane, and Reeside, 1936).

In southeastern Utah and southwestern Colorado the Todilto is separated from the Wingate by several intervening formations.

Limestone slabs of the Todilto, examined by Swain (1946) contained many specimens of the single species "Metacypris" todiltoensis Swain, a form recently transferred to Cytheridella Daday by Kietzke (1992). The described species came from a locality six miles north of Thoreau, New Mexico, on State Highway 56. It resembles species of "Metacypris" described by Jones (1886) from the Morrison Formation at Canyon City, Colorado.

Jurassic? Kayenta Formation, Arizona

The Kayenta Formation of northern Arizona and southeastern Utah consists of less than 100 feet (33 m) to 1200 feet (366 m) or more of sandstone and siltstone overlying the Wingate Sandstone and underlying the Navajo Sandstone. It is classified as Upper Triassic? by the U.S. Geological Survey, but the ostracodes obtained from the Kayenta (Kietzke and Lucas, 1994) suggest Lower Jurassic age. The following species were identified from the Kayenta in northeastern Arizona: Scabriculocypris n. sp., Darwinuloides magna (Jiang), and Darwinula sarytirmensis (Sharapova). The fauna suggested to the authors a correlation with the Sinemurian (Lower Jurassic) of China (Kietzke and Lucas, 1994).

Jurassic Morrison Formation, South Dakota, Wyoming, Montana, and Colorado

The ostracode fauna of the Upper Jurassic Morrison Formation was being monographed by Betty Kellett Nadeau in the mid-1900's but her work has apparently not been completed. The present writer received light-microscope photographs of her collection through the courtesy of the late F.M. Swartz. Some of the photographs are included in the present work. About 20 species-level taxa are represented in the Nadeau collection.

According to Sohn (1957, 1958) the "Morrison Formation" ostracodes described by Roth (1933) and Harper and Sutton (1935) were all from the Lower Cretaceous Lakota Sandstone.

The species Theriosynoecum wyomingense (Branson) is typical of the Morrison in Colorado, Montana, South Dakota and Wyoming (Sohn, 1982). The species Pseudocypridina piedmonti Roth was cited by Peck (1951b) from Morrison localities in the Black Hills area, South Dakota, as being typical of the Morrison. There is doubt as to whether the species also occurs in the Lakota Sandstone as maintained by Sohn (1979).

The Morrison Formation consists of green and red mudstones and shales, quartz sandstones, and thin nonmarine limestones, and ranges in thickness from about 22 feet (67 m)

in South Dakota to as much as 1500 feet (457 m) in Utah, but averages less than 300 feet (91 m) thick. It rests on the marine Sundance Formation, Upper Jurassic in northern areas and on other marine and nonmarine Jurassic strata in Utah. In the Front Range of Colorado it lies on Permian? and Triassic strata.

T. R. Jones (1886) recorded a fauna from presumed Morrison deposits near Canyon City, Colorado in the then-described "Atlantosaurus Beds". The specimens were supplied by the U. S. Geological Survey. In subsequent studies the ostracode-bearing layers were found to lie mainly below the dinosaurian strata. Jones recognized seven ostracode species: "Metacypris" forbesi Jones, "Metacypris" bradyi Jones, "Metacypris" whitei Jones, Darwinula leguminella (Forbes), "Cypris" purbeckensis Forbes, "Cytherideis" marshii Jones, and "Cytheridea" atlantosaurica Jones.

"M." forbesi and "M." bradyi, tentatively assigned to Theriosynoecum Branson, should receive further study according to Sohn (1982). They also resemble "Cytheridella" todiltensis (Swain), Kietzke (1992) from the Todilto Limestone member of the Morrison in New Mexico.

A comprehensive study of Morrison Ostracoda by Schudack, Turner and Peterson is reported to be in press (Schudack, 1997).

Jurassic Shongaloo Formation, Arkansas

The subsurface Shongaloo Formation, lower Cotton Valley Group of the northern Gulf of Mexico area (Swain and Anderson, 1993) consists in wells of about 1000 feet (330 m) of updip facies, nonmarine red and greenish gray shale and sandstone especially in southern Arkansas, and downdip facies, white sandstone and shelly limestone in northern Louisiana, eastern Texas and western Mississippi. In an updip well in Union County, Arkansas, Swartz and Swain (1946) recorded Cypridea sp. aff. C. wyomingensis Jones in the Shongaloo Formation. The marine facies of the Cotton Valley Group has yielded more than 80 species of Ostracoda (Swain and Anderson, 1993) many of which suggest brackish estuarine or lagoonal environments, but which are unlike Jurassic nonmarine species.

CRETACEOUS

Cretaceous Draney Limestone and Peterson Limestone, Wyoming and Idaho

The Lower Cretaceous Gannett Group (Mansfield and Roundy, 1916) crops out in Bannock County, southeastern Idaho and in Lincoln County, southwestern Wyoming and to the north. The Group is represented by five formations in ascending order: Ephraim Conglomerate,

Peterson Limestone, Bechler Conglomerate, Draney Limestone, and Tygee Sandstone. Ostracodes were described by Peck (1941) from the Peterson and the Draney.

The Peterson Limestone consists of interbedded nonmarine limestones and shales 200 feet (61 m) thick. Samples of the shales in southeastern Idaho and southwestern Wyoming yielded the following: Metacypris angularis Peck.

The Draney Limestone of the same region contained the following: Candona sp. indet., Cypridea anomala Peck, C. brevicornis Peck, C. compta Peck, C. diminuta Vanderpool, C. nitidula Peck, C. wyomingensis Jones and Cypris purbeckensis Forbes of Jones (1893).

Cretaceous "Lower Bear River Formation", Wyoming

A sequence of black shale and thin nonmarine limestone overlying the Gannett Group in southwestern Wyoming is, according to Peck (1941), indistinguishable from the Upper Cretaceous Bear River Formation but contains microfossils more closely related to the Lower Cretaceous Gannett Group. The following species were recorded from this so-called "Lower Bear River" by Peck (1941) from localities in Lincoln County, southwestern Wyoming: Cypridea anomala Peck, C. compta Peck, C. nitidula Peck and Metacypris persulcata Peck (=Dryelba persulcata (Peck) by Sohn (1982). All except the last species occur also in the Draney Limestone, upper Gannett Group, southeastern Idaho and southwestern Wyoming. None of the species is known to occur in the Upper Cretaceous Bear River Formation.

Cretaceous Kootenai Formation, Montana

The Lower Cretaceous Kootenai Formation of central and southwestern Montana yielded nonmarine Ostracoda described by Peck (1941). In central Montana the Kootenai consists of red, gray and green sandstone and shales, thin lenticular limestone and a workable coal bed. Plant fossils and poorly preserved freshwater invertebrates were recorded from these beds (Fisher, 1909). In southwestern Montana the Kootenai is represented by lower units of conglomerate, maroon, green and yellow sandy shale and buff sandstone, and upper thick nonmarine limestone with gastropods (Richards and Pardee, 1925).

The following species were recorded from the Kootenai by Peck (1941): Cypridea anomala Peck, C. ornata Peck, Longispinella longispina (Peck), Cypridea wyomingensis Jones, and Metacypris angularis Peck. Several of the species are common to the Kootenai and to the Draney and Peterson limestones of southwestern Wyoming and Idaho.

Cretaceous Lakota Formation and Minnewaste Limestone, South Dakota

Nonmarine Ostracoda have been recorded from the Lower Cretaceous Lakota Formation of southwestern South Dakota (Peck, 1951; Sohn, 1958, 1977, 1979). According to Sohn the ostracodes described by Roth (1933) and by Harper and Sutton (1935) as coming from the Upper Jurassic Morrison Formation of South Dakota are, in reality, from the Lower Cretaceous Lakota Formation. The Lakota ostracodes occur in dark shales about 20 feet (6 m) thick overlying the Jurassic Unkpapa Sandstone in the Hot Springs area, South Dakota.

The following species were identified by Peck (1951) from the Lakota: Longispinella longispina (Peck), Pseudocypridina inornata (Peck). Sohn (1979) recorded the following from the Lakota: Trapezpoidella rothi Sohn, Trapezoidella trapezoidalis (Roth), Limnocypridea morrisonensis (Roth), Pseudocypridina laeli Sohn, P. henrybelli Sohn, P. inornata (Peck) Longispinella asymmetrica Sohn, Pseudocypridina piedmonti Roth, "Cypridea" sp.

The Lakota Formation of southwestern South Dakota consists of up to 550 feet (167 m) of gray siltstone, sandy mudstone, claystone, limestone. The Minnewaste Limestone Member of the lower Lakota consists of light gray limestone, locally sandy and with black shales, thickness over 300 feet (91 m), and is overlain by the Fuson Shale.

The Minnewaste Limestone contains nonmarine ostracodes recorded by Peck (1951): Pseudocypridina inornata (Peck) and Ulwellia minuta Peck. The specimens came from limestones and interbedded shales of the Minnewaste.

Cretaceous Cloverly Formation, Wyoming

The Lower Cretaceous Cloverly Formation of central and northern Wyoming and southwestern Montana consists of non-marine shales, gray and varicolored and dense platy freshwater limestones. The Cloverly yielded the following ostracodes (Peck, 1951) near Lander, Wyoming: Cypridea quadrata Peck, C. obesa Peck, C. compta Peck, and C. anomala Peck.

Several of these species also occur in the lower Cretaceous Draney Limestones and "Lower Bear River" of southwestern Wyoming and in the Kootenai Formation of Montana (Peck, 1951).

Cretaceous "Newark Canyon" Formation, Nevada

A small collection of nonmarine Ostracoda was recorded from a shaly sequence in Pine Valley, Nevada, tentatively correlated with the Newark Canyon Formation, Lower Cretaceous (Aptian?) of central Nevada (Sohn, 1969). The exposure studied by Sohn was sampled at two locations in the center of W1/2 NW1/4 Sec. 21, T. 29 N., R. 53 E., close to the Eureka-Elko

County line, northeastern Nevada. The thickness of the shale was not recorded but one collection was obtained halfway of the shale slope and the other at the top of the slope.

The Newark Canyon Formation of the Eureka area, Nevada, consists of up to 1800 feet (548 m) of freshwater limestone, conglomerate, siltstone, sandstone and grit (Nolan et al., 1969). The age was previously determined as Aptian on the basis of mollusks and a fish (Sohn, 1969). Undescribed ostracodes in the type Newark Canyon are different from those in the Pine Valley occurrences studied by Sohn.

The following species were recorded from U.S.G.S. locality 27753: Cypridea (Cypridea) pecki Sohn, "Paracypridea" sp. and Petrobrasia? sp.

The second Pine Valley collection studied by Sohn, U.S.G.S. locality 27754 at the top of the shale slope, and presumably stratigraphically higher than 27753, yielded Cypridea (Bisulcocypridea) biscostata Sohn. As is discussed under the Sheep Pass Formation, east-central Nevada, the present writer suggests that this species in an Upper Cretaceous rather than a Lower Cretaceous form as it is an important element of the Upper Cretaceous-Paleocene Sheep Pass Formation. For this reason the age of U.S.G.S. locality 27754 may be late rather than early Cretaceous.

Cretaceous Strata, North Carolina, Arkansas, Texas and Louisiana

A thick sequence of intertonguing marine and nonmarine Lower Cretaceous and Upper Jurassic? mainly clastic strata is present beneath the coastal plain of North Carolina (Swain, 1952; Swain and Brown, 1964, 1972; Brown and others, 1972). The subsurface units of these ages were designated by the letters I, H, G, and F (Swain and Brown, 1972). Those containing Ostracoda were found to have faunal relationships to the Cotton Valley Group (Unit H), Coahuila Group (Unit H), and Trinity Group (Unit G) of the Gulf Coast region of the United States.

The non-marine facies of Unit H is represented in North Carolina by 80 to 1500 feet (457 m) of varicolored shale and sandstone with Cypridea (C.) sp., C. (C.) menevoides Swain and Brown, Potamocypris? sp., Cypridea (C.) dequeenensis Swain and Brown, C. (C.) cf. wyomingensis Jones, and Gomphocythere cf. silvana Martin. C. dequeenensis occurs in the lacustrine? De Queen Limestone, Pike County, Arkansas (Vanderpool, 1982; Swain and Brown, 1964) and the Glen Rose Formation of wells in NW Louisiana. Several other ostracodes that suggest brackish water conditions also occur in Unit H. i.e. Hutsonia spp., Taxiodiella sp., Fabanella sp. and Juvenix sp. The non-marine forms probably are of Lower Cretaceous age.

Unit G in the North Carolina subsurface consists of 160 to 1720 feet (524 m) of red, green and gray shale, massive sandstone and thin bioclastic limestone. The following non-marine ostracodes were identified in Unit G: Ilyocypris? tumida Swain, Cypridea (C.) diminuta Vanderpool, and C. (C.) cf. wyomingensis Jones. C. diminuta was recorded earlier from the Trinity Group, De Queen and Glen Rose Formations of Arkansas, Louisiana, and Texas (Vanderpool, 1928). Lower Cretaceous, Late Aptian-early Albian of Texas and Arkansas. The Atlantic and Gulf Coast Lower Cretaceous forms represent lacustro-fluvial assemblages.

Cretaceous "Washita Group", Florida

The Lower Cretaceous Washita Group equivalents of southern Florida consists of entirely subsurface oolitic limestone, dolomite, foraminiferal limestone, anhydrite and thin shale ranging from 825 feet (251 m) to 2169 feet (667 m) in known thickness (Swain, 1982). The strata are predominantly marine and brackish water but locally contain non-marine layers of probably sabhka origin, representing lakes and ponds that developed on exposed carbonate flats.

Two species, Limnocythere? sp. 1 and L.? sp. 2 were recorded from deep wells in Hendry and Pinellas Counties, Florida (Swain, 1982). Three probably brackish water species were recorded with the Limnocytheres: Eucytherura pinellensis Swain, Clithrocytheridea paucipunctata Swain, and C. holoreticulata Swain.

The presence of the miliolid foraminifer Nummoloculina heimi Bonet was used to establish the late Albian age of the deposits.

Cretaceous Bear River Formation, Wyoming

The Bear River Formation of southwestern Wyoming consists of dark gray non-marine shales and limestones up to 539 feet (164 m) thick. Ostracoda were recorded from the Bear River by Peck (1951b) and earlier by Jones (1893). According to Peck, some of the species assigned by Jones to the Bear River probably came from the Draney Limestone or other part of the Lower Cretaceous Gannett Group.

The following species occur in the Upper Cretaceous Bear River beds of the Evanston area, southwestern Wyoming (Peck, 1951b): Bisulcocypridea skeeteri (Peck), Pseudocypridina laevicula Peck, Looneyella monticula (Jones) and L. quadrispina Peck.

STRATIGRAPHIC DISTRIBUTION OF CENOZOIC NONMARINE OSTRACODA

PALEOCENE

Late Cretaceous, Paleocene, and Eocene? Sheep Pass Formation, Nevada

The lacustrine and fluvial Sheep Pass Formation was named from Sheep Pass Canyon, T. 10 N., R. 62 and 63 E., Egan Range, White Pine County, Nevada (Winfrey, 1958; Kellogg, 1964; Fouch, 1979) where it attains a thickness of 3325 feet (1014 m). Winfrey (1958) and Kellogg (1964) recognized several non-volcanic members in ascending order: member A, conglomerate and sandstone 835 feet (255 m); member B, argillaceous dolomitic, fossiliferous limestone 1072 feet (327 m) thick; member C, sandstone, siltstone and sandy limestone 621 feet (189 m) thick; member D siliceous, dolomitic freshwater limestone 647 feet (197 m); and member E, silty, clayey limestone, varicolored sandstone, and channel sandstone 150 feet (46 m) thick.

Ostracoda were collected from member B in Sheep Pass Canyon by M.B. Stam. The species include: Clinocypris? sp., Paracypridopsis? sp., and Cypridea (Bisulcocypridea) bicostata Sohn (Swain, 1987b).

A second exposure of the Sheep Pass Formation in Johnson Canyon, Grant Range, T. 8 N., R. 58 W., Nye County, Nevada where 195 feet (60 m) of the formation are exposed out of a total thickness of 460-490 feet (140-150 m) reported to occur in the Grant Range. In white to tan dolomitic limestone overlying Mississippian sandstone (Fouch, 1979) M.B. Stam collected ostracodes of the following species (Swain, 1987b): Herpetocypris winfreyi Swain, Paracypridopsis? fouchi Swain, Candoniella whitei (Scott and Smith), Cypridea (Bisulcocypridea) nyensis (Swain), and Darwinula sp. The Sheep Pass member B in Sheep Pass Canyon is suggested to be late Cretaceous or Paleocene (Fouch, 1979; Swain, 1987b), while the ostracode-bearing beds in Johnson Canyon appear to be younger, probably late Paleocene or early Eocene (Swain, 1987b) based on the distribution of Cypridea (B.) bicostata in Sheep Pass Canyon and of C. (B.) nyensis in Johnson Canyon. The former species was recorded by Sohn (1969) from Cretaceous rocks in Pine Valley, Nevada, west of Egan and Grant Ranges. The latter species was described by Swain (1964a, b) from the subsurface Paleocene of the Railroad Valley Oil Field, Nevada.

In an earlier study (Swain, 1964a), core samples from the Shell Oil Co. No. 1 Eagle Springs well, Nye County, Nevada, penetrated the Sheep Pass Formation. The following ostracodes were recorded: Scottia? eaglespringsensis Swain, S.? trapezoidalis Swain, and S.? sublanceolata Swain. All three forms are small size compared to the other Sheep Pass ostracodes, less than 0.5 mm in length and are smooth-surfaced. They occur in lacustrine calcarenite or calcirudite and possibly represent interstitial forms.

Paleocene Fort Union Formation, Montana

The Fort Union Formation, a sequence of mainly nonmarine shales, sandstone and lignite in Montana, Wyoming, North and South Dakota and Colorado varies from less than 500 feet (152 m) to more than 3000 feet (914 m) thick. At or near the base in North Dakota is the Cannonball Member of the Fort Union Formation which has a marine (Midway Paleocene) microfauna, including ostracodes (Swain, 1949). Ostracodes from the nonmarine Fort Union were described from localities in southeastern Montana and western North Dakota (Swain, 1949).

In Powder River County, Montana the Fort Union (Tongue River Member) yielded: Candona sp., Darwinula cf. stevensoni Brady and Robertson, Cypridea (Bisulcocypridea) arvadensis (Swain), Pseudoeucypris pagei (Swain), Herpetocypris? sp. and Ilyocyprimorpha tuberculata (Swain). In North Dakota the Tongue River Member yielded Candona sp. and Limnocythere sp. (Swain, 1949).

EOCENE

Eocene Green River, Colton, Wasatch and Flagstaff Formations, Utah, Colorado and Wyoming

The Green River Formation of northern Utah, northwestern Colorado, and southwestern Wyoming consists of up to 5000 feet (1525 m) of lacustrine limestone, marlstone (calcitic and dolomitic calcilutite), oil shale, fossiliferous limestone, sandstone, water laid tuff, siltstone, and clay shale (Bradley, 1931; Dane, 1954; Pickard, 1955; Swain, 1956, 1964). In ascending order, the Green River has been divided into members in the Uinta and Piceance Creek Basins, Colorado and Utah: Douglas Creek including transition beds with the underlying Colton Formation of red beds, Garden Gulch, Parachute Creek, and Evacuation Creek. Ostracodes are mainly concentrated in the lower two members which are of shallow freshwater lacustrine and fluvial origin. Ostracodes also occur sparingly in the upper, alkaline, deeper water members. The ostracodes were mainly shallow water littoral dwellers but some species apparently ranged into deeper water profundal habitats. The extent to which the shallow water forms were transported into the deeper waters is not presently known, but the selective occurrence of species in the latter habitat suggests that transport can not account for their presence there entirely.

The Green River ostracode fauna is characterized by few species but large numbers of several species. Several of the species are quite rare. Swain (1956, 1964a, b) recorded 24 species from the formation, including those in the underlying Colton-Green River transition beds: Hemicyprinotus watsonensis watsonensis Swain, H. watsonensis emaciata Swain, H.

reniformis Swain, Heterocypris riflensis Swain, H. whiteriverensis Swain, Pseudoeucypris pagei (Swain), P. pagei subsp., Procyprois ravenridgensis Swain, Cyprois ephraimensis Swain, Stenocypris cf. sinuata Müller, Scottia subquadrata Swain, Potamocypris williamsi Swain, Advenocypris umbonata Swain, Candoniella whitei (Scott and Smith), Candona artesensis Swain, C. rangliensis Swain, Pactolocypris centronotus (Swain), Cypridea (Bisulcocypridea) arvadensis (Swain), Ilyocyprimorpha binodosa (Swain), Cypridea (Bisulcocypridea) bisulcata (Swain), Limnocythere sp., Metacypris cordata Swain and M. sp.

As a whole the Green River ostracode fauna is typified by the presence of most of the species in the lower part of the formation and by their gradual disappearance upward, as environmental conditions changed.

The following assemblage zones were recognized by Swain (1964) in ascending order in the Green River Formation of northern Utah and northwestern Colorado.

The Procyprois ravenridgensis Zone represents the earliest ostracode zone encountered in the Uinta and Piceance Creek Basins, although some of the other ostracode species occur locally below the lowest record of P. ravenridgensis. It represents the Upper Colton (Wasatch) and lower Colton-Green River Transition Beds. The base of the zone was selected at the earliest occurrence of the nominal species at any given locality. The nominal species preferred muddy bottom habitats of lake or pond, with lower carbonate content than did some of the other species.

The Cypridea (Bisulcocypridea) bisulcata Zone, because of the distinctive appearance of the nominal species, is one of the stratigraphically most useful of the Green River Zones. The zone covers the interval consisting of the upper part of the Colton-Green River transition beds and the lower part of the Green River Formation (Swain, 1964). The species is only infrequently found in fluvial marginal belts or in highly organic lacustrine facies. All the ostracode species of the overlying zones occur in this zone, and Potamocypris williamsi is present with it in several places. The thickness of rock included in the zone varies from less than 150 feet (46 m) to over 400 feet (122 m).

The Pseudoeucypris pagei Zone lies principally in the Garden Gulch Member but also in the lower part of the Parachute Creek Member and less commonly in the uppermost part of the Colton-Green River transition beds. The large, strongly convex shell of P. pagei is easy to recognize. The thickness of rock included in the Zone varies from less than 100 feet (30 m) to about 400 feet (122 m). Hemicyprinotus watsonensis and Cyprois ephraimensis commonly occur with it, along with Potamocypris williamsi occasionally but the latter has not been found in the overlying zone.

The Cyprois ephraimensis Zone is based on a nominal species that occurs with the other species below but does not range as high as Hemicyprinotus watsonensis, the nominal species of the overlying zone. The greater relative height and the more pronounced flattening of this form compared to H. watsonensis and Candoniella whitei make it easy to recognize. The thickness of the rock included in the Zone varies from less than 100 feet (30 m) to about 800 feet (244 m).

The Lower Hemicyprinotus Zone lies in the middle part of the Green River Formation and roughly corresponds to the sandstone-bearing, possibly deltaic, section of the middle Green River. This zone is overlain in the marginal parts of the Piceance Creek and Uinta Basins by a Barren Zone. Toward the deeper parts of the basins the deltaic facies and the Barren Zone disappear and the Lower and Upper Hemicyprinotus Zones unite. H. watsonensis occurs by itself in the Lower Zone or in some places with Candoniella whitei. These two species are the most abundant and long-ranging of the Tertiary ostracodes in the Uinta and Piceance Creek Basins. The top of the Lower Hemicyprinotus Zone lies below the "Mahogany Ledge" oil shale in the middle Parachute Creek Member.

The Barren Zone includes that portion of the Green River Formation from a level 325 to 700 feet (99-213 m) below the Mahogany Ledge to the upper part of the formation. Although barren of ostracodes, the rocks contain impressions of Diptera (flies) larvae in great numbers on platy carbonate bedding planes (Bradley, 1931). That author also described and illustrated bacteria, fungi, microscopic algae, moss and fern spores, pollen, testate protozoans and fragments of arachnids and insects in thin sections of Barren Zone rocks.

The Upper Hemicyprinotus Zone has been recognized in the Upper Green River and lower Uinta Formations, especially where the two formations interfinger. In addition to H. watsonensis, Candoniella whitei occurs sparingly in the zone. The latter species was described from the upper Green River of the Bridger Basin, southwestern Wyoming (Scott and Smith, 1951).

An evolutionary study of Hemicyprinotus watsonensis (Swain, 1964a) demonstrated that the species gradually increased in average length and in degree of elongation from the base to the top of the Green River Formation. It is possible that a new species could be erected for the forms in the upper Green River. If so, the evolutionary process would have taken about 5 million years to form the new species, more or less the time span of the Green River Formation.

Taylor (1972) and Kaesler and Taylor (1971) recognized two biofacies in Green River deposits in Wyoming. Cypridea (Bisulcocypridea) bisulcata and Metacypris paracordata were mud dwellers, while Potamocypris williamsi, Pseudoeucypris pagei and Hemicyprinotus watsonensis were plant dwellers.

Eocene and Oligocene Elko Formation, Nevada

The Elko Formation in the vicinity of Elko County, Nevada is about 1585 feet (484 m) thick and consists of platy lacustrine oil shales, below, and tuffaceous siltstone and tuff in its upper half (Solomon et al., 1979). Potassium-argon dates in the upper Elko, based on biotite, range from 37.1 to 38.9 my. indicating late Eocene to early Oligocene age.

At the site of the abandoned Catlin Shale Products Co. oil shale works at the southwest edge of Elko, a sequence of oil shales about 55 feet (16 m) thick was studied for ostracodes (Swain, 1987b). The sequence represents member 2 of the Elko Formation as delineated by Solomon et al. (1979), which is about 72 feet (22 m) thick beginning 203 feet (62 m) above the base of the formation. Most of the specimens studied came from fissile to platy, siliceous, organic shale and oil shale just above the Catlin works. The specimens are very thin-shelled and badly flattened by tectonic compression, and although locally abundant, are often difficult or impossible to identify. Only three species were identified: Pontoniella catlinensis Swain, the dominant form, Amplocypris? sp., and Pactolocypris cf. centronotus (Swain). The latter species is similar to a form in the Green River Formation, middle Eocene, of Colorado and Utah (Swain, 1987b). The ostracode fauna suggests, by the kinds of species and the delicate nature of the shells, deposition in quiet water, below wave base.

Eocene or Oligocene Twin Bridges and Illipah Formations, Nevada

A lacustrine sequence of fossiliferous limestone and marl is exposed in the Twin Bridges area about 15 miles (24 km) south of Elko, Nevada. The sequence was informally called the Twin Bridges limestone by van Houten (1956) and was formalized as a formation by Swain (1987b). The exposed Twin Bridges Formation is about 630 feet (265 m) thick at the type locality (Sharp, 1939) but is said to be up to 1000 feet (305 m) thick in the Carlin-Pinon Range (Smith and Ketner, 1976) and 1100 feet (335 m) thick in the Adobe Range north of Elko (Silitonga, 1976; Solomon et al., 1979).

The middle part of the formation at Twin Bridges yielded abundant ostracodes and gastropods, both types of which are very thin-shelled and difficult to separate from matrix (Swain, 1987b, 1988, 1990b). The following ostracode species were recorded: Scottia? sp. aff. S. eaglespringsensis Swain, Pactolocypris cf. centronotus (Swain), Hemicyprinotus sp. aff. H. watsonensis Swain, Cyprois whiteflatensis Swain, Herpetocypris? winfreyi Swain, Procyprois twinbridgensis Swain, and Paracypridopsis fouchi Swain. Several of the preceding species resemble some of those in the Eocene Green River Formation of eastern Utah and western Colorado and in the Sheep Pass Formation (Late Cretaceous to Eocene) of eastern Nevada.

Additional Twin Bridges species are: Eucypris sp. aff. E. bobcatensis Swain and Cypricercus sp. aff. C. palisadensis, Swain. These resemble species from the Oligocene Indian

Well and the Miocene Humboldt Formations of central and northeastern Nevada. A final group of Twin Bridges species includes: the tropical? form Chlamydotheca dixieflatensis Swain, Cypriconcha sp., Timiriasevia heylmuni Swain, Cyclocypris ketneri Swain, and Pontoniella catlinensis Swain. Of these, the final one is also found in the Eocene-Oligocene? Elko Formation at Elko, Nevada.

The age of the Twin Bridges Formation is suggested to be late Eocene (Swain, 1987b), younger than the Sheep Pass Formation of eastern Nevada, but older than, or partly the same age as, the Elko Formation near Elko, Nevada. The Elko Formation has been dated radiometrically at 37-39 my (AAPG COSUNA Chart, 1985), early Oligocene or late Eocene.

The layers of the Twin Bridges that contain abundant ostracodes and gastropods are thin and are concentrated at certain horizons. The limestone is highly porous at some levels and yields a petroleum odor on fresh break. The strata appear to have been deposited in a large lake having abundant nutrients and of variable depth from littoral to profundal. Sources of coarse clastic material were distant from the lake basin but the siliceous nature of the limestone suggests a source of dissolved silica from nearby areas. Many if not most of the ostracode species recorded here may have come from profundal facies of the formation, judging from the fine grained laminated nature of the rock, but it can not be determined whether the specimens lived in the deeper waters or were washed out from shallow areas of the lake.

The Illipah Formation, possibly of Eocene age (Humphrey, 1960) in Illipah Creek Valley, south of US Highway 50, White Pine County, Nevada was sampled in tan lenticular sandstone (Swain, 1987b). Only fragmentary unidentifiable ostracodes were observed.

OLIGOCENE

Oligocene Indian Well Formation, Nevada

The Indian Well Formation as exposed in Elko County, central and north-central Nevada, consists of up to 3,300 feet (1005 m) of volcanic ash, tuff, agglomerate, and interbedded ashy, fluvial and thin lacustrine sands, clays and siltstones. The formation is underlain by the Eocene Elko Formation and is overlain by the Miocene Humboldt Formation. The Indian Well has been dated as Oligocene (33-37 my) radiometrically (Smith and Ketner, 1976).

A locality at Dixie Flats, Sec. 29, T. 31 N, R. 55 E., Elko County is mapped as Humboldt Formation (Smith and Ketner, 1976) but lies close to the Humboldt-Indian Wells boundary and its lithologic character of pale gray and white ashy siltstone and shale is similar to that of the Indian Well (Swain, 1987b). A collection from that locality yielded Cypricercus cf.

sacramentoensis Swain and C. cf. toanoensis Swain. The first species was described from the North Creek Formation (Oligocene and Miocene) of the Sacramento Pass region, White Pine County, Nevada and the second species from both the North Creek and the Humboldt Formations of northeastern Nevada. The latter species was also recorded from either the upper Indian Well or the lower Humboldt Formation in the Palisades section, Pine Valley, Eureka county, Nevada (Swain, 1986b). A sequence of mostly felsic ignimbrites in the Railroad Valley, east-central Nevada, locally termed the Garrett Ranch Group is roughly equivalent to the Indian Well Formation, and produces oil.

Oligocene Hells Bells Canyon Formation, Nevada

A sequence of interbedded water-laid tuffs, limestones, welded tuffs and basal ignimbrites up to 1000 feet (305 m) thick, in the Pahranagat area, Lincoln County, southern Nevada, was named the Hells Bells Canyon Formation by Dolgoff (1963). It is overlain by the Hiko Tuff, dated radiometrically as 17-19 my (Early Miocene). Samples of waterlaid tuff in the upper 400 feet (122 m) yielded Timiriasevia heylmuni Swain. This species was described from the late Eocene Twin Bridges Limestone of central Nevada. The species may range higher than previously known or the formation may be older than thought by Dolgoff. The volcanic nature of the Hells Bells deposit suggest Oligocene rather than Eocene age in this area.

Oligocene? Edwards Creek Tuff, Nevada

Platy ashy shale and siltstone mapped as Edwards Creek Tuff crop out near the head of the valley of Edwards Creek, T. 19 N., R. 38 E., Churchill County, Nevada. The formation is dated at 27 my, Oligocene (AAPG COSUNA Chart 1985). Samples of the siltstone collected by the writer and D.J. Bergstrom yielded Candonopsis sp. aff. C. shahejiensis Hou et al. (Swain, 1987c). This form resembles one in the Esmeralda Formation of west-central Nevada (Swain, 1985), and in the middle Oligocene of eastern China (Hou et al, 1978).

Oligocene? Titus Canyon Formation, Nevada and California

An exposure mapped as the Titus Canyon Formation of Oligocene? age (Stewart and Carlson, 1978) was sampled one mile (1.6 km) east-southeast of Nevada Highway 374 at Daylight Pass on the Nevada-California border (Swain, 1987c). The outcrop consists of about 5 feet (1.5 m) of gray, sublithographic, algal, rubbly lacustrine limestone and underlying mottled, argillaceous algal? marlstone. The limestones occur in a sequence of several hundred feet of white and pale greenish gray, laminated, indurated tuff. A non-sulcate form of the ostracode Cypridea? was identified in the marlstone (Swain, 1987c). If correctly identified, the non-sulcate Cypridea suggests early Paleogene or Mesozoic age rather than Oligocene. The volcanic nature of the enclosing deposits, however, is supportive of an Oligocene age.

Oligocene Pruett Formation, Southwestern Texas

A Tertiary rock succession 3000-4000 feet (915-1219 m) thick in Brewster County, southwestern Texas was named the Buck Hill Volcanic series by Goldich and Elms (1949). Three formations consisting largely of tuffs and lava flows are recognized in the Buck Hill, in ascending order: Pruett Formation, 900-1000 feet (274-305 m) thick; Duff Formation, 1000-1400 feet (305-427 m) thick; and Tascotal Formation, 1000 feet (305 m) thick, approximately. Samples containing non-marine ostracodes from thin lacustrine limestones were supplied by S.S. Goldich from two localities in the Pruett Formation in Agua Fria and Cathedral Mountains Quadrangles, Brewster County, Texas (Swain, 1987c).

All the ostracode specimens as well as associated gastropod and charophyte specimens are silicified. Both localities are of early Oligocene age, based on radiometric determinations obtained by Goldich.

A locality in freshwater limestone in Sheep Canyon, southeastern Cathedral Mountain Quadrangle yielded the following species: Cyclocypris cf. forbesi Sharpe, Candona sp. aff. rangliensis Swain, Darwinula cf. stevensoni (Brady and Robertson), Cytheridella alpinensis Swain, and C. sp. A second locality in the Agua Fria Quadrangle, northeastern Brewster County, Texas yielded only Cypricercus? sp.

The Cytheridellas are of interest in being the first representatives of the genus recorded from the North American Tertiary. Living members of the genus are characteristic of tropical South American and Central American slightly brackish water lakes. There is one unpublished occurrence known to the writer from Mud Lake, Florida in a sample supplied by the late R.C. Bright.

Oligocene Chadron Formation, Northwestern Nebraska

The Chadron Formation, White River Group, was named from Chadron, Dawes County, Nebraska. It occurs in western Nebraska, northeastern Colorado, Montana, South and North Dakota and eastern Wyoming. Near Adelia, northwest of Crawford, Sioux County, Nebraska, the Chadron is 143 to 209 feet (44-64 m) thick and consists of volcanic ash, silt and sand. It underlies the Brule Formation and unconformably overlies the late Cretaceous Pierre Shale or the Lance Formation.

A collection of non-marine Ostracoda obtained by Richard Konizeski at Helmsville, Nebraska, contained the following: Candona cf. candida (O.F. Müller), C. cf. distincta Furtos, C. cf. havanaensis Staplin, C. cf. ohioensis Furtos, C. cf. poseyensis Staplin, C. sp., Cyprois sp. 1, C. sp. 2, Pactolocypris cf. suborbicularis Swain.

The assemblage is of interest in that it contains several species resembling those in the Pliocene and Pleistocene of the central United States as well as one species typical of the Miocene Humboldt Formation of Nevada. The Humboldt also has several other species that suggest an Oligocene age. It was pointed out previously (Swain, 1986) that the lower part of the Humboldt and of equivalent Esmeralda Formations of Nevada may be Oligocene in age. The rest of the Chadron assemblage is unlike the Oligocene forms so far known from the Great Basin.

Oligocene? of Madison County, Montana

Samples of pale gray and white siliceous clay containing delicate thin shelled nonmarine Ostracoda was obtained by J. A. Dorr, Jr. from the Oligocene?, sec. 28, T. 8 S., R. 45 W., Madison County, Montana. The following were identified: Candona cf. simpsoni Sharpe, C. cf. cachensis Swain, Pontoniella sp. and Cypricercus cf. elegantulus Swain. C. simpsoni is known from Pleistocene to Recent, C. cachensis is from the Pliocene, Pontoniella sp. resembles a Pliocene form and C. elegantulus occurs in the Miocene or late Oligocene. No definite age assignment can be made for the collection at present.

Oligocene-Miocene Renova Formation, Montana

Lacustrine deposits of the Arikareean Renova Formation, Passamari Member, in the Ruby Basin, southwestern Montana, yielded a unique foraminiferal and ostracodal assemblage (Van Nieuwenhuise and He, 1986). Three species of the brackish water foraminiferal genus Elphidium are associated with nonmarine Ostracoda. The assemblage is suggested to have developed during the proposed maximum marine transgression in the late Oligocene and early Miocene. Introduction of the foraminifera may have been accomplished by migrating waterfowl or spawning salmon. Further publication of the data is planned by the authors.

STRATIGRAPHIC DISTRIBUTION OF NEOGENE PRE-HOLOCENE NON-MARINE OSTRACODA

MIOCENE

Oligocene and Miocene Humboldt and North Creek Formations, Nevada

As originally described by Sharp (1939), the Humboldt Formation comprised up to 8000 feet (2440 m) or more of highly variable interbedded lacustrine, fluvial and extrusive and pyroclastic igneous rocks in northeastern Nevada. It has been recognized for many years that an angular unconformity lies within Sharp's original Humboldt (Becker, 1969; Swain, Becker, and Dickinson, 1967). The field relationships were shown to the writer in 1955 by M. B. Stam, Gulf Oil Co. Smith and Ketner (1976) restricted Sharp's original Humboldt to the lower 2000 feet (610 m) or so of the original sequence on the outcrop and assigned the remainder of the Humboldt beds to the Palisades Canyon Rhyolite and the Hay Ranch Formations of Regnier (1960). The revised Humboldt is suggested by those authors to be Miocene based on vertebrate fossils and on radiometric dates (Smith and Ketner, 1976), while the overlying Palisades Canyon Rhyolite and the Hay Ranch Formation are Pliocene and early Pleistocene.

A subsequent study of the Humboldt Ostracoda Swain (1986b) included samples from 3400 feet (1037 m) of the formation in the Gulf Oil Co. No. 1 Thousand Springs well, Elko County, Nevada. Ostracoda were obtained from throughout the formation in that well, but differed in mode of preservation of the shells. Those collected above 1670 feet (509 m) are white and opaque, whereas those below that level are dark brown, translucent and filled with calcite. Many other localities also yielded Humboldt Ostracoda.

In east-central Nevada, east of Sacramento Pass, White Pine County, a succession of lacustrine, fluvial and pyroclastic rocks, possibly as much as 7000 feet (2133 m) thick, was named the North Creek Formation by Young (1960). Radiometric (K-Ar) dating of volcanic rocks of the North Creek Formation range from about 35 my (early Oligocene) to 20.2 my (middle Miocene).

TABLE 1. The following freshwater ostracode species were identified in the Humboldt and related formations, northeastern Nevada.

Family CYPRIDIDAE
Subfamily CYPRIDINAE
Genus **Eucypris**
Eucypris ornatoides elongata, Swain
Eucypris ornatoides ornatoides Dickinson and Swain
Eucypris bobcatensis, Swain
Eucypris microreticulata, Swain

Eucypris sp. 1

Eucypris sp. 2

Eucypris sp. 3

Eucypris sp. 4

Genus **Pseudoeucypris**

Pseudoeucypris sp.

Genus **Cypricercus**

Cypricercus toanoensis, Swain

Cypricercus hunterensis, Swain

Cypricercus palisadensis, Swain

Cypricercus sacramentoensis, Swain

Cypricercus elegantulus, Swain

Subfamily HERPETOCYPRIDINAE

Genus **Herpetocypris**

Herpetocypris sp.

Subfamily BRADYCYPRINAE

Genus **Bradycypris**

Bradycypris? medicinecreekensis, Swain

Subfamily CYPRINOTINAE

Genus **Cyprinotus**

Cyprinotus sp. aff. *C.* cinglaensis Brady

Cyprinotus? sp. 1

Cyprinotus sp. 2

Genus **Heterocypris**

Heterocypris blairensis Swain and Wagoner

Genus **Hemicyprinotus**

Hemicyprinotus? simplex, Swain

Hemicyprinotus? sp.

Genus **Pactolocypris**

Pactolocypris suborbicularis attenuatus, Swain

Pactolocypris cf. *suborbicularis suborbicularis* Swain

Pactolocypris biprojectus Swain

Pactolocypris laevus, Swain

Pactolocypris cancellatus pronodosus, Swain

Family Cypridopsidae

Subfamily CYPRIDOPSINAE

Genus **Cypridopsis**

Cypridopsis? sp.

Genus **Paracypridopsis**

Paracypridopsis curta, Swain

Subfamily DISOPONTOCYPRIDINAE

Genus **Pontoniella** Mandelstam

Pontoniella sp.

Family Notodromadidae

Genus **Procyprois**

Procyprois? sp.

Family CANDONIDAE

Subfamily CANDONINAE

Genus **Candoniella**

Candoniella sp. aff. *C. parca* Hou et al.

Candoniella pinecreekensis, Swain

Genus **Candona**

Candona devexoidea Dickinson and Swain

Candona sp. 1

Family PONTOCYPRIDIDAE

Genus **Clinocypris**

Clinocypris? sp.

Family ILYOCYPRIDIDAE

Genus **Ilyocypris**

Ilyocypris thousandspringsensis, Swain

Family CYPRIDOPSIDAE

Subfamily CYPRIDOPSINAE

Genus **Cypridopsis**

Cypridopsis? sp.

Genus **Paracypridopsis**

Paracypridopsis? curta, Swain

Superfamily DARWINULACEA

Family DARWINULIDAE

Genus **Darwinula**

Darwinula sp.

Superfamily CYTHERACEA

Family LIMNOCYTHERIDAE

Genus **Limnicytheropteron**

Limnicytheropteron winecupensis, Swain

Genus **Elkocythereis**

Elkocythereis minutidontis Dickinson and Swain

Family CYTHERIDEIDAE

Subfamily CYTHERIDEINAE

Genus **Cyprideis**

Cyprideis, n. sp.

Ostracoda were obtained from the middle? 142 feet (43 m) of the formation 5.85 miles (9.4 km) east of Sacramento Pass, and north of U.S. Highway 50, in lacustrine limestones interbedded with volcanics (Swain, 1986b).

Forty-two species and subspecies of freshwater Ostracoda were obtained from the Humboldt and related formations in northeastern Nevada (Table 1). In the Medicine Creek section, Elko County Eucypris bobcatensis is very abundant in porous limestone about 100 feet (30 m) above the base of the section but has not been found elsewhere. Cypricercus elegantulus, Limnicytheropteron wincupensis, and Pactolocypris suborbicularis are common in the basal 50 feet (15 m) of the Medicine Creek section.

Pactolocypris palisadensis occurs in the lowermost Humboldt or uppermost Indian Well Formation in the Palisade section, Eureka County, Nevada, but was found in only one other locality, the Wilkins Ranch No. 2 well, Elko County in Oligocene or Miocene beds. Cyprinotus sp. aff. C. cingalensis, a unique dorsally umbonate species, was found only in the middle Humboldt Formation in the Cobre, Nevada section.

Pactolocypris laevis marks the upper part of the Humboldt in several northeastern Nevada sections. One of the principal Humboldt species Cypricercus toanoensis does not occur as high in the formation as does P. laevis in the Medicine Creek, Winecup Ranch and Cobre areas, but lies in the upper part of the formation in the Wilkins Ranch No. 2 well, just beneath the Hay Ranch Formation. C. toanoensis also occurs in the North Creek Formation and in the lowermost Humboldt or Indian Well Formation at Palisades, Nevada. Cypricercus hunterensis typically does not range as high in the Humboldt as does C. toanoensis.

Eucypris ornatoides elongata of the Humboldt does not range as high as C. toanoensis and C. hunterensis. Eucypris microreticulata also has a restricted range within the Humboldt, not reaching the upper part of the formation.

Pactolocypris suborbicularis and P. biprojectus of the lower Humboldt occur in the Esmeralda Formation of Nye and Mineral Counties, Nevada. Heterocypris blairensis of the middle and lower Esmeralda is present in the middle and lower Humboldt.

In summary, the upper part of the Humboldt Formation in northeastern Nevada is represented by Pactolocypris laevis; the middle Humboldt by (in more or less descending order) Cypricercus toanoensis, C. hunterensis, Pactolocypris suborbicularis attenuatus, Eucypris ornatoides elongata, E. microreticulata and Heterocypris blairensis; the lower Humboldt by Pactolocypris suborbicularis, Cypricercus palisadensis, and Hemicyprinotus? simplex.

Radiometric ages of tuffs, latites and rhyolites in the Humboldt range from 9 to 15 my (AAPG COSUNA Chart, 1985), middle to late Miocene. The ostracodes suggest that part of the formation may be older than this. Both the Esmeralda Formation of west-central Nevada and the Humboldt are dated as Miocene, but only three ostracode species occur in both formations based on present knowledge. Whether the dissimilarity in the two faunas represents environmental or age differences can not be determined at present.

Miocene Esmeralda Formation, Nevada

The Esmeralda Formation (Turner, 1900) is a very thick sequence of lacustrine, fluvial, and pyroclastic rocks exposed in the Silver Peak area, southern Big Smoky Valley, Esmeralda County, west-central Nevada. The thickness of the type Esmeralda recorded by Turner, 14,800 feet (4512 m) may be excessive because of normal faulting in the Tertiary of this part of the Great Basin (Ferguson et al., 1953). Redesignation of the type area of the Esmeralda as the Alum area north of Silver Peak (Robinson et al., 1968; Albers and Stewart, 1972; Moiola, 1969) greatly restricts the concept of Turner's Esmeralda Formation to what was his upper Esmeralda.

Stratigraphic work by geologists of the then-existing Gulf Oil Company, particularly by Marshall Stam, showed the occurrence of numerous ostracode-bearing beds in the so-called Esmeralda Formation of Stewart and Ione Valleys north of the type area of the Esmeralda. The present writer had the opportunity to study these ostracode faunas through the courtesy of the Gulf Oil Company, Humboldt State University, California and with the support of National Science Foundation grant No. EAR830248, and to subsequently publish on them in association with several colleagues (Swain, Becker, and Dickinson, 1971; Swain and Wagoner, 1979, 1985; Swain, 1985, 1987c).

In the type area of the Esmeralda the following species were identified from the basal layers of the formation (Swain, 1985); Heterocypris blairensis, Cypricercus mineralensis, Huabeinia cf. unispinata, Candoniella sp., Cypricercus sp., and Xiyingia? cf. alta Hou et al. This faunule includes one species of the upper Esmeralda at Blair Junction, 25 miles (40 km) to the north and two species of the middle Esmeralda in Stewart and Ione Valleys about 100 miles

(161 km) north of Silver Peak. Two other species resemble species in the Oligocene of eastern China (Hou et al., 1978, 1988).

About 35 ostracode species have been recorded from the Esmeralda Formation and its equivalents in Esmeralda, Mineral, and Nye Counties, western Nevada. Of these the following are typical of the Esmeralda: *Pactolocypris cancellatus*, *P. pactolensis*, *P. biprojectus*, *Heterocypris blairensis*, *Kassinina microreticulata*, *Advenocypris concinnus*, *Procyprois gracilis*, *Hemicyprinotus ionensis*, *Eucypris fingerrockensis*, *Cypricercus mineralensis*, and *Dongyingia lariversi*. There is poor correspondence between the Esmeralda ostracode faunas of the type area near Silver Peak, Nevada and those of the outcrops farther north in the Blair Junction (Table 2), Stewart and Ione Valley (Table 3) as noted above. This lack of similarity may be the result of poor connections between various lake basins, a difference in age, or a combination of several factors. The Esmeralda lakes were shallow, mainly calcium-rich water bodies during the times when ostracodes lived in them, judged from the limestone and tufa deposits associated with them. Ostracode coquinas are common. The basal Esmeralda in the type area has the ostracodes in a sandy matrix. According to some writers (Nick, Clausen, and Buckheim, 1982) the Esmeralda lake of Stewart Valley, Mineral County began as a fresh-water body that became more alkaline or saline as time progressed. Other studies (Swain, 1985) show that apparent brackish-water ostracode species occur throughout the formation.

TABLE 2 - Distribution of Ostracode Species in the Esmeralda Formation, Blair Junction Section, Nevada (after Swain and Wagoner, 1985)

	1A	1B	1C	$1C_2$	$1C_3$	1D	1F	1H	1I	1J	1K	1M	1N	1O	1P	1Q	1R	2A	2C	Unit 4 4-1	Unit 4 4-2	Unit 4 4-7
Kassinina paucireticulata, S.&W.	x	x	x	x	x	x	x	x	x	x	x	x	x	x	x	x	x	x				
Kassinina microreticulata, S.&W.	xx̲	x	x	x	x	x									x							
Kassinina mesoreticulata, S.&W.		x	x	x	x	x	x	x	x	x	x	x	x	x	x	x						
Heterocypris blairensis, S.&W.		xx̲	xx̲	x	x̲	x	x̲	xx̲	x̲	xx̲	xx̲	xx̲	xx̲	xx̲	xx̲	x̲						
Pactolocypris cancellatus, S.&W.		x	x̲	x		x			x̲	xx̲		xx̲	xx̲	x̲	x̲	x̲						
Dogielinella coaldalensis, S.&W.				x										x̲								
Candona cf. paraohioensis Staplin				x																		
Candona parabretzi, S.&W.	x̲		x̲	x	x̲		x̲			x̲			x̲	x?	?̲	x̲	x̲	x	x̲?̲			?̲
Paracypria? sp.				x																		
Cypridopsella? esmeraldensis, S.&W.	x		x							x												
Bradycypris? sp.										x												
Ussuriocypris? subreniformis, S.&W.								x														
Cypricercus oblongatus, S.&W.	x̲	x	x		x	x̲	x̲	x̲	x̲	x	x̲	x̲	x̲	x̲	x̲		x̲		x̲	?̲		?̲
Herpetocypris stenoformis D.&S.				x																		
Pontoniella? sp.														x								
Pseudocandona? sp.					x									x								
Limnocythere sp.						x̲										x̲	?̲		x̲			
Darwinula? sp.																						x̲
Undifferentiated ostracodes																					x̲	

X-Identified by FMS from collections supplied by JLW
X̲ - Recorded in manuscript by JLW

TABLE 3

Table 3 List of Ostracoda recorded at localities in Stewart and Ione Valleys, Mineral and Nye Counties, Nevada (after Swain, 1985).

Species	Loc 5	Loc 3	Loc 1	Loc 4	Loc 2
Eucypris fingerrockensis, Swain	5	31	6	4	2
Hemicyprinotus ionensis, Swain		134	1		
Hemicyprinotus? sp.		5			
Ussuriocypris subreniformis		2			
Phacocypris cf. pisiformis		2			5
Phacocypris aff. tongbinensis		1		1	3
Scottia aff. subquadrata		1			
Cyprinotus? aff. reniformis				1	
Cypricercus minensis		6			
Cypricercus aff. minensis		2			
Cypricercus mineralensis		205		1	3
Heterocypris blairensis		6	40	7	
Advenocypris? concinnus		30			
Pactolocypris pactolensis		122	2	1	
Pactolocypris biprojectus	4	10			
Pactolocypris suborbicularis		3			
Kassinina microreticulata	3	1			
Huabeinia? aff. huidongensis		1			
Dongyingia? lariversi		3			
Xiyingia aff. elliptica					2
Pontoniella? sp.	5	2			
Procyprois? gracilis	1	24		8	
Procyprois? sp.		1			
Cypridopsella esmeraldensis		6			
Candoniella sp. 1	30	4		1	1
Candoniella sp. 2	1	3		4	
Candona cf. paraohioensis	2	3			
Candona cf. compressa				1	
Candona cf. subtriangularis	4				
Candona sp.	4				
Candonopsis? aff. shahejiensis				1	
Elkocythereis? aff. bramletti	6				
Limnocythere sp.	3				
Cytheromorpha? sp.	2				
Approximate interval in meters above base of lake bed sequence	30-45	244	275	305	425-455

The age of the Esmeralda Formation has been cited as Miocene (11-23 my) by K-Ar age of igneous rocks associated with it (Ekren et al., 1980), but the ostracodes include several species that suggest an Oligocene age (Swain, 1985). The formation may range in age from Oligocene to Miocene.

Miocene Horse Camp Formation, Nevada

The Horse Camp Formation (Moores, 1969; Moores, Scott and Lumsden, 1968) is a succession of conglomerates, fanglomerates, sandstones, and facies-related lacustrine marls, siltstones and volcanic ash up to 10,000 feet (3048 m) in Nye County, central Nevada. The formation overlies volcanic rocks of Oligocene age, 20-33 my.

In samples of lacustrine Horse Camp obtained by E.M. Moores, the following ostracodes were identified (Swain, 1987c): Eucypris fingerrockensis Swain, Stenocypris horsecampensis Swain, Potamocypris moorei Swain, Paracypridopsis? sp., and Darwinula sp. E. fingerrockensis suggests a correlation with the middle part of the Esmeralda Formation, Miocene, of western Nevada.

Miocene Horse Spring Formation, Nevada

The Horse Spring Formation was named by Longwell (1921) for a succession of lacustrine and fluvial deposits up to 3000 feet (915 m) thick in the Muddy Mountains, Clark County, southern Nevada. Ebanks (1965) also applied the name Horse Spring to lacustrine rocks along the south flank of Fossil Ridge, Gass Peak Quadrangle, north of Las Vegas, Nevada. A collection of ostracodes was supplied to the present writer by Ebanks in 1965 and radiometric dates of 15.2 to 15.9 my were later determined on tuffs lying below the ostracode beds (Ebanks, 1986, written communication). In 1986 a measured section 730 feet (223 m) thick of the Horse Spring beds was sampled on the south side of Fossil Ridge by the writer and D.J. Bergstrom (Swain, 1987c). Ostracode-bearing limestone, shale and ash were sampled at intervals of 10 feet (3 m), 210 feet (64 m), 356 feet (109 m), 710 feet (216 m), and 730 feet (223 m) above the base of the exposure. The Ebanks collection is believed to have come from limestones about 710 feet above the base of the exposure.

The following species were recovered at the 710 and 730 feet levels: Kassinina gasspeakensis Swain, Hemicyprinotus sp. aff. H. ionensis Swain, Eucypris cf. microreticulata Swain, and Cypricercus cf. mineralensis. Several of the species suggest a correlation with the Esmeralda Formation of western Nevada, and the Humboldt Formation of northeastern Nevada.

Although part of the Esmeralda and the Humboldt may be as old as Oligocene, the radiometric ages of the Horse Spring appear to confirm a Miocene age for that formation.

Miocene? New Pass Tuff, Nevada

An exposure of laminated white diatomite and volcanic ash mapped as New Pass Tuff sampled 1.05 mile (1.69 km) west of the Lander-Churchill County Line, Nevada on the north side of U.S. Highway 50 at New Pass Summit, near the north end of the Desatoya Range (Swain, 1987c). The nearly horizontal beds are dated at 22 my (AAPG COSUNA Chart, 1985), lower Miocene. Darwinula sp. a form not diagnostic as to age was identified from the locality.

Miocene? Truckee Formation, Nevada

The Truckee Formation was named for fluvial and lacustrine deposits up to 3000 feet (315 m) in western Nevada, south and east of Reno (Thompson, 1956). The type Truckee, along Truckee River is made up largely of diatomite beds (Axelrod, 1956) but freshwater limestones are common in the upper several hundred feet.

Samples of the middle Truckee diatomite in which ostracodes are preserved as internal and external molds were collected by Ira L. La Rivers in 1950, two miles WSW of Hazen, Churchill County, Nevada. The plant fossils of the middle Truckee are dated as Hemphillian (Upper Miocene-Lower Pliocene) by Axelrod (1956), but the middle Truckee is classified as upper Miocene (Clarendonian and early Hemphillian) on the AAPG COSUNA Chart (1985).

The following species were identified from the middle Truckee (Swain, 1990a): Eucypris cf. fingerrockensis Swain, Pactolocypris cancellatus (Swain and Wagoner), P. pactolensis Swain, Hemicyprinotus ionensis Swain, Dogelinella? coaldalensis Swain, and Limnicytheropteron sp. aff. L. winecupensis Swain. These species suggest a correlation with the Esmeralda Formation of western Nevada (Swain and Wagoner, 1985; Swain, 1985) and with the Humboldt Formation of northeastern Nevada (Swain, 1986b). Based on these comparisons the middle Truckee at the locality sampled is assigned a Miocene age.

PLIOCENE

Pliocene Cache Valley Formation, Idaho and Utah

The Cache Valley Formation was named by Peale (1879) for conglomerate, calcareous siltstone, sandstone, shale, and marl up to 1140 feet (348 m), exposed in Cache Valley, southeastern Idaho and adjacent northwestern Utah (Williams, 1948). According to Smith (1953) the Cache Valley Formation forms the upper part of the Salt Lake Group, overlying the gray tuffs, thin conglomerates and basal stromatolitic limestones of the West Spring Formation (Figure 2). The limestone of the West Spring about 1200 feet (366 m) thick was believed by Adamson, Hardy, and Williams (1955) to be part of the Cache Valley Formation.

A small ostracode collection of about 20 species from the Cache Valley Formation was obtained by the late R.C. Bright from the valley wall of Hoopes Creek, NW1/4, SE1/4, NE1/4 sec. 10, T. 12 S., R. 40 E., Franklin County, Idaho (Swain, 1986c). The locality represents an exposure of about 10 feet (3 m) of shell marl and white siltstone dipping gently southeast. The Pliocene gastropod Pliopholyx occurs in the exposure. In the Mink Creek area about 12 miles (19.2 km) southeast of the Hoopes Creek locality, Keller (1952) recognized in the Salt Lake Group an upper Mink Creek conglomerate unit about 3400 feet (1037 m) thick underlain by thin bedded white limestone and soft tuff 2200 feet (671 m) thick, probably representing the Cache Valley Formation. This sequence rests on Cambrian rocks. The Hoopes Creek collection is estimated by Bright to have come from within the tuff and limestone unit but can not be placed more closely.

The following species were obtained from the Hoopes Creek locality (Swain, 1986c): Cypricercus sp. aff. C. reticulatus Zaddach, Herpetocypris franklinensis Swain, Potamocypris cf. comosa Furtos, Cyclocypris sp. aff. C. serena (Koch), Cypria hoopescreekensis Swain, Candona diffusoidea Swain, C. dipunta Swain, C. patzcuaro Tressler, C. kingscreekensis Swain, C. sp. aff. C. elliptica Furtos, C. sp. aff. C. symmetrica Hou et al,. Candoniella elongatolactoides (Dickinson and Swain), Candona cachensis Swain, Fusocandona cachevalleyensis Swain, F. exotica Swain, Darwinula cf. stephensoni (Brady and Robertson), Neolimnocythere? prestonensis Swain, Limnocythere huntingdonensis Dickinson and Swain, Elkocythereis postdorsoalata Swain, and Cyprideis sp.

The dominance of candonid ostracodes, by comparison with other ostracode groups, suggests carbonate-rich, shallow, eutrophic lake habitats. There are several uniquely-shaped candonid species and these, together with limnocytherids and the Cyprideis in the collection suggest elevated levels of dissolved solids. The fauna is interpreted as having lived in a shallow, somewhat alkaline lake in a subtropical to temperate environment. As compared to the Pliocene Glenns Ferry ostracode fauna of south-central Idaho (Swain, 1986a) the present fauna suggests less alkaline conditions than those of the Glenns Ferry assemblage wherein there are even more exotic species than in the Hoopes Creek collection.

Two of the Cache Valley species occur in the Salt Lake Group of Cache and Box Elder Counties, Utah (Swain, 1947). One species occurs in the Hay Ranch Formation, northeastern Nevada (Dickinson and Swain, 1967). This relative lack of common species in other deposits of apparently similar age in the Great Basin has been noted in previous studies. The problem of whether these represent small but significant differences in age as well as in environmental differences is comparable to that of existing ostracode assemblages in modern Minnesota lakes and the writer is inclined to place emphasis on environmental rather than age factors in the problems mentioned above (Swain, 1995).

Pliocene Hay Ranch Formation ("Upper Humboldt" Formation), Nevada

The Humboldt Formation of Sharp (1939) in northeastern Nevada was redefined by Regnier (1960) to be represented by the Hay Ranch Formation (Pliocene and Pleistocene?) and the underlying restricted Humboldt Formation (Miocene). The Hay Ranch rests unconformably, with angular relationships, around the margins of the intermontane basins, but within the deeper parts of the basins the two may be concordant if not conformable. The Hay Ranch is of fluvial and lacustrine origin and includes pyroclastics as well as sills, dikes and lava flows locally.

The Hay Ranch Ostracoda were studied by Dickinson (1959), and Dickinson and Swain (1967) from four localities in northeastern Nevada. In the Gulf Oil Corp. No. 1 Wilkins Ranch well, two miles north of Wells, Nevada, the Hay Ranch beds are in excess of 7223 feet (2202 m) and consist of a lower sequence of calcareous tuff and an upper sequence of buff and cream colored marl and interbedded marly sandstone and conglomerate.

Twenty-one species of non-marine Ostracoda and two species of Cladocera, based on their ephippia, were recorded from the Hay Ranch. One ostracode genus and 19 species, and one genus and two species of Cladocera were newly described.

The species are: Cypricercus stami Dickinson and Swain, C. parafuscatus D. and S., Cypris osmondi D. and S., C. potamides D. and S., Prionocypris sp. aff. P. glacialis (Sars), Herpetocypris stenoformis D. and S., Eucypris ornatoides D. and S., E. punctatula D. and S., Caspiollina microstriata D. and S., Hemicyprinotus krafti D. and S., Heterocypris wilkinsensis D. and S., H. humboldtensis D. and S., Cypria nevadensis D. and S., Candona devexoidea D. and S., Candoniella elongatolactoides (D. and S.), Candona sp. aff. C. candida Müller, C. chasei D. and S., Limnocythere huntingdonensis D. and S., L. pterygoventrata D. and S., Elkocythereis bramletti D. and S., E. minutidonta D. and S., Daphnia humboldtensis D. and S., and Wilkinsia gondoliformis D. and S.

The ostracode succession in the Wilkins Ranch well suggests that four biozonal units can be recognized in the Hay Ranch beds, but other such complete well records or surface sections are not yet available to corroborate the units. They are as follows, in descending order:

Elkocythereis bramletti Unit: extends from 957 to 1194 feet (292 to 364 m) of depth. Elkocythereis bramletti, E. minutidontis and Limnocythere pterygoventrata are abundant and several other species are common in the unit.

Candona chasei Unit: extends from 1194 to 2677 feet (364 to 816 m) in the well. Candona chasei, C. devexoidea, Cypris potamides, Herpetocypris stenoformis, Heterocypris wilkinsensis, H. humboldtensis and Cypria nevadensis are found only in the unit.

Cypricercus parafuscatus Unit: extends from 2677 to 4252 feet (816 to 1296 m). C. parafuscatus, Limnocythere pterygoventrata, Elkocythereis minutidontis, E. bramletti and Wilkinsia gondoliformis are common and Candonas are rare or absent.

Limnocythere pterygoventrata Unit: extends from 4717 to 6860 feet (1438 to 2091 m). The nominal species is almost the only ostracode in the unit.

The Hay Ranch fauna has been recorded at several locations in northeastern Nevada (Dickinson and Swain, 1967; Swain, 1986b).

Pliocene Glenns Ferry Formation, Idaho

The Glenns Ferry Formation of southwestern Idaho (Malde and Powers, 1962) is up to 2000 feet (610 m) thick and consists of fluvial, lacustrine and flood plain sands, gravels, and silts, forming the middle part of the Idaho Group, Upper Pliocene.

A sample of the Glenns Ferry Formation was collected by the late R.C. Bright in 1954 from a road cut ("Big Cut") in sec. 14, T. 4 S., R. 1 W., Owyhee County, Idaho. Anderson (1965) referred the strata at this locality to his Oreana Formation, a unit subsequently referred to a lacustrine facies of the Glenns Ferry Formation (Swirydczuk et al., 1979). The age of the Big Cut occurrence is stated to be late Pliocene on the basis of fission-track dating and vertebrate remains (Smith, 1985). A large bivalve and gastropod fauna is known from the Glenns Ferry in the Oreana Quadrangle, and non-marine ostracodes representing 20 species have been described (Swain, 1986b). The ostracode-bearing beds are suggested to represent shore-face deposits of a large lake (Smith, 1985).

Anderson (1965) reported that non-marine ostracodes occur at several of his studied localities in the Glenns Ferry, from determinations made by the late D.J. Jones. These determinations include about 25 ostracode species representing the genera Metacypris?, Tuberocypris, Tuberocyproides, Paleocypridina, Darwinuloides, Ilyocypris?, Cyprois, Candona, Limnocythere, Darwinula, and Scottia as well as one or more unknown genera. The present writer was not able to verify the presence of all of these genera in the collection he studied (Swain, 1986a).

The following Glenns Ferry species were recorded (Swain, 1986a): Eucypris sp., Cypricercus stami Dickinson and Swain, Herpetocypris andersoni Swain, Paracypria declivoidea Swain, Pontoniella sp., Oreanocypris acutispinosus Swain, O. subtriangularis Swain, O. ascioformis Swain, Tuberocyproides oreanensis Swain, Candona catherinecreekensis Swain, Candoniella elongatolactoides (Dickinson and Swain), Fusocandonopsis owyheensis Swain, Candona sp., Malheurocypris variabilonodosa (Swain), Elkocythereis postdorsoalata Swain, Limnicytheropteron cephalotoidea Swain, L. typica Swain, Limnocythere sp. 1, L. sp. 2, Theriosynoecum? sp., Cytherideidae? sp. indet.

The rather high degree of ornamental complexity of several of the Glenns Ferry species suggests an alkaline or saline lake, by comparison with other non-marine faunas. The levels of dissolved solids are suggested to have been in the hundreds, rather than thousands, ppm. Modern Pyramid Lake, Nevada, with approximately 4700 ppm dissolved solids supports only a small living ostracode population. During earlier stages of Lake Lahonton, of which Pyramid Lake is a remnant, there were abundant ornate Limnocythere and Cyprideis, as well as abundant Candona (Swain, 1977). Presumably, the content of dissolved solids in those early Holocene stages of Lake Lahonton was much less than at present, and when such content reached a limiting level, the ostracodes largely died out.

In a study of Pliocene Ostracoda from the western United States, Forester (1990) recorded several species, all identified with living species, from the Glenns Ferry Formation dated by him at 3.7-3.3 Ma, at Hagerman, Sand Point, Flatiron Butte and Grandview, Idaho. The species at Hagerman include in a lower sample Candona caudata, Cytherissa lacustris and Limnocythere ceriotuberosa; and in an upper sample, Candona rawsoni, C. ohioensis, Limnocythere itasca, L. paraornata, Cypridopsis vidua, Heterocypris incongruens, Cyclocypris serena, Pelocypris sp. and "exotic limnocytherids". The so-called exotic limnocytherids were also recorded higher in the Hagerman section. Judged from his illustrations, several of the species named above were recorded under different names by Swain (1986a, 1987a). At Sand Point, Idaho only exotic candonids and limnocytherids were found. The exotic species were described and illustrated by Swain (1986a), at another locality in Idaho. At Flatiron Butte, Idaho, Cypridopsis okeechobei, Ilyocypris bradyi, Cyclocypris ovum (all modern species) and Heterocypris sp. occur in addition to others recorded above. At Grandview, Limnocythere sappaensis a cited living alkaline-water species occurs with Candona caudata, a freshwater species. Lower annual temperatures than today and higher precipitation in the Lower Glenns Ferry and more or less equal conditions to today are suggested by Forester (1990) to be indicated by the ostracodes. Because of uncertainty as to their identity, the species named by Forester have not been included in the Systematic section. Forester did not refer to Dan Jones' earlier work (in Anderson, 1965).

Miocene and Pliocene Salt Lake Group, Utah and Idaho

The name Salt Lake Group was applied by Hayden (1869) to white and light gray beds cropping out in Morgan and Salt Lake Valleys, north-central Utah. These beds were subsequently named the Norwood Tuff (Eardley, 1944). An early Oligocene age was determined for the Norwood deposits based on titanotheres and an artiodactyl, and formed the lower Salt Lake Group of other authors (Williams, 1948; Adamson et al., 1955; Keller, 1952; Smith, 1953). The Norwood Tuff, 50-2000 feet (15-610 m) thick is overlain unconformably, where present, by the Collinston Conglomerate (2500 feet) (762 m), the West Spring and Cache Valley Formations (up to 7674 feet) (2339 m), and the Mink Creek Conglomerate (up to 3435 feet) (1047 m) thick. The Cache Valley Formation in northern Utah contains ostracodes described elsewhere (Swain, 1947). The formation in southern Idaho also yielded a non-marine ostracode fauna (Swain, 1986c).

The western limit of the Salt Lake Group lies at least as far west as the Goose Creek uranium district (Mapel and Hail, 1959) southernmost Cassia County, Idaho and adjacent parts of Box Elder County, Utah and Elko County, Nevada. In the Goose Creek area the "Salt Lake Group" is about 2250 feet (686 m) thick and overlies strata about 900 feet (274 m) thick referred to the Payette? Formation. The latter consists of shale, ash, sandstone, conglomerate, thin carbonaceous shale and lignite. Mapel and Hail (1959) correlate the Tertiary rocks of the Goose Creek area with the Humboldt Formation of Sharp (1939) of north-central Nevada. The original Payette Formation represents lacustrine and pyroclastic deposits along Payette River, western Boise County, Idaho. The so-called Payette of the Goose Creek area may represent the Humboldt Formation (Miocene) of present day nomenclature and the Salt Lake Group of the Goose Creek area may correlate with the Hay Ranch Formation (Pliocene) (=Upper Humboldt) of northern Nevada (Swain, 1987a).

The ostracodes of the Salt Lake Group occur primarily, if not entirely in the Cache Valley Formation. In addition to previous records (Swain, 1947; 1986c) several other Salt Lake collections in northern Utah and southern Idaho were studied (Swain, 1987a). Most of these are from the Pliocene Cache Valley Formation, but several others can not be dated more precisely than late Miocene or Pliocene, undifferentiated. The species identified in the 1987 study are listed in Table 4. Several localities of problematical Salt Lake or Humboldt or a transition zone between the two formations in Box Elder County, Utah are also listed in Table 4.

TABLE 4

Table 4 Distribution of Ostracoda in Salt Lake Group and Humboldt Formation, western Utah and southeastern Idaho, studied in this paper (after Swain, 1987).

	1	2	3	4	5	6	7	8	9	10	11	12	13	14	Other
Eucypris cf. ornatoides									x		x				c
Cypris potamides												x			c
Herpetocypris cf. franklinensis										x					b
Cypricercus hunterensis						x	x								a
Cypricercus cf. toanoensis													x		a
Cypricercus stami														x	c
Hemicyprinotus oneidensis			x												a
Cypridopsis saltlakensis		x									x				
Cypria sp.									x						
Candona cachensis	x	x		x		x					x	x			
Candona paracaudata	x	x													b
Candona cf. compressa		x									x				e
Candona subtriangularis		x													e
Candona sp.													x		
Candona devexoidea														x	c,a
Tuberocypris acuminatus?		x								x					d
Tuberocypris quadralis		x		x											
Tuberocypris ovalis		x													
Tuberocyproides dipleura		x													
Limnicytheropteron unicornis		x		x											
Limnocythere? williamsi		x													
Limnocythere? dibolba		x													
Metacypris? clarkstonensis	x														
Cyprideis maladensis			x												
Charophytes								x							
Stratigraphic assignment: Salt Lake (SL), Humboldt (H) or Salt Lake, undifferentiated	SL	SL	SL	SL	SL	SL	H	SL	H	H	H	SL	SL	SL	

a, Humboldt Fm. (Miocene), NE Nevada (Swain, 1986); b, Cache Valley Fm. (Pliocene), Franklin Co., Nevada (Swain, 1987); c, Hay Ranch Fm. (Pliocene), NE Nevada (Dickinson and Swain, 1967); d, Glenns Ferry Fm., SW Idaho (Swain, 1986a); e, Esmeralda Fm. (Oligocene?-Miocene), SW Nevada (Swain, 1985).

Twenty-two species of freshwater ostracodes were identified in the 1987 study of which 16 probably represent some part of the Cache Valley Formation. Seven of the Salt Lake species have also been found in the Humboldt Formation of presumed Miocene age in northeastern

Nevada and northwestern Utah (Swain, 1986b). Three of the species also occur in the Esmeralda Formation (Miocene) of western Nevada (Swain and Wagoner, 1985; Swain, 1985). Three other species are also found in the Pliocene Glenns Ferry Formation of southern Idaho or in the Cache Valley Formation of southeastern Idaho (Swain, 1986a, 1987a). Three additional species occur in the Pliocene Hay Ranch Formation of northeastern Nevada (Dickinson and Swain, 1967). It appears that the Salt Lake Group ostracodes range from Miocene to early or middle Pliocene.

Pliocene in Beartrap Canyon Quadrangle, California

A succession of Pliocene and Lower Pleistocene non-marine sediments totalling 18,000 feet (6488 m) occurs in the Beartrap Canyon Quadrangle and the adjacent areas, northern Los Angeles County, California (Eaton, 1939). Eaton recognized the following divisions: First Division (Lower? Pliocene), 5000 feet (1524 m) thick; Second Division (Middle? Pliocene), 7000 feet (2134 m) thick; Third Division (Upper Pliocene?), 6000 feet (1829 m) thick; and Fourth Division (Lower? Pleistocene), about 3000 feet (915 m) thick.

A sample of the First Division, containing non-marine ostracodes was collected by Charles E. Bettinger, in 1949, in sec. 7, T. 7? N., R. 17 W., Los Angeles County, California in the upper 2500 feet of the First Division. The First Division consists of fluvial brown and black siltstone and sandstone, and lacustrine greenish gray sandstone, siltstone, and shale. The ostracode specimens are brown and translucent, generally unlike the chalky-textured or clear calcite preservation of the Neogene ostracodes from the Great Basin studied by the writer (Swain, 1990a).

Only three ostracode species were identified from the lacustrine upper part of the First Division (Swain, 1990a): _Eucypris_ cf. _microreticulata_ Swain, _Bradycypris_? cf. _medicinecreekensis_ Swain and _Candona_ cf. _paracaudata_ Swain. The first two species were recorded in the Humboldt Formation, Miocene, of northeastern Nevada (Swain, 1986b) and the third species from the Cache Valley Formation (Pliocene) of northern Utah (Swain, 1947, 1987b). Based on these limited data the First Division is only questionably Pliocene and could possibly be Miocene.

Pliocene at Localities in California, Arizona and Utah

The lacustrine ostracodes at Pliocene localities in northern California, central and southern Arizona and southern Utah were recorded by Forester (1990). At Alturas, California, Pliocene strata estimated at 4.3 Ma contained the modern species _Limnocythere ceriotuberosa_ and Candona caudata. The area was suggested to be similar to modern conditions of climate at that time. Earlier work on Pliocene ostracodes of California (Chapman, 1896) was not referred to.

At Verde Valley, central Arizona, estimated at 4.5 to 4.0 Ma, Pliocene strata yielded Candona inopinata, C. acuminata, Ilyocypris bradyi, Cypridopsis vidua, Heterocypris incongruens? (all modern species) and Potamocypris sp. in carbonate deposits. The assemblage suggested to the writer colder annual temperatures and greater precipitation than at present. At Ranch, southern Arizona (2.7-2.6 Ma), Candona inopinata?, Limnocythere herricki, Candona cf. renoensis, Limnocythere staplini and Cypridopsis salebrosa (modern species) occur with the so-called Flat Tire vertebrate fauna. The species suggest lower annual temperatures and similar precipitation as at present.

At Beaver Basin, Utah (2.4-2.01 Ma) Limnocythere sappaensis, L. sp. and Candona rawsoni together with a cladoceran ephippium referred to modern Moina (appears to be Wilkinsia gondoliformis Dickinson and Swain, 1967) suggest temperatures and precipitation similar to the present. Some of the ostracodes illustrated but not described by Forester appear to be those of Dickinson and Swain, 1967. The species of uncertain identity recorded by Forester are not included in the Systematic section of the present work.

Pliocene Merced Formation, California

The Merced Formation near Berkeley, California consists of more than 5000 feet (1524 m) of non marine sandstone, shale and impure limestone, as well as equivalent marine strata (Lawson, 1893). Chapman (1896) described nine species and subspecies of non marine Ostracoda from localities south and east of Berkeley in presumably lacustrine deposits, as follows: Cypria subangulata Chapman, Cypris procera Chapman, Cyclocypris californica Chapman, Herpetocypris merriamiana (Chapman), H. lata (Chapman), Cypridopsis pliocaenica Chapman, Candona candida depressa Chapman, C. lactea acuminata Chapman, and C. gracilis Chapman.

The deposits are believed to be Pliocene and Pleistocene?. Most of the nonmarine ostracodes occur in impure limestone or calcareous shale but at one locality a few specimens are in argillaceous sandstone.

Through the courtesy of D.E. Peterson, the writer was able to examine ostracods from the localities recorded by Chapman. The observed specimens were too badly compressed and fragmented to be identified with Chapman's species. The relatively large size of the Merced specimens, however, could be confirmed.

Pliocene Teewinot Formation, Wyoming

Pliocene fresh-water ostracodes from Jackson Hole, Wyoming were recorded by Sohn (1956). The fauna was collected by J.D. Love from the Teewinot Formation, the age of which was based on vertebrate fossils as middle Pliocene. Collections were made at four localities: (1) type section of Teewinot Formation, total thickness 407.4 feet (124.2 m), cut on west side of oiled highway, two miles northeast of junction with U.S. 187, NW1/4, NW1/4 sec. 18, T. 42 N., R. 115 W., Wyoming; collections were made 81.7 to 213 feet (24.9 to 64.9 m) above base; (2) supplementary section, Teewinot Formation, total thickness 172 feet, along an abandoned ditch west of Fish and Wildlife Service hay barn on west line NW1/4 sec. 36, T. 42 N., R. 116 W, Wyoming; collections made 95.5 to 147 feet above base; (3) middle facies Teewinot Formation, North Antelope section, west part of sec. 1 and east part of sec. 2, T. 43 N., R. 115 W., Wyoming, collected 36 feet (11 m) above base; (4) middle facies Teewinot Formation, secs. 21, 22, 28, 29. T. 42 N., R. 115 W., Wyoming; collected 69 feet (21 m) above base.

Sohn (1956) recorded the following 12 species in open nomenclature and one identified form: Ilyocypris? sp., Limnocythere? sp. 1, L.? sp. 2, L.? sp. 3. Lineocypris? sp., "Candona" sp. 1, "C." sp. 2, "C." sp. 3, "C." sp. 4, "C." sp. 5, "C." sp. 6, C. cf. cachensis Swain, and Cyprinotus sp. 1. Candona cachensis occurs in the Salt Lake Group of Utah (Swain, 1947, 1987a) and in the Cache Valley Formation (Pliocene) of Idaho (Swain, 1987a).

Pliocene Laverne and Rexroad Formations, Western Kansas

The nonmarine Pliocene strata of southwestern Kansas comprise the Laverne, Ogallala and Rexroad Formations, lower, middle, and upper Pliocene, respectively (Gutentag and Benson, 1962). The Laverne and Rexroad have yielded ostracodes in Meade County, Kansas.

The Laverne Formation consists of shallow lacustrine gray micaceous sandstone and shale about 80 feet (24 m) thick. Cyprideis littoralis Brady, Candona crogmaniana Turner and Limnocythere staplini? Gutentag and Benson were recorded from the Laverne at one locality. The Cyprideis probably indicates slightly saline or alkaline lake waters.

The Rexroad Formation in Meade County, Kansas consists of up to 80 feet (24 m) of lower sand and gravel, middle silt and carbonaceous beds and upper clay silt and caliche beds. Ostracodes were found in the upper limonitic, calcitic sandstones and are represented by Cyprideis littoralis Brady. A saline or alkaline shallow lake is suggested by the species.

According to Benson (1967) the tropical genus Chlamydotheca occurs in the Pliocene of southwestern Kansas, but further details have not been provided.

Pliocene and Pleistocene? Idaho Group, Oregon

Neogene lacustrine sandstones and siltstones in southeastern Oregon are represented by an older sequence, the Payette Formation, of Miocene age, which is interbedded with Columbia River basalts, and a younger sequence, the Idaho Group (Pliocene and Pleistocene?) interbedded with Snake River basalts (Ross and Forester, 1958). The Idaho Group, from which ostracodes were obtained at two localities in Oregon, is up to several thousand feet thick. It has been subdivided into several formations in nearby southern Idaho, and from one of these, the Glenns Ferry Formation, ostracodes have been described (Swain, 1986a).

A locality on U.S. Highway 20, located 15 miles west of Burns, Harney County, Oregon was sampled by the writer in 1960 at four levels below a lava flow. The following species were identified: *Candoniella* cf. *elongatolactoides* (Dickinson and Swain), *C*. cf. *protzi*, Hartwig, *Herpetocypris* cf. *stenoformis* Dickinson and Swain, and *Limnocythere* cf. *posterolimba* Delorme (Swain, 1990a). The first and third species were recorded from the Hay Ranch Formation, Pliocene, of northern Nevada. The other two species occur in Pleistocene and Holocene deposits, for example, both are found in Holocene raised beach deposits of Lake Lahonton, Nevada (Swain, 1977).

The Idaho Group, undifferentiated, at a locality in Malheur County, Oregon, NE$^1/_4$, SW$^1/_4$, sec. 9 T. 22 S., R. 46 E., was sampled by L.R. Kittleman in 1965. The following species were identified (Swain, 1990a): *Candona kittlemani* Swain, *Fusocandona cachvalleyensis* Swain, *Malheurocypris* sp., and *Neolimnocythere*? *prestonensis* Swain. The collection is believed to be middle or late Pliocene based on the presence of two species of the Cache Valley Formation, middle or late Pliocene of Idaho (Swain, 1947, 1987a).

PLEISTOCENE

Pleistocene in Illinois and Adjacent Areas

The Pleistocene ostracodes of Illinois were recorded in an extensive study by Staplin (1963a, b) from deposits ranging in age from Nebraskan to late Wisconsin Stages (Table 5).

The early Pleistocene *Nebraskan* Glacial Stage ostracodes were obtained from the Sankoty Sand, sec. 12, T. 15 N., R. 9 E., Hennepin Quadrangle, Bureau County, Illinois. The fauna was suggested to represent an assemblage typical of slightly alkaline lake water: *Candona leightoni*, *C*. spp., *Limnocythere varia*?, *L*. *herricki*, *Cyclocypris ovum*, *C*. *forbesi*, *Cypria obesa*, *Cypridopsis vidua*, and *Potamocypris* sp.

Table 5 Pleistocene Section of Illinois and Lake Michigan Basin, Mankato Division Above Calumet I by J. L. Hough (after Staplin, 1963).

Stage	Substage	Moraines	Lake Stages	Lake Level (feet)
	Recent			580
			Algoma	
			Nipissing	605
			Chippewa	230
	Mankato*		Algonquin	605
			Toleston	605
		Valders	Calumet II	620
			Two Creeks (Bowmanville)	580-
		Port Huron	Calumet I	620
			Glenwood	640
		Lake Border	(Glenwood Compressed)	640
		System	Glenwood	640
	Cary*	Tinley		
			Early Lake Chicago	
Wisconsin		Valparaiso		
		Rockdale		
		Minooka		
		Marseilles		
		Chatworth		
		Farm Ridge		
		Cropsey	Lake Illinois deltas	
	Tazewell*	Normal		
		Metamora		
		Bloomington		
		Leroy		
		Shelbyville		
	Iowan		Loess	
	Farmdale		Loess	
Sangamon			Silt	
Illinoian*			Loveland silt, loess	
Yarmouth*				
Kansan*				
Aftonian*				
Nebraskan			Sankoty*	

* Units from which ostracodes were recovered.

The Aftonian? Interglacial Stage Ostracoda were found in silts in sec. 3, T. 6 N., R. 5 E. Glasford Quadrangle, Fulton County, Illinois. The fauna represents that of a cool alkaline lake with little vegetation, according to the author and may be slightly post-Aftonian in age. The fauna consists of Candona wanlessi, C. swaini (=C. rawsoni), C. cf. lactea, Cytherissa lacustris, Limnocythere chippewaensis, and Cyclocypris forbesi.

The Kansan Glacial Stage is represented by several faunules studied by Staplin. In the Arenzville Quadrangle, Cass County, sec. 27, T. 17 N, R. 11 W. in bluish gray calcareous silt, the following were found: Eucypris trapeziformis, Cyclocypris ovum, C. sharpei, C. forbesi, Cypridopsis vidua, Candona truncata, C. simpsoni, C. spp., Limnocythere verrucosa and L. scotti. The assemblage is stated to be typical of lakes with weedy or grassy shores.

Samples of late Kansan or of early Yarmouthian interglacial Stage representing the Sappa Silt (including the Pearlette Ash Bed) from Iowa, Nebraska and Kansas yielded ostracodes (Staplin, 1963). The Sappa Silt is believed to have been a fluvial deposit formed at the close of the Kansan glacial Stage. The faunas are indicative of clear alkaline waters with larger Candonas than in their Illinois counterparts. Ilyocypris in some samples suggests current action during deposition of the containing deposits.

Samples of the Sappa Silt from sec. 11, T. 2 N., R. 20 W., Harlan County, Nebraska yielded: (a) below ash beds, Ilyocypris gibba, I. bradyi, Candona cf. lactea, C. simpsoni, C. swaini (=C. rawsoni), C. sp., C. wanlessi; and (b) above ash beds, C. sappaensis, C. lactea, C. sp., Limnocythere sappaensis, Cypria maculata? Cyclocypris sharpei and Herpetocypris? sp.

Sappa Silt samples from SE$^1/4$ sec. 15, SW$^1/4$ sec., T. 98 N., R. 48 W., Lyon County, Kansas contained Ilyocypris bradyi, Candona sappaensis, C. cf. lactea, C. simpsoni, C.(P.) marengoensis, Limnocythere trapeziformis, L. sappaensis, Cyclocypris forbesi, C. ovum, Cypridopsis vidua and Cyprinotus sp.

In Meade County, Kansas, the Sappa Silt at the Cudahay Packing Co. ash mine, 7 miles north of Meade, Kansas yielded, in the lower 3 feet of clay: Ilyocypris bradyi, Candona (Pseudocandona) marengoensis, C. (Pseudocandona?) punctata, C. thomasi, C. caudata, C. simpsoni, C. illinoiensis, C. sp., Cypricercus tuberculatus, Cypria turneri, Cycylocypris? fragilis, C. laevis?, and Cypridopsis vidua. The overlying silt contained Ilyocypris bradyi, I. gibba, Candona willmani, Candona spp., and Cypridopsis vidua.

In western Kansas, SW$^1/4$ sec. 25, T. 14 S., R. 11 W., the Grand Island Member of the Meade Formation yielded Candona paraohioensis, C. cf. lactea, C. simpsoni, and Cypria maculata?. The overlying Sappa Silt, below the Pearlette Ash, yielded Ilyocypris gibba, Candona (P.) marengoensis, C. acuta, C. spp., Cypria maculata?, Limnocythere sp., Potamocypris smaragdina, Cyclocypris ovum, and Physocypria pustulosa.

Yarmouth Interstage. In Fulton County, Illinois, SE$^1/4$ NW$^1/4$ sec. 8, T. 5 N., R. 4 E., samples of Yarmouth marl, 11 feet (3.6 m) below Illinoian till, contained Candona cf. lactea, C. spp., Limnocythere verrucosa, Cyclocypris forbesi, Physocypria pustulosa, Cypria sp. and Cypridopsis vidua. In Yarmouth silt at that locality there were Candona simpsoni, C. sp., Limnocythere verrucosa and Cyclocypris sp.; many more unidentified candonids were also present.

Yarmouth silt in SW$^1/4$ NE$^1/4$ SW$^1/4$ sec. 29, T. 16 N., R. 12 W., Morgan County, Illinois yielded only Candona truncata and Cyclocypris sharpei. The indicated environment was that of a shallow weedy lake with rich plant or forest border. A sample of Yarmouth loessal? silt at Petersburg, Menard County, Illinois contained Candona thomasi. The indicated environment was that of a small temporary pond.

Illinoian Glacial Stage (Table 6). Silt of this age in SE^1/4 SW^1/4 sec. 5, T. 24 N., R. 4 W., 2.5 miles east of Pekin, Tazewell County, Illinois contained: Candona ohioensis, C. paraohioensis, C. candida, C. simpsoni?, Limnocythere verrucosa, Cyclocypris forbesi, C. ovum, Cypridopsis vidua, and Cyclocypris fragilis.

In Fulton County, Illinois, SE^1/4, NW^1/4, sec. 33, T. 7 N., R. 5 E., Illinoian silt yielded Candona simpsoni, C. sp. Limnocythere verrucosa and Cyclocypris ovum. The indicated environment is that of a weedy, grass-bordered, slightly alkaline lake. Also in Fulton County NW^1/4 SE^1/4 sec., 8, T. 5 N., R. 4 E., laminated Frederick silt and sand contained Candona cf. lactea and C. sp.

In Calhoun County, Illinois, SW^1/4 NW^1/4 SW^1/4 sec. 9, T. 8 S., R. 2 W., gray, calcareous silt lying below leached loess yielded: Candona sigmoides, C. candida, C. spp., C. rawsoni, C. crogmaniana, C. scopulosa, C. bretzi, C. wanlessi, C. cf. lactea, C. simpsoni, C. leightoni, C. eriensis, Cypria obesa, C. turneri, Eucypris? sp., Cyclocypris ovum, Cypridopsis vidua, and Limnocythere reticulata. The indicated environment is that of permanent clear, alkaline lake with muddy bottom and marginal aquatic plant growth. In the NW^1/4 SW^1/4 sec. 22, T. 9 S., R. 2 W., Calhoun County, calcareous Illinoian clay contained Candona pearlensis, C. cf. lactea and Cyclocypris sharpei. The first species also occurs in the Cary? deposits of Story County, Iowa. A cool climate is the suggested environment.

In Johnson County, Illinois, SW^1/4 SE^1/4 NE^1/4 sec. 12, just north of Brussels, the Brussels silt contained Candona illinoiensis, C. candida, C. ohioensis, C. thomasi, C. eriensis?, Cypricercus reticulatus, Cyclocypris sharpei, and Cypridopsis vidua. Indicated environment was that of a flood plain lake.

A sample of Illinoian Loveland? loess at Knoxville, Marion County, Iowa yielded Candona thomasi, C. simpsoni, Limnocythere reticulata, Cyclocypris obesa, C. sharpei, and Cypridopsis vidua. The indicated environment was that of a temporary alkaline pond with moderate vegetation.

Wisconsin Glacial Stage (Table 7). a. Mississippi River Area of northwestern Illinois. These are valley train deposits having probably cold alkaline waters. In Rock Island County, Illinois, SE^1/4 SE^1/4 SW^1/4 sec. 28, T. 17 N., R. 2 W., post-Sangamon-pre-Tazewell silts have Candona cf. lactea.

Tazewell substage silts in SE^1/4 SE^1/4 NE^1/4 sec. 11, T. 14 N., R. 5 W., Mercer County, Illinois contained Candona rawsoni, C. cf. lactea and Limnocythere sp. In NW^1/4 NW^1/4 sec. 21, T. 21 N., R. 3 W., Whiteside County in the Garden Plain Esker, silts contain Candona simpsoni, C. cf. lactea, C. spp., Eucypris sp. and Limnocythere sp.

Mankato Substage silts in SW^1/4 SE^1/4 SE^1/4 sec. 19, T. 19 N., R. 3 E., Whiteside County, Illinois yielded Candona rawsoni and C. candida. Also in Whiteside County, SE^1/4

NE1/4 sec. 4, T. 21 N., R. 7 E., blue gray sandy silt yielded: Ilyocypris bradyi, I. gibba, Candona sp., Limnocythere sp., Cypridopsis vidua, Cyclocypris forbesi and Physocypria pustulosa. The indicated environment was of strong current action.

Mankato silts in NW1/4 sec. 20, T. 28 N., R. 1 E., Jo Daviess County, Illinois contained: Candona cf. lactea, C. rawsoni, C. fluviatilis, C. wanlessi?, Limnocythere varia, L. sp. and Potamocypris minuta?. The indicated environment is one with current action. A nearby location in Galena, Illinois yielded: Candona cf. lactea, C. bretzi?, Limnocythere varia, Ilyocypris gibba and Potamocypris sp. Further, in Jo Daveiss County NW1/4 SE1/4 sec. 24, T. 28, N., R. 1 W. at rear of power plant at south edge of Galena, Mankato silts contained: Candona cf. lactea, C. rawsoni, C. wanlessi, C. caudata, Cypridopsis vidua, Potamocypris minuta, Ilyocypris bradyi, and Limnocythere varia. In sec. 24, T. 27 N., R. 1 E., four miles south of Galena, Illinois across road from Tri-State Zinc Mine, were found: Ilyocypris bradyi and Candona spp. In SW1/4 sec. 22, T. 26 N., R. 2 E. Jo Daveiss County tenacious silts contained: Candona cf. lactea, Cyclocypris sharpei, and Cypridopsis vidua.

b. Illinois River Valley. The Tazewell Substage is well represented by ostracodes in this area. The indicated environment is that of cool alkaline ponds and lakes mainly in Bloomington slackwater silts. In La Salle County, Illinois, center N1/4, SW1/4 sec. 25, T. 16 N., R. 11 E., Lake Illinois deltaic sands, transported specimens of Candona rawsoni, C. cf. lactea, and C. caudata? were found. In Fulton County, Illinois, center east line SW1/4 sec. 4, T. 4 N., R. 3 E., massive pink Bloomington silt has Candona rawsoni, C. cf. lactea, Limnocythere varia, L. reticulata, L. ornata wabashensis, Ilyocypris shafferi (known only from Bloomington silt deposits), Cypricercus tuberculatus, Cyclocypris forbesi, Cypridopsis vidua, and Potamocypris variegata. In another section of Bloomington silts, NE1/4 NE1/4 sec. 8, T. 4 N., R. 3 E., Fulton County, Illinois the cold water ostracodes Candona rawsoni, C. cf. lactea and C. distincta occur.

Samples of Bloomington silts in NW1/4 NE1/4 sec. 11, T. 4 N., R. 3 E., Fulton County, Illinois contained: Candona rawsoni, C. cf. lactea, C. simpsoni, C. eriensis?, C. crogmaniana, C. distincta, C. caudata, C. havanaensis, C. wanlessi, Ilyocypris shafferi, Limnocythere sharpei, L. sp., and Cypridopsis vidua.

In the SW1/4 NW1/4 sec. 24, T. 2 N., R. 1 E., one mile northeast of Browning, Fulton County, Illinois, Bloomington slackwater silts yielded Candona rawsoni, C. cf. lactea, C. simpsoni, C. poseyensis?, C. distincta, C. bretzi?, C. wanlessi, C. thomasi, C. sp., C. acuta, C. shawneensis?, Cypricercus tuberculatus, Limnocythere varia?, Cypridopsis vidua, Cyclocypris forbesi, C. sharpei, Cypria maculata? C. obesa, and Cypricercus? sp.

In Pike County, Illinois, SE1/4 SE1/4 SW1/4 sec. 3, T. 7 S., R. 2 W., Tazewell silts contained Candona thomasi, C. cf. lactea, C. truncata, C. spp., Cyclocypris sharpei, C. ovum, and Cypridopsis vidua. The indicated environment in the absence of Candona rawsoni and the presence of C. truncata suggests warmer or less alkaline water than is typical of this area.

c. Mississippi River Area, southwestern Illinois. Tazewell silt deposits (upper terrace of Shaw and Savage, 1912), SW cor. SE1/4 NE1/4 sec. 4, T. 8 S., R. 2 W., Jackson County yielded Candona simpsoni, C. truncata?, C. minuticandida, C. (Pseudocandona?) marengoensis, C. (P.) albicans, Cypricercus tuberculatus. The indicated environment is that of cool alkaline water. Also in Jackson County, Illinois SE1/4 NW1/4 sec. 4, T. 9 S., R. 1 W., 2.6 miles north of Carbondale on U.S. 51, Tazewell silts contained Candona simpsoni, C. caudata, C. eriensis?, C. acuta? and Cypria maculata. The indicated environment is the same as that of the preceding.

d. Chicago area and Lake Michigan Basin. In NE1/4 NW1/4 NW1/4 sec. 24, T. 2 N., R. 7 E., Kane County, north of Gilberts, Tazewell yellow lake clays contained Candona fluviatilis, Cypridopsis vidua and Cypricercus? sp. In SE1/4 SW1/4 SW1/4 sec. 31, T. 36 N., R. 13 E., on 183rd Street 0.1 mile E. of Harlem Ave., Cary age Lake Tinley clay yielded Candona cf. lactea, C. rawsoni and Limnocythere chippewaensis? At Blue Island, sec. 5, T. 37 N., R. 14 E., Ashland Ave. and west 138th Street, east end of clay pit, Glenwood or Calumet stage of Lake Chicago, Cary or Mankato Substage, sand, four feet above till contained: Candona rawsoni, C. simpsoni, C. caudata, Cypria maculata, Limnocythere verrucosa and Darwinula stevensoni. Silt below this sand contained, in addition to those above: Candona cf. lactea, C. elliptica alta, C. wanlessi, C. acuta, C. truncata, C. eriensis?, Limnocythere varia, L. trapeziformis, L. sharpei, L. pseudoreticulata, Cytherissa lacustris, Ilyocypris bradyi, Cyclocypris? fragilis, C. forbesi, C. laevis, Physocypria pustulosa, Cypridopsis vidua and Cyprinotus glaucus. The indicated environment is a slightly alkaline, cool, quiet, shallow lake with moderate vegetation.

At Calumet Lake, SE1/4 SW1/4 sec. 35, T. 37 N., R. 14 E., Chicago Brickyard in Cary or Mankato sandy silt (Toleston) and clay (Glenwood) were found: Cyclocypris forbesi, Candona rawsoni, C. cf. lactea, C. spp. and Cyprinotus glaucus? At Englewood, SW1/4 SE1/4 SE1/4 sec. 33, T. 39 N., R. 13 E. in Brickyard clay pit, in Toleston or post-Toleston stage of Lake Chicago, were found: Candona cf. lactea, C. simpsoni, C. thomasi, and C. caudata. At Sag Bridge, SE cor. sec. 17, T. 37 N., R. 12 E. in late or post Mankato fossiliferous marl were found: Candona elliptica alta, C. truncata, C. distincta, C. simpsoni, C. (Pseudocandona) punctata, Darwinula stevensoni, Limnocythere verrucosa, Potamocypris smaragdina, Cypria maculata?, Physocypria pustulosa, Cyclocypris ovum, C. laevis?, C. forbesi and Cypridopsis vidua.

In Lake Michigan, cores taken in the deepest part of the northern basin of the lake in Chippewa low water stage contained Candona houghi and Limnocythere chippewaensis.

e. Ohio and Wabash Rivers drainage areas, southeastern Illinois and Indiana. The deposits are Mankato and Cary? slackwater sediments with large and varied ostracode assemblages. In sec. 13, T. 9 S., R. 10 E., Gallatin County, Illinois, 4.9 miles north of Shawneetown, mile 852.7 of Ohio River, Mankato sandy silt contains Candona simpsoni, C. fluviatilis, C. spp., Darwinula stevensoni, Cypricercus reticulatus, Limnocythere ornata wabashensis, Reticulocypris tuberculata, Cypria anterosinuata, C. maculata?, Potamocypris minuta, P. variegata, Cypridopsis vidua, and Physocypria pustulosa. The indicated environment is that of warmer than presently in northern Illinois and less alkaline, partly because candonoids are less prominent than in the latter area.

Also in Gallatin County, Illinois, sec. 26, T. 9 S., R. 9 E., 2.2 miles (3.5 km) west of Shawneetown, in Mankato or late Cary sandy silt there occur: Ilyocypris gibba, I. bradyi, Candona cf. lactea, C. fluviatilis, C. pustulosa, C. shawneensis, C. rawsoni, Limnocythere sharpei and Potamocypris sp. The indicated environment is that of water with slight current.

In Posey County, Indiana, T. 5 S., R. 13 W., just south of railroad track at Solitude, in Mankato or late Cary marly silt, were found: Candona cf. lactea, C. rawsoni, C. simpsoni, C. distincta, C. caudata?, C. truncata, C. acuta, C. thomasi, Cypria obesa, Limnocythere scotti. L. ornata wabashensis, Cypricercus reticulatus, Cyclocypris forbesi, C. sharpei, and Cypridopsis vidua.

Near Mt. Vernon, Indiana, 1.5 to 2 miles (2.4-3.2 km) east on Route 62, in basal 2 foot Mankato to late Cary silt are found: C. (Pseudocandona) punctata, C. poseyensis, Cyclocypris forbesi, and Limnocythere sp. The succeeding 3-foot marly silt contains a rich fauna: Candona ohioensis, C. eriensis?, C. distincta, C. truncata, C. poseyensis, C. sigmoides, C. acuta, C. caudata, C. simpsoni, C. paraohioensis, C. minuticandona, C. willmani, C. crogmaniana, C. (Pseudocandona) albicans, C. (P.) marengoensis, C. (P.) punctata, Paracandona euplectella, Limnocythere verrucosa, L. sharpei, Cypria maculata?, C. obesa, C. mediana, C. anterosinuata, Cyclocypris? fragilis, C. forbesi, C. ovum, Cypricercus reticulatus, Cypridopsis vidua and Physocypria pustulosa. The indicated environment is that of slightly alkaline water, moderate aquatic vegetation and a range in depth of water.

A peat bed, 8 inches (20 cm) thick overlying the preceding marly silt represents development of a marsh containing the following not found in the underlying silt: Candona rawsoni, C. thomasi, Limnocythere ornata wabashensis, Potamocypris smaragdina, Darwinula stevensoni and Cyclocypris laevis? Also near Mt. Vernon, Indiana, SW1/4 sec. 24, T. 7 S., R. 14 W., a Mankato or late Cary silt yielded Candona truncata, C. distincta, C. willmani, C. wabashensis, C. thomasi, C. spp. Cypricercus reticulatus, Limnocythere scotti, Cypria obesa, and Cyclocypris sharpei. The indicated environment is a shallow marshy pond.

In Madison County, Ohio, "Orleton Mastodon Site," two miles west of Plumwood, Ohio, the following were recorded in a 22-foot (6.7 m) thick marl bed: Candona ohioensis, C. paraohioensis, C. candida, C. fluviatilis, C. simpsoni, C. thomasi, Limnocythere verrucosa, L. ornata wabashensis, Potamocypris variegata?, Cyclocypris? fragilis, C. forbesi, C. laevis?, C. ovum and Cypridopsis vidua. An overlying 13-foot peaty "muck" bed had the following additional forms: Candona truncata, and Potamocypris smaragdina. The indicated environment is a small alkaline lake 3 to 10 feet (0.9-3.2 km) deep succeeded by a weed-choked lake passing to a bog.

West of Port Stanley, Ontario along the shore of Lake Erie, a Tazewell silt yielded: Candona candida, C. truncata, C. simpsoni, C. spp., Limnocythere trapeziformis, L. verrucosa, Cyclocypris ovum, C. forbesi, Potamocypris variegata and Cypridopsis vidua. Cary-age silt in the same exposure yielded: Candona ohioensis, C. paraohioensis, C. simpsoni, Cyclocypris forbesi, Limnocythere verrucosa and Cypridopsis vidua.

Candona rawsoni was recorded in Wisconsin-age deposits at the Fairlawn, Ohio mastedon site (Teeter, 1970).

Table 6 a. Ostracoda characteristic of the Illinoian stage in Illinois;
b. Ostracoda restricted to Tazewell and Mankato Substages, Wisconsin stage in Illinois (after Staplin, 1963).

a. Lake deposits:

-*Candona ohioensis*
-*Candona candida*
-*Candona paraohioensis*
-*Candona rawsoni*
-*Candona* cf. *C. lactea*
-*Candona leightoni*
**Candona pearlensis*
**Candona illinoiensis*
**Candona* cf. *C. marengoensis*
Limnocythere verrucosa
Cypria obesa
Cypria turneri
Cypridopsis vidua
Cyclocypris spp.

b. Tazewell:

Candona havanaensis
Illyocypris shafferi
Limnocythere sp. indet. A.

Mankato:

Candona houghi
Candona pustulosa
Candona elliptica alta
Ilyocypris shaweentownensis
Cypria anterosinuata
Reticulocypris tuberculata
Potamocypris minuta

Pond deposits:

-*Candona thomasi*

Candona simpsoni

Cyclocypris sharpei

Cypria obesa

Cypridopsis vidua

*Confined to the Illinoian in Illinois.

-Rare or absent in Recent waters of Illinois.

Table 7 Ostracoda species of Cary - pre-Recent lakes in Lake Michigan - Chicago area, after Staplin, 1963.

Species	Lake Chicago Stages Cary-Mankato (nearshore)	 Algonquin-pre-Recent (basinal)	Pre-Recent (small lake)
Candona rawsoni	abundant	abundant	absent
Candona cf. *C. lactea*	abundant	common	absent
Cytherissa lacustris	occasional	common	absent
Candona acuta	occasional	occasional	----------*
Darwinula aurea	occasional	----------*	occasional
Candona houghi	absent	abundant	absent
Limnocythere chippewaensis	absent	common	absent
Candona crogmaniana	absent?	common	----------*
Limnocythere verrucosa	occasional	absent	occasional
Candona truncata	occasional	absent	common
Cypria spp.	present	absent	present
Cyclocypris spp.	present	absent	present
Physocypria spp.	present	absent	present
Limnocythere varia	present	absent	present
Limnocythere trapeziformis	present	absent	present
Limnocythere pseudoreticulata	present	absent	present
Candona elliptica alta	present	absent	present

* Should be present.

Pleistocene Matayuma-age Sediments, Illinois

A silt deposit about 20 m thick in Hancock County, west-central Illinois is suggested to be Matayuma age, about 730,000 to 830,000 years B.P. based on the remnant magnetic signature and two mammal species (Miller, et al., 1994). A fauna of 117 taxa includes insects, mollusks, ostracodes, mammals and plants. The fauna, magnetic signature and isoleucine epimerization suggest that these are the oldest Pleistocene deposits known in Illinois. The suggested climate is that now found in northeastern Iowa.

Only two genera of ostracodes were recovered: Physocypria spp. and juveniles of Heterocypris sp. The latter suggests occurrence in springs, seeps and bodies of water receiving ground water discharge.

Pleistocene Pre-Illinoian Sediments, Indiana

Sediments of early Pleistocene, pre-Illinoian age, about 40 m thick, near Connersville, Fayette County, southeastern Indiana, yielded a suite of mollusks, ostracodes and fish (Miller et al., 1993). Reversed paleomagnetic orientation and the extent of conversion of L-isoleucine to D-alloisoleucine of the samples suggests an age greater than 730,000 years (the time of the Brunhes-Matayuma paleomagnetic reversal), but probably younger than Pliocene.

The deposits are suggested to have formed in a cool-water lake, 8-16° C, that may have developed in a buried valley system. The deposits may be correlative with the Blackford Member of the Jessup Formation.

Three species of ostracodes were identified from a silt layer in the lower part of the exposure: Cytherissa lacustris, Candona caudata and Cyclocypris ampla. C. lacustris is typical of cool and deep lakes with low dissolved solids; C. caudata occurs in streams and deep water lakes; and C. ampla is found in cold fresh water lakes and in somewhat saline northern prairie lakes.

Pleistocene (Kansan), Kansas and Nebraskan?, Nebraska

Early Pleistocene (Kansan) clays, silts, and gravels at two localities in north-central Kansas were studied for ostracodes and other freshwater fossils by Gutentag and Galli-Oliver (1969). In a quarry at Abilene, Kansas, NW1/4 NE1/4 SW1/4 sec. 26, T. 13 S., R. 2 E., Dickinson County, an 8-foot (2.4 m) unit of gravel and silt, interbedded, yielded: Candona cf. lactea, C. nyensis (=C. rawsoni), C. renoensis, C. simpsoni, C. truncata, Cypricercus cf. reticulatus, and Limnocythere reticulata. An overlying 6-foot (1.8 m) clay unit, underlying the Pearlette Ash equivalent yielded Candona cf. lactea and Ilyocypris gibba. The environment indicated at these localities is that of aggrading streams, shallow lakes with mud substrates and grass and woods along the borders.

At Chapman, Kansas, SE1/4 SE1/4 sec. 28, T. 12 S., R. 4 E. in another quarry, 34 feet (10.4 m) of Kansan age silts, clays and gravels yielded: Candona acuta, C. simpsoni, and Cypridopsis vidua. The environment indicated here is that of a permanent moving, aggrading stream and shallow water lakes bordered by woods, grass and damp ground (Gutentag and Olivier, 1969).

In a road cut exposure along East Turkey Creek, SW1/4 SW1/4 SW1/4 sec. 26, T. 4 S., R. 2 E., six miles (9.6 km) south of Abilene, Kansas, Benson (1967) recorded the following species below the Pearlette Ash: Candona simpsoni, Physocypria pustulosa, Cypricercus tuberculatus, and Cypridopsis vidua. Above the Pearlette Ash the following were recorded: Candona renoensis, Ilyocypris gibba, Physocypria pustulosa, Cypricercus tuberculatus, Cypridopsis vidua, Cyprinotus sp. and an uncertain cyprid.

Candona rawsoni was obtained by Gutentag (1972) from the Sand Draw locality (Nebraskan?) in north-central Nebraska, indicating a cool-climate episode (Stout, 1975).

Pleistocene Bone Bed Site, Benton County, Missouri

A small collection of non-marine Ostracoda was obtained by J.E. King (1973) from a bone bed site at Boney Spring, northwestern Benton County, Missouri. The locality is one of the late Pleistocene peat and clay (± 16,500 years BP) beneath 4.3 m of Holocene.

The collection includes the following species: Cypridopsis? sp., immature, Candona cf. bretzi, C. (Pseudocandona?), n. sp., C. (P.?) cf. marengoensis, Limnocythere cf. varia and Cytheridella? sp. immature.

Candona bretzi was recorded by Staplin (1963) from Illinoian to post-Mankato deposits of Illinois. Limnocythere varia was recorded from Tazewell to Mankato deposits of Illinois.

Pleistocene Wisconsin-age Sediments, Missouri

The distribution of Candona rawsoni and C. patzcuaro in cores of Wisconsin-age sediments of east-central and southeastern Missouri was recorded by Stout (1975). Stout considered the two nominal forms to be end members of a polytypic series and to comprise one species, C. patzcuaro Tressler, a senior subjective synonym. The end members are distinguished on the basis of the degree of angularity of the posterior margin of the shell.

The colder climate "phene" C. rawsoni and the warmer climate "phene" C. patzcuaro are about equally abundant in core material dated at about 18,920 yrs. BP and suggest a combination of cold glacial meltwaters from the north and local warm water drainage into a slackwater lake on the ancient Mississippi River flood plain. Intermediate forms between the two end members also occur, along with abundant forms of the Candona lactea and C. wanlessi types. The latter are suggested by the author to be pre-adult instars of the nominal species.

In other parts of the cores one or another of the two end members predominates, suggesting relative increase or decrease in the cold meltwater elements in the preserved faunas.

This paper presents a philosophical approach to fossil nonmarine ostracode speciation that is provocative but needs further study and documentation.

Pleistocene, Wrenshall Lake Clays, and Lake Itasca Marl, Minnesota

Lake clays of late Pleistocene (Mankato?) glacial Lake Nemadji, near Wrenshall, northeastern Minnesota, yielded two species of ostracodes (Swain, 1995). The clays were once dug for brick manufacture. An exposure of about 10 m of red and gray clay just east of Wrenshall, Carlton County, Minnesota contained Cypricercus reticulatus wrenshallensis Swain and C. sp. The coloration of the clays and the lack of contained organic matter suggest oligotrophic, possibly cold water conditions in the lake.

A late Pleistocene marl at Lake Itasca, Minnesota, yielded, in a Bison skull, Cypria cf. maculata Hoff.

Pleistocene Rita Blanca Lake and Other Lacustrine Deposits, Texas

Three genera of non-marine Ostracoda were recorded in samples from early Pleistocene deposits of Rita Blanca Lake, near Channing, Hartley County, Texas (Benson, 1969). The ostracodes occur in large numbers in sediments of the ancient lake, but are generally delicate, poorly preserved and difficult to separate from the matrix.

The following were recorded: Limnocythere sp., resembling group of L. sanctipatricii Brady and Robertson and L. staplini Gutentag and Benson; Candona sp. resembling C. crogmaniana Hoff; and Cyprideis sp. resembling C. salebrosa ven den Bold. The indicated environment is an alternation of slightly saline waters with Limnocythere in light colored varve layers and fresher waters with Candona in darker colored varve layers (Benson, 1969).

Candona patzcuaro, a tropical to mild climate species was recorded in Texas, in the Pleistocene, where it occurs with C. lactea types of shells (Wise, 1962).

Late Pleistocene Archeological Sites, Texas and New Mexico

A species of Chlamydotheca, C. llanoensis Pierce, was described by Pierce (1988) and recorded at several localities in Texas and New Mexico. The species was described from the Blackwater Draw Formation, Lubbock Lake Archeological Site, Lubbock County, Texas. The fossils occur in Stratum 2 at that site which has been dated at 10,060 ±70 to 10,530 ±90 years before present. The species was also recorded at Blackwater Draw Archaeological Site, Roosevelt County, New Mexico in a fluviatile? facies associated with the Indian cultural level; in a spring deposit associated with skeletal Mammuthus sp. and lithic flakes; and in fluvial sediments dated at 17,000 years BP at pluvial Lake Brownfield, Terry County, Texas. At all of the localities, the associated gastropods are of fluvial types.

According to Pierce, C. llanoensis briefly colonized the southern High Plains during a late Pleistocene warming trend. It disappeared when the climate changed during the early Holocene.

Other ostracode species occurring with C. llanoensis are: Candona nyensis Gutentag and Benson (=C. rawsoni Tressler), Ilyocypris bradyi Sars, and occasionally Candona caudata Kaufmann and Cypridopsis vidua (O.F. Müller).

Pleistocene Lake Bonneville Group, Utah

The Pleistocene lacustrine and fluvial deposits of the Lake Bonneville Basin, Utah were studied for ostracodes by Lister (1975). The sediments were sampled in two core holes near Salt Lake City: (1) Saltair core hole SE1/4 sec. 25, T. 1 N., R. 3 W., Salt Lake County, at elevation of 4205 feet (1282 m) and depth of 200 m; and (2) Section 28 core hole SW1/4 sec. 28, T. 1 N., R. 2 W., elevation 4218 feet (1286 m), depth 236 m.

The Lake Bonneville Group included two formations that represent two major lacustrine episodes: Little Cottonwood Formation and overlying Draper Formation. According to Lister, most of the ostracodes he studied came from unnamed strata older than the Little Cottonwood Formation.

In a discussion of Bonneville Basin lacustrine sediments Oviatt, Sack and Currey (1994) refer to unidentified ostracodes among the fossils in the sediments.

Table 8 Summary of Environmental Preferences of Recent Representatives of Species of Ostracodes Found in Great Salt Lake Basin Sediments. From Various Sources, Including Delorme (1970a, b, c, d; 1971), after Lister (1975).

SPECIES	HABITAT
Candona acutula	Shallow water, lakes, nearctic.
C. camuramarginis	Temporary ponds, nearctic.
C. caudata	Streams and deeper parts of lakes, cosmopolitan.

SPECIES	HABITAT
C. adunca	------
C. pronopa	------
C. rawsoni	Ponds and lakes, tolerant of moderately high amounts of dissolved minerals, nearctic.
Cyclocypris ampla	Wide tolerance, ponds and lakes, nearctic.
C. serena	Wide tolerance, ponds and lakes, nearctic.
Physocypria pustulosa	Lakes and ponds, not found in very cold water, holarctic.
Cyprinotus glaucus	Lakes, not found in very cold water, holarctic.
Eucypris serrata	Lakes and ponds, not found in very cold water, holarctic.
Cypridopsis vidua	Lakes and ponds, in weeds, cosmopolitan.
Potamocypris granulosa	Lakes and ponds, holarctic.
P. pallida	Springs, holarctic.
P. smaragdina	Lakes and ponds, not found in arid regions, holarctic.
P. unicaudata	Lakes and ponds, holarctic.
Ilyocypris biplicata	Streams and lakes, holarctic.
I. gibba	Streams and lakes, holarctic.
Pelocypris tuberculata	Playa lakes, nearctic, possibly neotropical.
Cyprideis salebrosa	Tolerant of saline water, lakes and ponds, neotropical and nearctic.
Limnocythere ceriotuberosa	Lakes, nearctic.
L. friabilis	Lakes, nearctic.
L. illinoiensis	Cool lakes, nearctic.
L. itasca	Cool lakes, nearctic.
L. paraornata	Streams, nearctic.
L. parascutariense	Streams and lakes, nearctic.
L. pseudoreticulata	Lakes and permanent ponds, nearctic.
L. staplini	Lakes, tolerant of moderate to high salinities, nearctic.
Darwinula stevensoni	Large, cool lakes, cosmopolitan.

Thirty-one nonmarine ostracode species were identified by Lister from the lower Lake Bonneville Group. Detailed stratigraphic ranges of the species within the cores was not provided. A list of the recorded species and their inferred environmental preferences is given in Table 8. Several of the species appear to be confined to the Pleistocene based on present knowledge: Candona prenopa Lister, C. camuramarginis Lister, C. adunca Lister, Cypridopsis sp. of Lister, Limnocythere pseudoreticulata Staplin, and L. staplini Gutentag and Benson. The rest of the species occur in living assemblages.

Pleistocene of Lake Cochise Area, Arizona

The Wisconsin and pre-Wisconsin lake clays and sands of Lake Cochise basin, near Willcox, Arizona were studied for Ostracoda by Cameron and Lundin (1977). The lake sediments which are at least 330 feet (100 m) thick, were deposited in Sulphur Springs Valley, a typical structural trough in the southern part of the Great Basin. Other Tertiary and Quaternary alluvial sands, gravels, silts and muds lie below the lake sediments. Radiocarbon dates on the Lake Cochise sediments show that they are late Pleistocene.

The sediments of Lake Cochise occur in two principal facies: a littoral coarse-grained sand-mud-gravel facies that extends about 0.8 km basinward from the marginal beach ridge; and a profundal black to greenish gray clay facies of the central lake basin. In the latter, "abrasion pH" values (made by grinding the sediment in distilled water) ranged from 8.7 to 9.6 and averaged 9.2. Eh values ranged from -106 to -306 mv. The ostracodes recorded from the beach localities and from the central lake locality are listed in Table 9. The beach facies is dominated by Candona simpsoni, C. wanlessi and Cypridopsis vidua. The offshore facies has Limnocythere robusta and L. pterygoventrata as the most abundant.

In addition, Cameron and Lundin (1977) recognized three informal zones in the central lakes core. Zone 1 of Wisconsin age occurs from 6 m to 14.8 m in depth in the core and represents full pluvial conditions. The presence of Limnocythere pterygoventrata and L. ceriotuberosa and the monocotyledon Ruppia represent alkaline and/or saline waters. Ostracodes are abundant in this interval. Zone 2, from 14.8 to 24.7 m was a transitional interval between full pluvial conditions of Zone 1 to interpluvial conditions of Zone 3. There are fewer ostracodes, more barren intervals, the lake was smaller and with more nearshore facies and ostracode species. Moderately saline waters are indicated by the presence of L. ceriotuberosa. Zone 3 from 24.7 to 42 m in the core was an interpluvial interval with shallow water, more nearshore species and poor nutrient conditions. Limnocythere staplini suggests increased salinity during the interval.

Table 9 Abundance of ostracode species of Lake Cochise, Arizona core and beach ridge sediments (after Cameron and Lundin, 1977).

Species	Core		Beach Ridge	Mud	Beach Ridge	Sand
	No. of Specimens	%Fauna	No. of Specimens	% Fauna	No. of Specimens	% Fauna
Cyprinotus glaucus	8	0.0	-	-	32	0.4
Cypridopsis vidua	5	0.0	-	-	997	11.3
Potamocypris granulosa	5	0.0	-	-	349	4.4
Physocypria pustulosa	1	0.0	-	-	170	2.1
Candona devexoidea	-	-	-	-	4	0.1
Candona pearlensis	-	-	-	-	48	0.6
Candona patzcuaro	1	0.0	-	-	290	3.6
Candona simpsoni	-	-	-	-	1702	21.4
Candona truncata	-	-	-	-	28	0.4
Candona wanlesi	8	0.0	-	-	4213	52.8
Limnocythere robusta	4868	55.0	98	27.1	3	0.0
Limnocythere pterygoventrata	3575	40.0	252	72.4	7	0.1
Limnocythere staplini	75	0.8	-	-	-	-
TOTALS	8846	99.2	353	100.0	8070	99.7

Pleistocene Gubik Formation, Alaska

The Gubik Formation of the Arctic coastal plain, northern Alaska consists of late? Pleistocene deposits covering more than 20,000 square miles. The formation ranges in thickness from a feather edge to more than 150 feet (48 m). It consists of both marine and nonmarine facies on the north and south respectively, and includes coarse and fine clastics, peat and ice. The nonmarine facies yielded 12 species of ostracodes (Swain, 1963): Potamocypris pelludicus Swain, Prionocypris glacialis Sars, P. pigra (Fischer), Candona candida (Müller), C. lactea Baird, C. hyalina Brady and Robertson, Pseudocandona ikpikpukensis Swain, Candonopsis cf. kingsleyi (Brady and Robertson), Ilyocypris gibba Ramdohr, I. bradyi Sars, Limnocythere sancti-patricii Brady and Robertson, and Cytherissa simplissima (Swain). The marine facies of the Gubik is represented by 31 species of nearshore Arctic types.

E.M. Brouwers (Cypris Newsletter for Ostracodologists, 1992, p. 30) has indicated that at least part of the Gubik is Pliocene based on ostracodes, but a list of species has not been published.

Pleistocene Lake Lahonton, Nevada

Pleistocene sediments estimated to be about 250,000 years old, lying just above the Miocene Esmeralda Formation, were cored by the U.S. Bureau of Reclamation at Marble Bluff Dam, two miles south of Pyramid Lake, Nevada (Swain, 1977). Two indeterminate species of ostracodes, Candona sp. and Limnocythere sp. were recorded from depths of 12 to 22.7 m in gray silt and sand.

A large Holocene assemblage of ostracodes, no longer living in Pyramid Lake was recorded (Swain, 1977) from raised terraces and shore of the lake. Subsequently, a species tentatively identified as Cyprinotus salinus was submitted by A.J. Cohen to the present writer as an example of a species now living in the lake.

Pleistocene, Owens Lake, California

Lacustrine Ostracoda were identified in Pleistocene sediments up to 800,000 years old of Owens Lake, California (Carter, 1997). The species include Limnocythere ceriotuberosa, L. sappaensis, L. friabilis, L. cf. bradburyi, Candona aff. caudata, Cytherissa lacustris (major species), and L. itasca, L. aff. paraornata, L. platyforma and Cytheromorpha sp. (minor species). Several high-latitude glacial and interglacial episodes occur in the sequence, with glacial terminations at 180 Ka, 225 Ka, 340 Ka and 438 Ka as suggested by the alternations in the distribution of the species and their salinity tolerances. The data are close to but differ in detail from other evidence (i.e., spectral analysis mapping) of previous workers.

Suprageneric Classification Of Fossil Nonmarine Ostracoda In The United States

The classification used here is taken, with modifications, from that of Hartmann and Puri (1974).

Order Leperditicopida Scott, 1961

Family Isochilinidae Swartz, 1949

Order Palaeocopida Henningsmoen, 1953

Suborder Kloedenellinocopina Scott, 1961

Superfamily Kloedenellacea Ulrich and Bassler, 1908
Family Geisinidae Sohn, 1961 **
Order Podocopida Müller, 1894
Suborder Cypridocopina Jones (emend. Kozur, 1972)
Superfamily Cypridacea Baird, 1845
Family Cyprididae Baird, 1845
Subfamily Cypridinae Baird, 1845
Subfamily Dolerocypridinae Triebel, 1961
Subfamily Notodromadinae Kaufmann, 1900
Subfamily Cyproidinae Hartmann, 1963
Subfamily Scottinae Bronstein, 1947
Subfamily Cyprinotinae Bronstein, 1947
Subfamily Eucypridinae Bronstein, 1947
Subfamily Bradycypridinae Hartmann and Puri, 1974
Subfamily Herpetocypridinae Kaufmann, 1900
Subfamily Herpetocyprellinae Bronstein, 1947
Subfamily Disopontocypridinae Mandelstam, 1956 ***
Subfamily Pelocypridinae Triebel, 1962
Subfamily Huabeinidinae Li and Lai, 1978 (see Hou et al.,1988)
Family Candonidae Kaufmann, 1900
Subfamily Candoninae Kaufmann, 1900
Subfamily Thalassocypridinae Hartmann and Puri, 1974
Family Cyclocyprididae Kaufmann, 1900 ****
Family Paracyprididae Sars, 1923 ****
Family Ilyocyprididae Kaufmann, 1900
Family Cyprideidae Martin, 1940
Family Trapezoidellidae Sohn, 1979
Family Pontocyprididae Müller, 1894
Family Cypridopsidae Kaufmann, 1900
Subfamily Cypridopsinae Bronstein, 1947
Subfamily Cyprettinae Hartmann, 1963
Suborder Darwinulicopina Sohn, 1988
Superfamily Darwinulacea Brady and Norman, 1889
Family Darwinulidae Brady and Norman, 1889
Family Carbonitidae Swain, 1972
Superfamily Darwinuloidacea Molostovskaja, 1990

Family Darwinuloididae Molostovskaja, 1979
Superfamily unknown
Family Suchonellidae Mischina, 1972
Suborder Cytherocopina Gründel, 1967
Superfamily Cytheracea Baird, 1850
Family Cytheridae Baird, 1850
Subfamily Cytherinae Baird, 1850
Family Cytherideidae Sars, 1925
Subfamily Cytherideinae, Sars, 1925
Family Neocytherididae Puri, 1957 *
Family Limnocytheridae Klie, 1938
Subfamily Limnocytherinae Klie, 1938
Subfamily Metacypridinae Danielopol, 1965
Family Hutsoniidae Swain and Anderson, 1993
Family Dryelbidae Sohn, 1982
Family Cytherissinellidae Kashevarova, 1958
Order Myodocopida Sars, 1866
Suborder Myodocopina Sars, 1866
Superfamily Cypridinacea Baird, 1850
Family Cypridinidae Baird, 1850 **

* Questionable identification in nonmarine habitat
** Questionable nonmarine habitat
*** Referred to Candonidae-Candoninae by Hartmann and Puri, 1974
**** Subfamily of Candonidae by Hartmann and Puri, 1974

SYSTEMATIC PALEONTOLOGY

The following abbreviations are used in the generic descriptions: RV, right valve; LV, left valve; L, length; H, height.

Several generic and subgeneric reassignments of species have been made herein. The writer has preferred to use new names for Pliocene and older species rather than to expand the concept of extant forms to include characters not recorded in living species. Several mid-Tertiary and older generic assignments are candidates for reconsideration when additional material becomes available, particularly those based on modern genera, for example: <u>Bradycypris</u> Sars, <u>Cypricercus</u> Sars, <u>Candonopsis</u> Vavra, <u>Cyprinotus</u> Brady, <u>Cyclocypris</u>

Brady and Norman, Cypria Zenker, Cypridopsis Brady, Cypris O. F. Müller, Eucypris Vavra, Heterocypris Claus, Herpetocypris Brady and Norman (Sars), Ilyocypris Brady and Norman, Potamocypris Brady and Scottia Brady and Norman. The present report represents only a beginning in the study of this stratigraphically useful group of fossils.

Genus Advenocypris Schneider, 1956

(Cyprididae-Cypridinae?)

?Guangbeinia Huang and Shui, in Hou et al., 1978, p. 69. (see Hou et. al. 1988)

Advenocypris? concinnus Swain

Plate 1, figures 1-3

Potamocypris sp. aff. P. fulva Brady, Swain, 1964b, p. 176, pl. 2, figs. 4-6.

Advenocypris? concinnus Swain, 1985, p. 425, pl. 1, fig. 5; pl. 4, figs. 23, 24; pl. 5, figs. 7, 8.

Subtriangular-reniform, strong mid-dorsal overreach of RV by LV; smooth; L, holotype, 0.97; H, 0.62 mm.

Occurrence. Miocene, Esmeralda Formation, Black Spring Quadrangle, Nye County, Nevada; 2.85 (4.5 km) west of Black Spring (Swain, 1985).

Advenocypris umbonata Swain

Plate 1, figure 4

Potamocypris n. sp. and var. Swain, 1956, p. 136, pl. 3, figs. 11a-d.

Advenocypris umbonata Swain, 1964a, p. 273, pl. 43, figs. 3a-c.

Subtriangular, umbonate LV, strong dorsal and ventral overlap of RV by LV; smooth; L, holotype, 1.00; H, 0.62 mm (Swain, 1964a).

Occurrence. Eocene, Colton-Green River transition beds, 315 feet (96 m) above base, Cypridea bisulcata Zone, east end of Raven Ridge, Rio Blanco County, Colorado (Swain, 1964a).

Genus Amplocypris Zalanyi, 1944

(Cyprididae-Cypridinae?)

Amplocypris? sp.

Plate 1, figure 5

Amplocypris? sp., Swain, 1987c, p. 222, pl. 2, fig. 17.

Subtrapezoidal, highest anteriorly, dorsum convex, venter nearly straight; anterior broadly rounded, posterior narrowly rounded, strongly extended below, subtruncate above; LV larger than RV; compressed; smooth; L, 1.3; H, 0.8 mm (Swain, 1987c).

Occurrence. Eocene or early Oligocene, Elko Oil Shale, Catlin Works, Elko, Nevada (Swain, 1987c).

Genus Bairdiocypris Kegel, 1932

(Bairdiocypridae)

Bairdiocypris morrisonensis Roth, 1933

(see Limnocypridea morrisonensis Sohn).

Bairdiocypris trapezoidalis Roth, 1933

(see Trapezoidella trapezoidalis Sohn).

Genus Bradycypris Sars, 1924

(Cyprididae-Bradycypridinae)

Bradycypris? medicinecreekensis Swain

Plate 1, figures 10-13

Bradycypris? medicinecreekensis Swain, 1986, p. 204, pl. 2, figs. 15, 16, pl. 3, fig. 8, pl. 4, figs. 5, 6. Swain, 1991.

Subtrapezoidal, strongly convex in edge view, finely pitted; LV larger than RV, overlapping along dorsal slopes; hinge a rabbet furrow in LV into which fits edge of RV; L, holotype 1.01; H, 0.63 mm (Swain, 1976).

Occurrence. Miocene, Humboldt Formation, "Bobcat" limestone and conglomerate near base of formation, Medicine Creek section, Winecup Ranch, Elko County, Nevada; Gulf Oil Co. No. 1 Thousand Springs well, 1260-1270 feet (384-393 m). Elko County, Nevada (Swain, 1986). ?Pliocene lake beds, Beartrap Quadrangle, Los Angeles County, California (Swain, 1991).

Bradycypris? sp.

Plate 1, figure 14

Bradycypris? sp., Swain and Wagoner, 1985, p. 135, pl. 5, fig. 5.

Subovate-subtriangular, LV larger than RV, overlapping RV along dorsal slopes; surface a fine bertillion meshwork, thus differing from B?. medicinecreekensis Swain; L, 1.03; H, 0.60 mm (Swain and Wagoner, 1985).

Occurrence. Miocene, Esmeralda Formation, Blair Junction section, Esmeralda County, Nevada (Swain and Wagoner, 1985).

Genus Candona Baird, 1845

(Candonidae-Candoninae)

Candona acuta Hoff

Plate 1, figures 17, 18; plate 2, figures 1, 1a, b

Candona acuta Hoff, 1942, p. 96, pl. 5, figs. 65, 66; pl. 6, figs. 70-72. Staplin, 1963, p. 762, pl. 91, figs. 16, 19.

Female subtrapezoidal-reniform, highest posteriorly; anterior rounded, extended below, posterior truncate below and above, extended below; smooth to faintly roughened; inner lamellae narrow. Male subpyriform-reniform, larger than female, venter strongly convex. L, female, 1.03-1.18; H, 0.54-0.60 mm; L, male 1.28-1.32; H, 0.65-0.69 mm (Staplin, 1963).

Occurrence. Pleistocene, Kansan Stage to Mankato Substage, Wisconsin Stage of Illinois, Kansas; Recent Illinois (Staplin, 1963; Hoff, 1942). (see plate description).

Candona acutula Delorme

Plate 8, figures 25,26

Candona acutula Delorme, 1967, p. 358, Lister, 1975, p. 5, pl. 1, figs. 1-4, TF 2.

Subreniform to subtriangular, highest postmedially; posterior margin in males broader than in females; narrower than anterior in both; posterior margin markedly truncated; smooth; inner lamellae broad anteriorly. L, female 1.1; H, 0.57; L, male, 1.18; H, 0.68 mm. (Lister, 1975).

Occurrence. Pleistocene, Lake Bonneville Group, Saltair core hole, SE$^{1}/_{4}$ sec. 25, T. 1 N., R. 3 W., Salt Lake County, Utah (Lister, 1975).

Candona adunca Lister

Plate 8, figures 27,28

Candona caudata Kaufmann, Benson and McDonald, 1963, p. 12, pl. 2, fig. 11, TF 8, (fide Lister, 1975).

Candona adunca Lister, 1975,p. 10, pl. 1, figs. 20, 21, TF 8.

Elongate sublanceolate, subreniform, highest postmedially; posterior distinctly pointed, strongly extended below; smooth; inner lamellae relatively broad terminally; L, female, 1.47; H, 0.68 mm; males unknown (Lister, 1975).

Occurrence. Pleistocene, Lake Bonneville Group, Saltair core hole SE$^{1}/_{4}$ sec. 25, T. 1 N., R. 3 W.; Sec. 28 core hole, SE$^{1}/_{4}$ sec. 28, T. 1 N., R. 2 W., Salt Lake County, Utah (Lister, 1975).

Candona (Pseudocandona?) albicans (Brady)

Plate 1, figures 15,1 6

Candona albicans Brady, 1864, p. 61, pl. 4, figs. 8-10. Staplin, 1963, p. 763, pl. 93, figs. 21,

22.

Candona parallela Müller, 1900, p. 25, pl. 5, figs. 5, 6, 18, 19, 23-25.

Pseudocandona albicans (Brady), van Frausum and Wouters, 1990, p. 507.

Pseudocandona? albicans (Brady), Swain, 1995, p. 37, pl. 4, figs. 33, 34, 42.

Female subtrapezoidal, highest one-fourth of length from posterior end; dorsum straightened medially; venter slightly concave; anterior narrow, extended below; posterior broadly rounded; surface finely pitted medially, pustulose terminally and ventrally; interior surface weakly pustulose; inner lamellae broad anteriorly, narrow elsewhere; adductor muscle scar a loosely packed group of six spots. L, 0.8; H, 0.45 mm (Swain, 1995). Placed in Pseudocandona because of its shape and surface ornamentation.

Occurrence. Late Pleistocene, Tazewell and Mankato Substages, Illinois and Indiana; Pleistocene, Scotland. Recent, North America, Europe (Staplin, 1963).

Candona artesensis Swain

Plate 2, figures 2, 3

Candona n. sp., Swain, 1956, p. 136, pl. 3, figs. 13a, b.

Candona artesensis Swain, 1964a, p. 274, pl. 43, figs. 4a-e.

Subelliptical-subreniform, highest one-third of length from posterior end, dorsum straightened medially, LV larger than RV; curvature of ends nearly equal; smooth: L, 0.84; H, 0.45 mm (Swain, 1964).

Occurrence. Eocene, Colton-Green River transition beds and lower Green River Formation, Uinta and Piceance Creek Basins, Colorado and Utah, and at Fossil, Lincoln County, Wyoming.

Candona bairdioides (Jones and Kirkby)

Plate 3, figures 36, 37

Carbonita? bairdioides Jones and Kirkby, 1879, p. 38, pl. 3, figs. 24, 25

Candona bairdioides (Jones and Kirkby), Cooper, 1946, p. 65, pl. 8, figs. 3-10

Subelliptical, highest postmedially, dorsum straightened; posterior extended below, truncate above; LV larger than RV, overlapping it except ventrally; smooth; L, 1.28; H, 0.63 mm (after Cooper, 1946)

Occurrence. Pennsylvanian, Virgilian Series, Cohn to Omega Cyclothems, central Illinois (Cooper, 1946).

Candona bretzi Staplin

Plate 1, figure 21

Candona bretzi Staplin, 1963, p. 765, pl. 93, figs. 7,8.

Female elongate, reniform, highest medially, dorsum gently convex, venter concave, anterior narrower than posterior, both ends rounded; valve surface smooth; inner lamellae narrow, radial canals short and numerous; LV overlaps RV on all margins; hinge rabbeted, adductor scar with five or six spots, lying slightly anteromedially; L, 0.93-1.10; H, 0.48-0.55 mm (Staplin, 1963).

Occurrence. Pleistocene, Illinoian to post-Mankato Substage, Wisconsin Stage of Illinois.

Candona cachensis Swain

Plate 1, figures 19, 20, 22a, b, 23, 24.

Candona cachensis Swain, 1947, p. 520, pl. 76, figs. 7-10; Swain, 1987a, p. 122, pl. 1, fig. 19, pl. 2, fig. 5; Swain, 1986c, p. 373, pl. 3, figs. 13a, b; Sohn, 1956, p. 119, pl. 1, figs. 24-26.

Elongate subreniform, highest postmedially, dorsum convex, venter concave, anterior broadly rounded, posterior less broad, both ends extended below; LV larger than RV; surface smooth, internal surface with scattered normal pore pits; LV hinge a furrow to receive edge of RV; inner lamellae broad anteriorally, with deep vestibule, narrow elsewhere; adductor scar an anteromedian group of seven spots; two frontal spots; L, 1.23; H, 0.67 mm (Swain, 1947).

Occurrence. Pliocene, Salt Lake Group and Cache Valley Formation of northern Utah and southern Idaho (Swain 1987a, 1986c); Pliocene, Teewinot Formation; Jackson Hole, Wyoming (Sohn, 1956).

Candona camuramarginis Lister

Plate 8, figures 29, 30

Candona renoensis Gutentag and Benson, Delorme, 1970c, p. 1116, fide Lister, 1975.

Candona camuramarginis Lister, 1975, p. 9, pl. 1, figs. 18, 19, TF 7.

Subquadrate-subreniform, highest postmedially; height more than half of length; posterior much broader than anterior; smooth; anterior inner lamella very broad anteriorly, distinctly sinuous; L, 0.88, H, 0.51 mm (Lister, 1975).

Occurrence. Pleistocene, Lake Bonneville Group, Saltair and Section 28 core holes, SE$^{1}/_{4}$ sec. 25, T. 1 N., R. 3 W., SW$^{1}/_{4}$ sec 28, T. 1 N., R. 2 W., Salt Lake County, Utah (Lister, 1975).

Candona candida (O.F. Müller)

Plate 1, figure 25; Plate 2, figure 4,5

Cypris candida O. F. Müller, 1776, p. 199; 1785, p. 62, pl. 6, figs. 7-9.

Monoculus candidus (Müller), Jurine, 1820, p. 176, pl. 19, figs 7-8.

Candona lucens Baird, 1835, p. 100, pl. 3, fig. 5.

Cypris pellucida Koch, 1837, sp. 5.

Cypris lucida Koch, 1838, sp. 18.

Candona candida (Müller), Baird, 1845, p. 153.

Eucandona candida (Müller), Swain, 1961, p. 235, figs 1 a-g.

Candona candida (Müller), Staplin, 1963, p. 765, pl. 91, figs, 12-14.

Female subreniform-subtriangular, highest postmedially, dorsum strongly convex, slightly truncated medially, venter concave; LV larger than RV; surface smooth, but faintly reticulate under high magnification; hinge rabbet-grooved in LV; normal pore canals arranged in rows on either side of adductor muscle scar, scattered elsewhere; inner lamellae narrow; L, 1.01-1.07; H, 0.59-0.68 mm (Staplin, 1963).

Occurrence. Pleistocene (Illinoian-Wisconsin) of Illinois, Ohio. Recent and Pleistocene of Europe, Asia and North America, including Alaska (Gubik Formation).

Candona sp. aff. C. candida (O.F. Müller) Dickinson and Swain

Plate 1, figure 26

Candona sp. aff. C. candida (O.F. Müller) , Dickinson and Swain, 1967, p. 343, pl. 37, figs. 12, 15.

Elongate-reniform, highest postmedially, dorsum convex, venter concave, ends more or less equally rounded, the posterior a little narrower; smooth?; more elongate than C. candida; L, 1.38; H, 0.73 mm (Dickinson and Swain, 1967).

Occurrence. Pliocene, Hay Ranch Formation (=Upper Humboldt) of northern Nevada; Miocene, Humboldt Formation of northern Nevada (Dickinson and Swain, 1967).

Candona candida depressa Chapman

Plate 7, figure 7

Candona candida var. depressa Chapman, 1896, p. 98, pl. 3, fig. 19.

Elongate, subtrapezoidal-subreniform, highest postmedially; dorsum convex, venter concave; anterior broader than posterior, the latter subacuminate, both ends extended below; compressed, smooth; L, 1.69; H, 0.94 mm (after Chapman, 1896).

Occurrence. Pliocene, Merced Formation, Wildcat Canyon, in hills east of Berkeley, California (Chapman, 1896).

Candona catherinecreekensis Swain

Plate 1, figures 27-30

Candona catherinecreekensis Swain, 1986a, p. 96, pl. 2, figs. 6-11.

Female elongate, subtrapezoidal-subreniform, highest in posterior third; dorsum gently convex, angulated posteriorly, venter gently concave; anterior rounded, extended below, posterior bluntly pointed, strongly extended below, truncate above. Male elongate-reniform, venter concave, with small protuberances anteromedially, ends broadly rounded, extended below. LV larger than RV; smooth except for widely scattered normal pore pits; hinge of LV a rabbet groove to receive RV edge; adductor scar an anteromedian group of five tiny spots and a more anteroventral frontal spot; inner lamellae relatively narrow; L, 1.69; H, 0.88 mm. C. cachensis Swain lacks the anteroventral protuberances of this species (Swain, 1986).

Occurrence. Pliocene, Glenns Ferry Formation, Owyhee County, Idaho.

Candona caudata Kaufmann

Plate 1, figures 31, 32; Plate 2, figure 6; Plate 8, figure 31 - 33

Candona elongata Brady and Norman, 1889, p. 100, pl. 10, figs. 14-27.

Candona acuminata Brady and Norman, 1889 (non-Fischer, 1854), p. 104, pl. 9, figs. 9, 10; pl. 10, figs. 5,6.

Candona caudata Kaufmann, 1900, p. 365, pl. 24, figs. 16-20.

Candona caudata Kaufmann. Gutentag and Benson, 1962, p. 33, pl. 2, fig. 11, TF 8; pl. 26, figs. 17-23. Staplin, 1963, p. 767, pl. 91, figs. 10, 11. Benson, 1967, p. 232, TF 12-5.

Candona novacaudata Benson and McDonald, 1963, p. 15, pl. 2, figs. 1-4.

Female, elongate-reniform-lanceolate, highest postmedially; posterior in adults pointed, hook-like, extended below; smooth; vestibules along free margins; L, 1.10-1.27; H, 0.51-0.60 (Staplin, 1963).

Occurrence. Pleistocene (Illinoian-Wisconsin Stages), Nye Sink deposits, NW1/4 SW1/4 sec. 15, T. 6 N., R. 25 E. CM, Beaver County, Oklahoma; Wisconsin Stage Terrace deposits SE1/4 SW1/4 sec. 8, T. 52 N., R. 36 W., Platte County, Missouri (Gutentag and Benson, 1962). Pleistocene (Wisconsin of Illinois) (Staplin, 1963). Pleistocene, Sappa Formation, Osborne County, Kansas, USNM 649614, loc. U (Benson, 1967). Wisconsin Stage, Illinois (Staplin, 1963). Lake Bonneville Group, Saltair and sec. 28 core holes, SE1/4 sec. 25. T. 1 N., R. 3 W., SW1/4 sec. 28, T. 1 N., R. 2 W., Salt Lake County, Utah (Lister, 1975).

Candona caudata Kaufmann, Benson and McDonald 1963
(see Candona adunca Lister)

Candona chasei Dickinson and Swain
Plate 2, figures 7-9
Candona chasei Dickinson and Swain, 1967, p. 343, pl. 36, figs. 15-17, pl. 37, figs. 1-3.

Subreniform, highest postmedially, longest near venter; posterior much more narrowly rounded than anterior; LV larger than RV; smooth to very weakly punctate; L, 1.22; H, 0.65 mm (Dickinson and Swain, 1967).

Occurrence. Pliocene, Hay Ranch Formation ("Upper Humboldt"), Gulf Oil Co., Wilkins Ranch No. 1 well, Elko County, Nevada, 1194-2468 feet (364-752 m) (Dickinson and Swain, 1967).

Candona clauseni Swain
Plate 2, figures 10, 11
Candona clauseni Swain, 1987c, p. 434, pl. 1, figs. 17-20.

Male elongate, subquadrate-subreniform, highest one-third from posterior end; dorsum nearly straight; posterior more narrowly rounded than anterior; LV larger than RV; smooth; L, 1.10; H, 0.57 mm. Female relatively higher, with more pointed posterior end than males; L, 1.06; H, 0.60 mm (Swain, 1987c).

Occurrence. Miocene, middle Esmeralda Formation, Black Springs measured section, Nye County, Nevada; 125, 360, 380 feet (38, 104, 116 m) above base of section; lower Esmeralda Formation, Cedar Spring measured section, Nye County, Nevada, 80 and 100 feet (24 and 30 m) above base of section (Swain, 1987c).

Candona cf. C. compressa (Koch)
Plate 2, figures 12, 13
Cypris compressa Koch, 1838, sp. 17.
Cypris setigera Jones, 1850, p. 25, pl. 3, figs. 3a-c.
Candona compressa (Koch), Liljeborg, 1853, p. 129, pl. 26, figs. 1-3.
Candona cf. compressa (Koch), Swain, 1947, p. 521, pl. 76, figs. 3-6; Swain, 1985, p. 433, pl. 4, figs. 14, 15; Swain, 1987c.
Candona pubescens Koch, 1837; Vavra, 1891, p. 43, TF 9.;
Candona follax Müller, 1900, p. 27, pl. 4, figs. 2, 11-15, 17.
Candona fossulensis Hoff, Winkler, 1960, p. 925, pl. 122, figs. 10-13.
Candona leightoni Staplin, 1963, p. 776, pl. 94, figs. 20, 21 (fide Delorme 1970).

Ovate-subreniform, highest one-third from posterior end; posterior broader than anterior; LV appreciably larger than RV; smooth; inner lamellae of moderate width; L, 1.27; H, 0.67 mm (Swain, 1947).

Occurrence. Pliocene, Salt Lake Group, sec. 15, T. 13 S., R. 2 W., Cache County, Utah (Swain, 1947). Miocene, Esmeralda Formation, Pactolus Road, loc. 4, Nye County, Nevada (Swain, 1985); Humboldt Formation, loc. 11, Nevada (Swain, 1987c).

Candona crogmaniana Turner

Plate 2, figures 14,15; Plate 8, figure 34

Candona crogmaniana Turner, 1894, p. 20, pl. 8, figs. 24-33.

Candona crogmani Turner, Turner, 1895, p. 300, pl. 71, figs. 24-33, pl. 81, figs. 4, 5.

Candona crogmaniana Turner, Gutentag and Benson, 1962, p. 35, pl. 2, fig. 10; TF 9, Staplin, 1963, p. 768, pl. 91, figs. 15, 17, 18, 20, 21. Lister, 1975.

Female large, elongate-subtriangular, highest postmedially; posterior pointed, truncate above; smooth, vestibules along free margins; L, 1.14-1.72; H, 0.58-0.71 mm (Staplin, 1963).

Occurrence. Pleistocene late Kansan to early Yarmouthian Stages, above Pearlette Ash, SE$^1/_4$ NE$^1/_4$, sec. 1, T. 25 S., R. 7 W., Reno County, Kansas. Lower Pliocene, Laverne Formation, W$^1/_2$ sec. 29, T. 34 S., R. 30 W., Meade County, Kansas (Gutentag and Benson, 1962). Pleistocene, Illinoian Stage-Recent, of Illinois (Staplin, 1963). Lake Bonneville Group, Saltair core hole, SE$^1/_4$ sec. 25, T. 1 N., R. 3 W., Salt Lake County, Utah (Lister, 1975).

Candona devexoidea Dickinson and Swain

Plate 2, figures 16 - 18

Candona devexoidea Dickinson and Swain, 1967, p. 343, pl. 37, figs. 10, 11, 13, 14. Swain, 1986b, p. 212, pl. 2, fig. 13. Swain, 1987c, p. 124. Cameron and Lundin, 1977, p. 349, pl. 1, fig. 22.

Female elongate, subtriangular-subreniform, highest two-thirds from anterior end; dorsum highly arched, truncated posteriorly; posterior acuminate; LV only slightly larger than RV; smooth; inner lamellae broad anteriorly and postventrally; L, 1.17; H, 0.60 mm. Male arcuate elongate-reniform; L, 1.79; H, 0.66 mm (Dickinson and Swain, 1967).

Occurrence. Pliocene, Hay Ranch Formation ("Upper Humboldt"), one mile south of Elko, Nevada on Highway 46; Gulf Oil Co. No. 1 Wilkins Ranch well, Elko County, Nevada, 1194-1268 feet (364-752 m) (Dickinson and Swain, 1967). Miocene, Humboldt Formation, Gulf Oil Co., No. 2 Wilkins Ranch well, Elko County, Nevada, 3610-3640 (1100-1109 m) (Swain, 1986). Pliocene, Salt Lake Group, loc. 14, Utah (Swain, 1987). Late Pleistocene Lake Cochise deposits, Willcox Playa, Cochise County, Arizona (Cameron and Lundin, 1977).

Candona diffusoidea Swain

Plate 2, figures 19-21

Candona diffusoidea Swain, 1987a, p. 369, pl. 1, figs. 13-15.

Elongate, sublanceolate-subreniform, highest postmedially; posterior narrower than anterior; LV larger than RV; valves compressed, smooth; adductor scar with four spots and two frontal spots; L, 1.15; H, 0.53 mm (Swain, 1987a).

Occurrence. Pliocene, Cache Valley Formation, Hoopes Creek loc., Franklin County, Idaho (Swain, 1987a).

Candona dipunta Swain

Plate 2, figures 22, 23

Candona dipunta Swain, 1987a, p. 370, pl. 2, figs. 7, 8.

Elongate-reniform, highest postmedially; dorsum straightened medially; venter concave with slight projection posteriorly; posterior subtruncate above and with angular flange-like extension below; smooth; inner lamellae broad anteriorly; adductor group with four or five spots; L, 1.03; H, 0.60 mm (Swain, 1987a).

Occurrence. Pliocene, Cache Valley Formation, Hoopes Creek loc., Franklin County, Idaho (Swain, 1987a).

Candona distincta Furtos

Plate 2, figures 24-27

Candona distincta Furtos, 1933, p. 478, pl. 8, fig. 15; pl. 9, figs. 13-16; pl. 12, figs. 21, 22.

?Candona distincta Furtos, Staplin, 1963,p. 769, pl. 93, figs. 5-6; Benson, 1967, p. 222, TF 6-1-4, p. 229, TF 10-5.

Subtriangular-reniform (female) to elongate subreniform (male), highest postmedially, compressed, posterior equal to or narrower than anterior; smooth to weakly reticulate; vestibule of moderate width along free margins; adductor scars faint; L, female, 0.9-1.00; H, 0.46-0.54; L, male, 0.95-1.02; H, 0.53-0.57 mm (Staplin, 1963).

Occurrence. Pleistocene, late Cary or Mankato Substage and Tazewell Substage of Illinois (Staplin, 1963). Pleistocene, Sappa Formation, Osborne County, Kansas (Benson, 1967). ?Lower Oligocene, Chadron Formation, Helmville, Nebraska (coll. by R. Konizeski, unpublished). Recent of Ohio (Furtos, 1933).

Candona elliptica alta Staplin

Plate 3, figures 1, 2

Candona elliptica alta Staplin, 1963, p. 770, pl. 94, figs. 1, 2.

Thin-shelled, elongate, elliptical-subreniform; male higher than female; highest postmedially, ends subequal; smooth, inner lamellae wide anteriorly; L, female, 0.83-0.89; H, 0.39-0.40 mm (Staplin, 1963).

Occurrence. Pleistocene, Mankato Substage, Wisconsin Stage, Chicago area, Illinois (Staplin, 1963).

Candona sp. aff. C. elliptica Furtos

Plate 2, figures 28a, b, 29

Candona sp. aff. C. elliptica Furtos, 1933, Swain, 1987a, p. 372, pl. 3, figs. 11, 12a, b; pl. 5, fig. 4.

Elongate, subelliptical-subquadrate, highest three-fifths from anterior end; posterior more narrowly rounded than anterior; LV larger than RV, valves compressed; smooth; inner lamellae of moderate width; adductor scar a compact group of six spots; two frontal spots (Swain, 1987a).

Occurrence. Pliocene, Cache Valley Formation, Hoopes Creek locality, NW$^{1}/_{4}$ SE$^{1}/_{4}$ NE$^{1}/_{4}$ sec. 10, T. 12 S., R. 40 E., Franklin County, Idaho (Swain, 1987a).

Candona eriensis Furtos

Plate 3, figure 12

Candona eriensis Furtos, 1933, p. 483, pl. 8, figs. 11-14; pl. 12, figs. 8-10.

Candona eriensis? Furtos, Staplin, 1963, p. 771, pl. 93, figs. 17, 20.

Elongate, subtrapezoidal-subreniform, highest postmedially, truncate postdorsal slope; posterior narrower than anterior; smooth; inner lamellae of moderate width, narrower ventrally; adductor group may have sixth spot; L, female, 1.10 to 1.25; H, 0.50 to 0.61; L, male, 1.12 to 1.32; H, 0.55 to 0.67 mm (Staplin, 1963).

Occurrence. Pleistocene, Illinoian Stage, Mankato Substage, Wisconsin Stage of Illinois (Staplin, 1963). Recent of Ohio (Furtos, 1933).

Candona fluviatilis Hoff

Plate 3, figure 13

Candona fluviatilis Hoff, 1942, p. 60, pl. 1, figs. 7-9; pl. 2, figs. 10-11; Gutentag and Benson, 1962, p. 36, pl. 1, fig. 12; Staplin, 1963, p. 772, pl. 94, fig. 7.

Trapezoidal, highest medially, dorsum straightened; anterior broader than posterior, the latter truncate above; reticulate surface with pentagonal interconnecting narrow ridges; LV overlaps RV greatest along mid-venter; inner lamellae narrow; L, female, 0.68 to 0.90; H, 0.33 to 0.44 mm, male unknown (Staplin, 1963).

Occurrence. Pleistocene, late Kansan Stage or early Yarmouthian Stage, above Pearlette Ash, $SE^1/_4$ $NE^1/_4$ sec. 1, T., 25 S., R. 7 W., Reno County, Kansas; Wisconsin Stage terrace deposits, $SE^1/_4$ $SW^1/_4$ sec. 8, T. 52 N., R. 36 W., Platte County, Missouri (Gutentag and Benson, 1962). Pleistocene, Tazewell Substage, Wisconsin Stage to Recent of Illinois (Staplin, 1963). Recent of Illinois (Hoff, 1942).

Candona? gracilis Chapman

Plate 1, figures 8, 9

Candona gracilis Chapman, 1896, p. 99, pl. 3, figs. 23-25.

Very elongate-subelliptical, highest medially; dorsum weakly convex, venter concave, widest postmedially; smooth; ends nearly equally rounded, slightly extended below; L, 0.71; H, 0.22 mm (after Chapman, 1896).

Occurrence. Pliocene, Merced Formation, Wildcat Canyon in hills east of Berkeley, California (Chapman, 1896).

Candona havanaensis Staplin

Plate 3, figures 14, 15

Candona havanaensis Staplin, 1963, p. 773, pl. 93, figs. 10, 11.

Elongate, subelliptical-subreniform, highest medially; posterior narrower than anterior; smooth; LV overlaps RV on free margins; inner lamellae broadest anteriorly; L, female 0.80 to 1.00; H, 0.45 to 0.50; L, male 1.00 ; H, 0.50 (Staplin, 1963).

Occurrence. Pleistocene, Tazewell Substage, Wisconsin Stage, Illinois River Valley, Illinois (Staplin, 1963). ?Oligocene, Chadron Formation, Helmville, Nebraska (R. Konizeski, coll.).

Candona houghi Staplin

Plate 3, figures 16-18

Candona houghi Staplin, 1963, p. 773, p. 92, figs. 13-20.

Subtriangular-subtrapezoidal-subreniform, highest postmedially, truncate postdorsally; smooth to finely reticulate at high magnification; slight LV overlap of RV; inner lamellae broadest anteriorly; L, female 1.18 to 1.30; H, 0.61 to 0.69 mm; males not known (Staplin, 1963).

Occurrence. Pleistocene, late Mankato Substage, Wisconsin Stage to pre-Recent, Lake Michigan cores (Staplin, 1963).

Candona hyalina Brady and Robertson

Plate 3, figures 19-21

Cypris acuminata Zenker, 1854, p. 74 (part), pl. 2, fig. D3.

Candona hyalina Brady and Robertson, 1870, p. 18, pl. 19, figs. 5-8; pl. 5, figs. 4-11. Swain, 1963, p. 804, pl. 96, figs. 1a-c, TF 3a.

Subreniform-subtriangular, highest postmedially; dorsum with steep posterior slope; posterior subacuminate; valves moderately convex; smooth; inner lamellae broad anteriorly; about 30 radial canals; L, 1.1; H, 0.58 mm (Swain, 1963).

Occurrence. Pleistocene, Gubik Formation, non-marine facies, Arctic Coastal Plain, northern Alaska, Ikpikpuk Lake and Teshekpuk Lake Shot Holes from surface to 80 feet (24 m) (Swain, 1963). Recent, western Europe (Brady and Robertson, 1870).

Candona (Pseudocandona) ikpikpukensis Swain

Plate 7, figures 1 - 4

Pseudocandona ikpikpukensis Swain, 1963, p. 805, pl. 95, figs. 5a, b; pl. 96, figs. 10a, b, TF 3b.

Candona ikpikpukensis (Swain), Delorme, 1968, p. 868, pl. 4, figs. 52, 53; 1970, p. 1110, TF 163, 164.

Elongate reniform-trapezoidal, highest medially to postmedially; dorsum gently convex, straightened medially, venter concave; anterior broader than posterior, the latter subacuminate, extended below; LV slightly large than RV; compressed; surface with closely spaced pits; posteriorly there are six or more weak, widely spread narrow ridges that extend postventrally from large smooth anteromedian spot; a few faint ridges in anterior half; LV hinge a groove into which fits edge of RV; L, holotype, 1.11; H, 0.58 mm (Swain, 1963).

Occurrence. Pleistocene, Gubik Formation, non-marine facies, Arctic coastal plain, Alaska, Ikpikpuk Lake shot holes, line 6, shot point 8, depth 30-40 feet (9-12 m) and other Ikpikpuk Lake shot holes, averaging 30 feet (9 m) below surface (Swain, 1963).

Candona illinoiensis Staplin

Plate 3, figures 22, 23

Candona illinoiensis Staplin, 1963, p. 775, pl. 94, figs. 17-19.

Subquadrate, highest about one-third of length from posterior end, dorsum straightened, sinuous; unequal ended, posterior the broader; smooth; inner lamellae narrow; L, 0.85; H, 0.47 mm (Staplin, 1963).

Occurrence. Pleistocene, Illinoian Stage, Brussels Formation, of Illinois; late Kansan Stage, Sappa Silt, of Meade County, Kansas (Staplin, 1963).

Candona? kentuckyensis Scott and Summerson

Plate 3, figure 24

Candona kentuckyensis Scott and Summerson, 1943, p. 672, pl. 1, fig. 18.

Subrectangular, highest postmedially, dorsum and venter long, straight, subparallel; LV larger than RV; pitted? or partly smooth?; L, 1.07; H, 0.47 mm (Scott and Summerson, 1943), (atypical of Candona).

Occurrence. Lower Pennsylvanian, Kanawha Series Formation, along US. highway 119, just north of Molus, Harlan County, Kentucky (Scott and Summerson, 1943).

Candona kingscreekensis Swain

Plate 3, figures 25a, b, 26

Candona kingscreekensis Swain, 1986c, p. 371, pl. 2, figs. 12, 13, 15; pl. 3, figs. 7a, b.

Males subtriangular-subreniform, females elongate-subtriangular, highest postmedially, posterior pointed, more so in males than in females; LV considerably larger than RV, postdorsal slope truncated; smooth; inner lamellae broad anteriorly; adductor scar with five spots; two frontal spots; L, male, 1.43; H, 0.85 mm (Swain, 1986c).

Occurrence. Pliocene, Cache Valley Formation, Hoopes Creek loc., NW$^1/_4$ SE$^1/_4$ NE$^1/_4$ sec. 10, T. 12 S., R. 40 E.,Franklin County, Idaho (Swain, 1986c).

Candona kittlemani Swain, 1990

Plate 3, figures 27, 28

Candona kittlemani Swain, 1990a, p. 151, pl. 1, figs. 23, 28a, b.

Subtrapezoidal-subreniform, highest postmedially, dorsum straightened; posterior subacuminate below, truncate above; LV larger than RV; finely venose ornamentation, partly accentuated by filamentous algae? in type material; inner lamellae broad anteriorly; L, 0.97; H, 0.57 mm (Swain, 1990).

Occurrence. Pliocene, Idaho Group, sec. 9, T. 22 S., R. 46 E., Malheur County, Idaho (Swain, 1990).

Candona cf. lactea Baird

Plate 3, figure 29 - 31

Candona lactea Baird, 1850, p. 255, pl. 18, figs. 25-27. Brady, 1868, p. 382, pl. 24, figs. 55-58. Brady, Crosskey and Robertson, 1874, p. 134, pl. 1, figs. 14-16. Müller, 1912, p. 147.

Candona cf. lactea Baird, Staplin, 1963, p. 775, pl. 91, fig. 1 Benson, 1967, p. 232, TF 12-2. Swain, 1963, p. 804, pl. 95, fig. 2, pl. 96, figs. 2a, b, 5, 6a-c, TF 2d.

Candoniella lactea (Baird), Sun and Feng, 1997, Cypris, V.15, p. 82.

Subelliptical-subreniform, highest anteromedially, ends subequal, the posterior slightly the narrower; smooth; inner lamellae narrow; elliptical in dorsal view, more so than in most Candona; L, 0.48-0.79, H, 0.28-0.42 mm (Staplin, 1963). (May be confused with instars of other Candonas). Note that some Chinese workers place the specimen in Candoniella.

Occurrence. Pleistocene, Aftonian? Stage, Fulton County, Illinois; Illinoian Stage, Tazewell County, Illinois; Wisconsin Stage, Mankato Substage, Jo Davies County, Illinois; Tazewell Substage, La Salle County, Illinois, Fulton County, Illinois, Mercer County, Illinois; Cary Substage, Lake Tinley Clay, Chicago, Illinois; Cary or Mankato Substage, Calumet Lake, Chicago, Illinois, Gallatin County, Illinois, Posey County, Illinois; Cary or Mankato Substage, Story County, Iowa; Late Kansan, early Yarmouthian Substage, Sappa Silt, Harlan County, Nebraska; Lyon County, Iowa; late Kansan Meade Formation, Grand Island Member, sec. 25, T. 14 S., R. 11 W., Kansas; Mankato or late Cary Substage, western Indiana (Staplin, 1963). Pleistocene, Gubik Formation, Arctic coastal plain, northern Alaska, occurs in 21 of Ikpikpuk Lake shot holes, in Party 47 shot holes and in Teshekpuk shot holes, averaging 40 feet below ground surface in upper-most occurrences (Swain, 1963). Recent of Great Britain, Ireland, Netherlands.

Candona lactea acuminata Chapman

Plate 7, figures 8, 9

Candona lactea var acuminata Chapman, 1896, p. 99, pl. 3, figs. 20-22.

Elongate subelliptical-subreniform, highest medially, anterior broader than posterior and as seen in edge view, acuminate; dorsum gently convex, venter concave; smooth; L, 1.02; H, 0.60 mm (after Chapman, 1896).

Occurrence. Pliocene, Merced Formation, Wildcat Canyon, San Pablo Ridge and Bollinger Canyon, in hills east of Berkeley, California (Chapman, 1896).

Candona leightoni Staplin

Plate 3, figures 32, 33

Candona leightoni Staplin, 1963, p. 776, pl. 94, figs. 20, 21.

Candona compressa (Koch), Delorme, 1970, p. 1104.

Subquadrate, highest about one-fourth of length from posterior end; dorsum straightened; smooth to faintly reticulate at high magnification; LV overlaps RV on all margins;

inner lamellae broad anteriorly; L, female, 0.9 to 0.97; H, 0.48 to 0.53 mm; L, male, 0.94 to 1.00; H, 0.52 to 0.56 mm (Staplin, 1963).

Occurrence. Pleistocene, Nebraskan (Sankoty Sand) and Illinoian Stages of Illinois; Wisconsin Stage, Cary? Substage of Story County, Iowa (Staplin, 1963). Recent, Canadian Prairie lakes (Delorme, 1970).

Candona ? magnitata (Holland), Scott
Plate 3, figures 34, 35
Plate 4, figures 1, 2
Whipplella magnitata Holland, 1934, p. 346, pl. 25, figs. 2a-c.
Candona magnitata (Holland), Scott, 1944, p. 145, pl. 23, figs. 9-13.

Elongate-ovoid, highest postmedially?; dorsum more convex than venter; posterior end more pointed than anterior; RV overlaps LV midventrally; narrow inner flanges at ends; adductor scar circular, anteromedian; L, 1.04; H, 0.56 mm (Scott, 1944) (Atypical of Candona because of convex venter).

Occurrence. Permian, Nineveh Formation, Pennsylvania and West Virginia (Holland, 1934; Scott, 1944).

Candona (Pseudocandona?) marengoensis (Klie)
Plate 4, figure 3, 4
Candona marengoensis Klie, 1931, p. 337, 341, TF 13-16. Staplin, 1963, p. 777, pl. 94, figs. 10, 16. Benson, 1967, p. 235, TF 13-2.
Pseudocandona marengoensis (Klie), Swain, 1995, p. 37, pl. 5, figs. 1-5.

Small, thin, subquadrate, highest medially, straightened dorsum, slightly concave venter; posterior broader than anterior, truncate above; surface finely reticulate, with polygonal pattern; valves subequal, left the larger; inner lamellae narrow (Staplin, 1963). Quadrate outline and reticulate surface ally it with Pseudocandona. Adductor scar may have sixth spot (Swain, 1995); L, 0.39-0.84; H, 0.22 to 0.48 mm (Staplin, 1963).

Occurrence. Pleistocene, Illinoian Stage to Wisconsin Stage, Mankato or Cary Substage of Illinois and western Indiana; late Kansan Stage of Iowa and Kansas (Staplin, 9163). Sappa Formation, Osborne County, Kansas (Benson, 1967). Recent and Subrecent of Europe and U.S.A (Klie, 1931, Swain, 1995).

Candona (Pseudocandona?) sp. aff C. (P.) marengoensis Klie
Plate 7, figure 12

Elongate subreniform, highest slightly anteromedially; dorsum convex, venter concave; anterior slightly narrower than posterior; coarsely reticulate except anterior third which is nearly smooth in specimen at hand; L, 0.50; H, 0.31 mm; specimen coll. by J. E. King.

Occurrence. Pleistocene, bone bed at Boney Spring, Benton County, Missouri, coll. by J. E. King.

Candona minuticandida Staplin

Plate 4, figure 5

Candona minuticandida Staplin, 1963, p. 778, pl. 93, fig. 9.

Small, sturdy, subquadrate, highest postmedially; dorsum straightened, venter weakly sinuate; smooth; anterior narrower than posterior; inner lamellae broad anteriorly; L, 0.46 to 0.58; H, 0.26 to 0.33 mm (Staplin, 1963).

Occurrence. Pleistocene, Tazewell Stage and Mankato or late Cary Substage of Wisconsin Stage of Illinois (Staplin, 1963).

Candona ohioensis Furtos

Plate 4, figures 10-12

Candona ohioensis Furtos, 1933, p. 475, pl. 9, figs. 19, 20; pl. 10, figs. 8-12. Staplin, 1963, p. 778, pl. 93, figs. 23-25.

Large, elongate subreniform (males) and sublanceolate-reniform (females); among the largest of American Candona, highest postmedially; posterior bluntly pointed and truncate above in females, rounded in males; venter concave; smooth; inner lamellae broad anteriorly and postventrally; L, female 1.67 to 1.93; H, 0.75 to 0.90; L, males 1.70 to 1.83; H, 0.80 to 0.90 mm (Staplin, 1963).

Occurrence. Pleistocene, Illinoian Stage of Illinois; late Wisconsin Stage of Ontario and Ohio; Cary? Substage of Wisconsin Stage of Iowa; late Cary or Mankato Substage of Wisconsin Stage of western Indiana (Staplin, 1963). ?Lower Oligocene, Chadron Formation, Helmville, Nebraska, coll. by R. Konizeski (unpublished). Recent and Subrecent of northern United States (Furtos, 1933; Smith, 1991; Swain, 1995).

Candona pagei Swain, 1949

(see Pseudoeucypris pagei (Swain))

Candona? paleostraca Tasch

Plate 4, figures 13-15

Candona paleostraca Tasch, 1963,p. 1249, pl. 174, figs. 9-12.

Subovate-subreniform "lozenge-shaped", highest postmedially; anterior narrower than posterior; RV slightly overlaps LV along midventer; surface with closely set pores; adductor scar anteromedian, circular; L, 0.85; H, 0.59 mm (Tasch, 1963). Generic position uncertain.

Occurrence. Permian, Wellington Formation, Elmo VIII, bed 10, NE$^1/_4$ NW$^1/_4$ sec. 21, T. 16 S., R. 2 E., Dickinson County, Kansas (Tasch, 1963).

Candona parabretzi Swain and Wagoner

Plate 4, figures 16-18

Candona parabretzi Swain and Wagoner, 1985, p. 138, pl. 4, figs. 7, 8, 10.

Elongate subreniform, highest two-fifths from posterior end; presumed males flattened dorsally and ventrally and have more pointed posterior end than females; LV only a little larger than RV; shell more strongly convex and with blunter ends in edge view than typical of genus; surface with faint reticulate, diamond-shaped pattern, formed by rows of transverse furrows, seen at high magnification (X600); inner lamellae narrow; L, female 1.30; H, 0.73 mm (Swain and Wagoner, 1985).

Occurrence. Miocene, Esmeralda Formation, Blair Junction section, Esmeralda County, Nevada (Swain and Wagoner, 1985).

Candona paracaudata Swain

Plate 4, figures 19-25

Candona paracaudata Swain, 1947, p. 521, pl. 26, figs. 11-13. Swain, 1987a, p. 122, pl. 1, figs. 16-18, pl. 2, figs. 1, 2. Swain, 1990, p.150 , pl. 1, figs. 1-5.

Elongate-subovate to subreniform, highest in posterior third; dorsum straightened medially; posterior subacuminate ventrally; LV markedly larger than RV; smooth; inner lamellae broad anteriorly; L, 1.18; H, 0.67 mm (Swain, 1947).

Occurrence. Pliocene, Salt Lake Group, sec. 15, T. 13 N., R. 2 W., Cache County, Utah. Salt Lake Group, lc. 1, Utah (Swain, 1987a). Pliocene, Beartrap Quadrangle, Los Angeles County, California (Swain, 1990).

Candona paraohioensis Staplin

Plate 4, figures 26, 29

Candona paraohioensis Staplin, 1963, p. 779, pl. 92, figs. 10-12. Swain and Wagoner, 1985, p.138 , pl. 4, fig. 6. Swain, 1985, p.433 , pl. 2, fig. 13.

Candona pseudohioensis Swain and Wagoner, 1985 (error), p.147 , pl. 4, fig. 6.

Candona paraohioensis Swain and Wagoner, Swain, 1985 (error), p.433 , pl. 2, fig. 13.

Elongate reniform, highest postmedially, posterior equal to or more broadly rounded than anterior; smooth; dorsum more truncate in females than in males; inner lamellae broad anteriorly and postventrally; L, female 1.03 to 1.20; H, 0.53 to 0.62; L, males 1.12 to 1.18; H, 0.6 to 0.63 mm (Staplin, 1963).

Occurrence. Pleistocene, Illinoian Stage of Illinois; Mankato Substage of Wisconsin Stage of Wabash-Ohio River confluence area; Cary? Substage of Wisconsin Stage of Ontario; Late Kansan Stage of western Kansas (Staplin, 1963). Miocene, Esmeralda Formation, Blair Junction section, Esmeralda County, Nevada (Swain and Wagoner, 1985). Esmeralda Formation, Black Spring road section, 2.85 miles (4.5 km) west of Black Spring and Simon Mine area, 3.3 miles (5.3 km) east of Finger Rock Wash, Nye County, Nevada (Swain, 1985). Recent and Subrecent of north central United States (Smith, 1991).

Candona patzcuaro Tressler

Plate 4, figures 30-32

Candona patzcuaro Tressler, 1954, p. 139, TF 4-8. Delorme, 1970, p. 1113, TF 198-213. Swain, 1986c, p. 370, pl. 2, figs. 10, 11; pl. 3, figs. 1, 5, 8, 14; pl. 4, figs. 1, 2. Cameron and Lundin, 1977, p. 338.

Candona hipolitensis Tressler, 1954, p. 142, figs. 13-15 female of C. patzcuaro, fide Delorme, 1970, p. 1113.

Candona marchica Hartwig, 1899, Swain, 1955, p. 609, pl. 59, figs. 3a, b, TF 33, 33a.

Candona rawsoni Tressler, Stout, 1975 (junior subjective synonym).

Male elongate-subreniform, female elongate subquadrate-subreniform, highest one-third of length from posterior end; concave venter with protuberance one-fourth from anterior end in males; posterior in females truncated and with weak marginal flange on LV; exterior smooth, interior finely pustulose; LV larger than RV, overlapping it; inner lamellae broadest anteriorly; adductor scar with elongate sixth spot above group of five spots; L, female, 1.10; H, 0.64; L, male 1.42; H, 0.83 mm (Swain, 1986c). Has weaker posterior marginal flange on LV than C. rawsoni Tressler (a boreal species) but is very close to it in shell features.

Occurrence. Pliocene, Cache Valley Formation NW$^{1}/_{4}$ SE$^{1}/_{4}$ NE$^{1}/_{4}$ sec. 10 T.12 S. R.40 E., Franklin County, Idaho (Swain, 1986a). Recent, southern North America (Tressler, 1954). Pleistocene, Lake Cochise deposits, Arizona (Cameron and Lundin, 1977).

Candona pearlensis Staplin

Plate 4, figure 33; Plate 5, figures 1, 2

Candona pearlensis Staplin, 1963, p. 780, pl. 92, figs. 1-6. Cameron and Lundin, 1977, p. 349, pl. 1, fig. 9.

Elongate- subquadrate, subreniform, highest one-third of length from posterior end; dorsum straightened in adults; posterior narrower than anterior; smooth to finely reticulate under high magnification; male higher and mainly larger than female; L, male, 0.9 to 1.01; H, 0.5 t0 0.57 mm; L, female, 0.81 to 1.03; H, 0.40 to 0.54 mm (Staplin, 1963).

Occurrence. Pleistocene, Illinoian Stage of Illinois; Cary Substage, Wisconsin Stage of Iowa (Staplin, 1963). Late Pleistocene, Lake Cochise deposits Willcox Playa, Cochise County, Arizona (Cameron and Lundin, 1977).

Candona planidorsata Cooper

Plate 3, figures 38, 39

Candona planidorsata Cooper, 1946, p. 66, pl. 8, figs. 11-17.

Elongate subquadrate, highest postmedially, hinge nearly straight, about half of shell length, venter nearly straight; anterior narrower but more rounded than posterior, both extended below, posterior truncate above; LV larger than RV but along venter RV overlaps LV; smooth; L. 1.06; H., 0.56 mm (after Cooper, 1946)

Occurrence. Pennsylvanian, Des Moinesian, Summum Cyclothem, central Illinois (Cooper, 1946).

Candona poseyensis Staplin

Plate 5, figures 3, 4, 8

Candona poseyensis Staplin, 1960 in Winkler, p. 926, pl. 122, figs. 13-17; Staplin, 1963, p. 781, pl. 94, figs. 11, 12, 16.

Elongate elliptical, subreniform, highest postmedially; ends subequal in curvature; smooth; hinge margin straightened; male slightly higher than female; L, female, 0.81 to 0.90; H, 0.40 to 0.46; L, male, 0.93 to 0.94; H, 0.49 to 0.50 mm (Staplin, 1963).

Occurrence. Pleistocene, Mankato or late Cary Substage, Wisconsin Stage of western Indiana; lake silts, Lake Nipissing age, of Michigan (Staplin, 1963). ?Oligocene, Chadron Formation, Helmville, Nebraska (coll. by R. Konizeski).

Candona pronopa Lister

Plate 8, figure 35

Candona pronopa Lister, 1975, p. 8, pl. 1, figs. 10, 11, TF 6.

Subtrapezoidal-subreniform, highest postmedially; height about half of length; dorsum nearly straight to slightly sinuous; posterior strongly truncate above, extended and narrowly rounded, below; smooth; inner lamellae of moderate width; radial canals in part funnel-shaped; L, 0.83; H, 0.43 mm (Lister, 1975).

Occurrence. Pleistocene, Lake Bonneville Group, Saltair and sec. 28 core holes, SE1/4 sec. 25, T. 1 N., R. 3 W., sec. 28, SW1/4 sec. 28, T. 1 N., R. 2 W., Salt Lake County, Utah (Lister, 1975).

Candona cf. protzi Hartwig

Plate 5, figures 5 - 7

Candona protzi Hartwig, 1898, p. 476, fig. 2.

Candona cf. protzi Hartwig, Swain, 1977, pl. 1, figs. 2a, b. Swain, 1990, p. 150, pl. 1, figs. 12, 13, 17.

Elongate reniform; strongly concave venter, highest postmedially; ends rounded; L, 1.26; H, 0.66 mm (Swain, 1990).

Occurrence. Pliocene, Idaho Group, 15 miles (24 km) west of Burns, Oregon (Swain, 1990). Subrecent, raised beach sediments of Pyramid Lake, Nevada (Swain, 1977). Recent, Europe (Hartwig, 1898).

Candona pseudohioensis Swain and Wagoner 1985

(see Candona paraohioensis, Staplin)

Candona (Pseudocandona?) punctata (Furtos)

Plate 5, figure 9

Candona punctata Furtos, 1933, p. 485, pl. 13, figs. 2-8; Staplin, 1963, p. 781, pl. 94, fig. 9

Short, subtriangular, highest just posterior to middle; dorsum subumbonate, venter slightly concave; ends narrow, the posterior more so than anterior; punctate; frontal muscle scars very small; inner lamellae broad anteriorly and postventrally; L, 0.8 to 0.95; H, 0.45 to 0.53 mm (Staplin, 1963).

Candona (Pseudocandona?) pustulosa (Staplin)

Plate 5, figure 10

Candona pustulosa Staplin, 1963, p. 782, p. 94, fig. 8.

Subtrapezoidal, subreniform, highest anteromedially; dorsum straightened, venter slightly concave; coarsely pustulose, with intervening areas reticulate as in P.? fluviatilis; inner lamellae broadest anteriorly; adductor scar a compact 5-spot group, surmounted by a sixth spot; L, 0.60 to 1.00; H, 0.32 to 5.0 mm (Staplin, 1963).

Occurrence. Pleistocene, Mankato Substage, Wisconsin Stage, Gallatin County, Illinois (Staplin, 1963). Holocene (sub-Recent) of Minnesota (Swain, 1995).

Candona rangliensis Swain

Plate 5, figures 11, 12; aff. 13, 14

Candona rangliensis Swain, 1964a, p. 774, pl. 43, figs. 6a-c. Swain, 1987c, p.433 , pl. 5, figs. 13, 14.

Elongate subreniform, highest medially; dorsum truncate anteromedially; LV larger than RV; finely pitted; posterior truncate above, extended below; L, 1.0; H, 0.48 mm (Swain, 1964).

Occurrence. Eocene, Colton-Green River transition beds, east end of Raven Ridge, Rio Blanco County, Colorado (Swain, 1964). Oligocene, Pruett? Formation, Buck Hill volcanic series, Alpine Quadrangle, Brewster County, Texas (Swain, 1987c).

Candona rawsoni Tressler

Plate 4, figures 6-9

Plate 5, figures 31-33

Plate 8, figure 36 - 38

Candona rawsoni Tressler, 1967, p. 420, TF 5-11.

Candona sp., Bronstien, 1930, p. 140, pl. 4, figs. 17-21.

Candona obtusa Bronstein, 1947, p. 252, 321, TF 157 (non-Candona weltneri obtusa Müller, 1900).

Candona nyensis Gutentag and Benson, 1962, p. 37, pl. 2, fig. 1-33, TF 101.

Candona swaini Staplin, 1963, p. 785, pl. 91, figs. 4-9.

Candona patzcuaro Tressler, Stout, 1975 (senior subjective synonym)

Elongate reniform (males) to truncate-reniform (females), highest postmedially; ventral margin deflexed anteriorly in LV; posterior margin truncate below, slightly concave above in females; smooth; normal pores numerous; LV overlaps RV on all margins; inner lamellae broad anteriorly (Staplin, 1963). Has stronger posterior flange on LV of females than C. patzcuaro Tressler (a tropical species), but the two species are very similar.

Occurrence. Pleistocene, Illinoian to Wisconsin, Nye Sink deposits, NW$^1/_4$ SW$^1/_4$ sec. 15, T. 6 N., R. 25 E., Beaver County, Oklahoma; Illinoian to Wisconsin Stage, Odee Formation, NE$^1/_4$ sec. 35, T. 34 S., R. 29 W., and NW$^1/_4$ SE$^1/_4$ sec. 1, T. 34 S., R. 29 W., Meade County, Kansas; Sangamonian to early Wisconsin Stage, NW$^1/_4$ NE$^1/_4$ sec. 8, T. 33 S., R. 27 W., Meade County, Kansas; late Kansan to early Yarmouthian Stage above Pearlette Ash SE$^1/_4$ NE$^1/_4$ sec. 1, T. 25 S., R. 7 W., Reno County, Kansas; Kansan Stage, below Pearlette Ash, Canyon, Texas; Illinoian Stage sec. 10, T. 27 N., R. 24 W., Harper County, Oklahoma (Gutentag and Benson, 1962; Benson, 1967 as Candona nyensis). Pleistocene, Aftonian Stage to late Mankato Substage, Wisconsin Stage, of Illinois; Recent, Lake Michigan cores (Staplin,

1963). Lake Bonneville Group, Saltair and sec. 28 core holes SE1/4 sec. 25, T. 1 N., R. 3 W., SW1/4 sec. 28, T. 1 N., R. 2 W., Salt Lake County, Utah (Lister, 1975). Recent and Subrecent, northern United States and Canada (Tressler, 1967; Delorme, 1970 ; Smith, 1991; Swain, 1995).

Candona renoensis Gutentag and Benson

Plate 5, figure 15, 16 a, b

Candona renoensis Gutentag and Benson, 1962, p. 39, pl. 1, figs. 1, 2, TF 11. Benson, 1967, p. 222, TF 6-7, 9, 11. Not Delorme, 1970c, p. 1116, (=C. camuramarginis Lister, 1975).

Subtrapezoidal, subreniform, highest postmedially, female less high and less inflated than male; LV slightly larger than RV; smooth; inner lamellae broadest anteriorly; adductor scar with six spots; L, female 0.99; H, 0.53 mm; L, male 1.01; H, 0.61 mm (Gutentag and Benson, 1962).

Occurrence. Pleistocene, Sangamon Stage, Cragin Quarry, Big Springs Ranch, SW1/4 sec. 17, T. 32 S., R. 28 W., Meade County, Kansas; late Kansan or early Yarmouthian Stage, above Pearlette Ash, SE1/4 NE1/4 sec. 1, T. 25 S., R. 7 W., Reno County, Kansas (Gutentag and Benson, 1962). Yarmouthian Stage, Dickinson County, Kansas (Benson, 1967).

Candona rogersi Jones, 1862

(see Darwinula rogersi (Jones))

Candona salteriana Jones

Plate 5, figures 36, 37

Candona? salteriana Jones, 1862, p. 122, pl. 5, figs. 13, 14. Cooper, 1946, p. 66, pl. 8, figs. 21- 25.

Carbonita roederiana Jones and Kirkby, 1890, p. 138, TF 5-6b.

Elongate subovate, highest postmedially, dorsum gently convex, venter nearly straight; ends extended below, anterior narrower than posterior; LV overlaps RV except ventrally where RV overlaps LV; smooth; L. 1.35; H., 0.72 mm (after Cooper, 1946).

Occurrence. Pennsylvanian, Des Moinesian Series, Summum Cyclothem; Virgilian Series, Bogota Cyclothem, central Illinois (Cooper, 1946).

Candona sappaensis Staplin

Plate 6, figures 1, 2

Candona swaini sappaensis Staplin, 1963, p. 786, pl. 91, figs. 8, 9. (C. swaini Staplin, 1963c= C. rawsoni Tressler by Delorme, 1970).

Relatively larger and more "robust" than C. rawsoni; subreniform (male), subreniform-truncate (female); L, 1.14 to 1.41; H, 0.62 to 0.81 mm (Staplin, 1963).

Occurrence. Pleistocene, late Kansan, Sappa Silt, Lyon County, Iowa; Harlan County, Nebraska; above and below Pearlette Ash (Staplin, 1963).

Candona scopulosa Furtos

Plate 5, figures 17, 18

Candona scopulosa Furtos, 1933, p. 479, pl. 8, figs. 7-10, pl. 12, figs. 11-13. Staplin, 1963, p. 782, pl. 92, figs. 25-28.

Elongate subquadrate, subreniform, highest in posterior third; dorsum straightened, venter concave; posterior broader than anterior; smooth, radial canals numerous; LV slightly overlaps RV; inner lamellae broadest anteriorly; L, 0.92-1.99; H, 0.49-0.50 mm (Staplin, 1963).

Occurrence. Pleistocene, Illinoian and Tazewell Stages of Illinois (Staplin, 1963). Recent, Lake Erie (Furtos, 1933).

Candona shawneensis Staplin

Plate 5, figures 19, 20

Candona shawneensis Staplin, 1963, p. 783, pl. 92, figs. 7-9.

Subtrapezoidal-reniform smaller and more pointed posteriorly in female than in male, highest postmedially, dorsum straightened, venter concave; smooth; LV slightly overlaps RV; inner lamellae broadest anteriorly; L, female, 0.82 to 0.93; H, 0.42 to 0.50; L, male, 1.02, H, 0.60 mm (Staplin, 1963).

Occurrence. Pleistocene, Mankato or late Cary Substage, Wisconsin Stage, near Shawneetown, Illinois (Staplin, 1963).

Candona sigmoides Sharpe

Plate 5, figure 21

Candona sigmoides Sharpe, 1897, p. 455, pl. 45, figs. 4-11. Staplin, 1963, p. 784, pl. 92, figs. 21-24.

Subquadrate to subelliptical-reniform, highest postmedially; dorsum straightened, venter concave; smooth, inner lamellae widest anteriorly; L, 0.98 to 1.10; H, 0.51 to 0.59 mm (Staplin, 1963).

Occurrence. Pleistocene, Illinoian Stage of Illinois; Mankato or late Cary Substage, Wisconsin Stage of Ohio-Wabash Rivers confluence area (Staplin, 1963). Recent, Illinois (Sharpe, 1897).

Candona simpsoni Sharpe

Plate 5, figures 22 - 24

Candona simpsoni Sharpe, 1897, p. 452, pl. 46, figs. 1-6. Staplin, 1963, p. 784, pl. 91, figs. 2, 3. Benson, 1967. Cameron and Lundin, 1977, p. 349, pl. 1, figs. 1, 2.

?Candona reflexa Sharpe, 1897, p. 457, pl. 47, figs. 1-3 (fide Staplin, 1963).

?Candona exilis Furtos, 1933, p. 483, pl. 12, figs. 14-17 (fide Staplin, 1963).

Elongate, reniform-subelliptical, highest postmedially, dorsum straightened, posterior narrowly rounded; smooth, inner lamellae narrow; L, 0.57 to 0.81; H, 0.27 to 0.40 mm (Staplin, 1963).

Occurrence. Pleistocene, Nebraskan Stage, Bureau County, Illinois; Illinoian Stage, Sankoty Sand, Tazewell, Fulton and Calhoun Counties, Illinois; Tazewell Substage, Whiteside, Fulton, and Jackson Counties, Illinois; Cary or Mankato Substage, Wisconsin Stage, Lake Tinley clay, Chicago, Illinois; late Wisconsin of Ohio; Tazewell of Ontario; Mankato or late Cary Substage of Indiana; Cary Substage of Iowa; late Kansan of Nebraska, Kansas and Iowa; Recent of Illinois, Ohio, Michigan (Staplin, 1963). Sappa Formation, Osborne County, Kansas (Benson, 1967). Recent of Minnesota, South Dakota and Canada. Late Pleistocene, Lake Cochise deposits, Willcox Playa, Cochise County, Arizona (Cameron and Lundin, 1977).

Candona subovata Jones

Plate 1, figures 6, 7

Candona subovata Jones, 1893, p. 388, pl. 15, figs. 16a, b

Elongate, subreniform, highest postmedially; dorsum gently convex, venter weakly concave; anterior narrower than posterior, widest postmedially; smooth; L, 1.2; H, 0.65 mm (after Jones, 1893).

Occurrence. Upper Cretaceous, Montana Formation, Coalville, Utah (Jones, 1893).

Candona subreniformis Jones

Plate 7, figures 5, 6

Candona subreniformis Jones, 1893, p. 388, pl. 15, figs. 14a, b

Elongate subanceolate-subreniform, highest postmedially; dorsum convex, venter concave; anterior narrower than posterior, widest postmedially; smooth; L, 1.2; H, 0.6 mm (after Jones, 1893).

Occurrrence. Upper Cretaceous, Montana Formation, Coalville, Utah (Jones, 1893).

Candona subtriangularis Swain

Plate 5, figures 25-30

Candona subtriangularis Swain, 1947, p. 522, pl. 76, figs. 1, 2. Swain, 1985, p. 434,pl. 4, figs. 18, 19. Swain, 1987c, p. 123, not figured. Swain, 1990, p. 151, pl. 1, figs. 24-27.

Elongate subtriangular-subreniform, highest three-fifths from anterior end; dorsum arched; ends narrowly rounded, posterior bluntly pointed; LV larger than RV; smooth; inner lamellae broad; postventral cup-like pocket in inner lamellae; L, 0.90; ;H, 0.51 mm (Swain, 1947).

Occurrence. Pliocene, Salt Lake Group, sec. 15, T. 13, N., R. 2 W., Cache County, Utah (Swain, 1947). Miocene, Esmeralda Formation, Simon Mine area, loc. 5, Nye County, Nevada (Swain, 1985) (identification now believed questionable). Pliocene, Salt Lake Group, loc. 2 (Swain, 1987c). Pliocene, Idaho Group, Malheur County , Oregon (Swain, 1990).

Candona swaini sappaensis Staplin

(see Candona sappaensis)

Candona sp. aff. C. symmetrica Hou et al.

Plate 6, figures 3, 4

Candona sp. aff. C. symmetrica Hou et al., 1978, Swain, 1987a, p. 373, pl. 3, figs. 3, 4; pl. 4, fig. 5. Not listed by Hou et al., (1988) in their revision of Bohai Bay species.

Elongate sublanceolate-subreniform, highest submedially; posterior narrower than anterior; LV larger than RV and overlaps around entire periphery; valves compressed, smooth; L, 0.77; H, 0.33 mm (Swain, 1987a).

Occurrence. Pliocene, Cache Valley Formation, Hoopes Creek loc., Franklin County, Idaho (Swain, 1987a).

Candona thomasi Staplin

Plate 5, figures 34, 35

Candona thomasi Staplin, 1963, p. 787, pl. 94, figs. 3, 4.

Subelliptical-subreniform to subtrapezoidal-subreniform, highest in posterior third; anterior narrower than posterior, dorsum straightened; smooth; inner lamellae narrow; L, 0.70 to 0.76; H, 0.38 to 0.40 mm (Staplin, 1963).

Occurrence. Pleistocene, Yarmouth Stage to Mankato Substage, Wisconsin Stage of Illinois; Loveland Loess? and Cary Substage, Wisconsin Stage of Iowa (Staplin, 1963).

Candona truncata Furtos

Plate 6, Figures 5, 7

Candona truncata Furtos, 1933, p. 480, pl. 9, fig. 105; pl. 11, figs. 7, 8. Staplin, 1963, p. 787, pl. 93, figs. 13-16, 18, 19. Benson, 1967, p. 228, TF 10-7. Cameron and Lundin, 1977, p. 349, pl. 1, fig. 21.

Subtriangular-subtrapezoidal, highest postmedially and exceeds half of length, strongly truncate above and extended below posteriorly in females; more rounded in males; smooth; inner lamellae fairly broad; horizontal adductor muscle scar spot at top of group atypically rounded; L, 0.97 to 1.05; H, 0.57 to 0.64 mm (Staplin, 1963).

Occurrence. Pleistocene, Aftonian? to Illinoian Stage and Mankato Substage, Wisconsin Stage of Illinois; late Wisconsin and Recent of Ohio; Tazewell Stage of Ontario (Staplin, 1963). Pleistocene, Sappa Formation, Osborne County, Kansas (Benson, 1967). Late Pleistocene, Lake Cochise deposits, Willcox Playa, Cochise County, Arizona (Cameron and Lundin, 1977). Recent of Michigan, Ohio, Minnesota, North Dakota (Furtos, 1933; Smith, 1988; Swain, 1995).

Candona wabashensis Staplin

Plate 6, figures 8, 9

Candona wabashensis Staplin, 1963, p. 788, pl. 92, figs. 29, 30.Subtrapezoidal-subreniform, highest postmedially; male dorsum long and straight, female dorsum gently curved to straight medially; female posterior pointed, extended below, truncate above; male posterior more rounded; smooth; LV slightly larger than RV; inner lamellae broad anteriorly; L, male, 1.01 to 1.10; H, 0.56 to 0.60 mm; L, female, 0.90 to 0.99; H, 0.47 to 0.52 mm (Staplin, 1963).

Occurrence. Pleistocene, Mankato or late Cary Substage, Wisconsin Stage of Posey County, Indiana; Mankato or late Cary of Story County, Iowa (Staplin, 1963).

Candona wanlessi Staplin

Plate 6, figures 10, 11

Candona wanlessi Staplin, 1963, p. 789, pl. 93, figs. 3, 4. Cameron and Lundin, 1977, p. 349, pl. 1, figs. 11, 12.

Subelliptical-subreniform, highest medially to postmedially, ends nearly equally rounded; fine reticulate at X300; LV slightly larger than RV; inner lamellae narrow; L, 0.71 to 0.88; H, 0.39 to 0.49 mm; male and female shells appear to be similar (Staplin, 1963).

Occurrence. Pleistocene, Illinoian? to Wisconsin Stages of Illinois; late Kansan Stage of Nebraska (Staplin, 1963). Late Pleistocene, Lake Cochise deposits, Willcox Playa, Cochise County, Arizona (Cameron and Lundin, 1977).

Candona whitei Scott and Smith, 1951 (part)
(see Hemicyprinotus watsonensis, Swain)

Candona willmani Staplin
Plate 6, figures 12, 13
Candona willmani Staplin, 1963, p. 790, pl. 93, figs. 1, 2.

Elongate subtriangular-subreniform, highest in posterior third; dorsum arched; ends rounded, extended below; surface with minute pits seen at high magnification; inner lamellae narrow; L, female, 0.79 to 0.81; H, 0.44 to 0.47 mm; L, male, 0.84 to 0.94; H, 0.48 to 0.53 mm (Staplin, 1963).

Occurrence. Mankato of late Cary Substage, Wisconsin Stage of Wabash-Ohio River confluence area; late Kansan of Meade County, Kansas (Staplin, 1963).

Candona, n. sp., Swain, 1956
(see Candona artesensis Swain, 1964a).

Candona? sp., Swain
Plate 6, figures 14, 15
Candona sp., Swain, 1949, p. 175, pl. 32, figs. 11-13.

Elongate subtriangular-subreniform, highest medially; posterior slightly narrower than anterior; smooth; inner lamellae narrow; radial canals closely spaced; L, 1.0; H, 0.52 mm (Swain, 1949) (Atypical of Candona).

Occurrence. Paleocene, Fort Union Formation, Tongue River Member, NW1/4 sec. 31, T. 9 S., R. 49 E., Moorhead Coal District, Powder River County, Montana, 125 feet (38 m) above Anderson Coal (Swain, 1949).

Candona sp., Benson
Plate 6, figures 16
Candona sp., group of C. crogmaniana Hoff, Benson, 1969, p. 109, pl. 21, figs. 3, 5, 7.

Elongate subreniform to subtrapezoidal, highest postmedially, truncated postdorsally (female); smooth; narrow inner lamellae; L, 1.46; H, 0.66 mm (Benson, 1969).

Occurrence. Early Pleistocene, Rita Blanca lake deposits, Hartley County, Texas (Benson, 1969).

Candona sp., Swain

not figured

Candona sp., Swain, 1985, p. 434, not figured.

Elongate-reniform, venter attains greatest concavity anteromedially; narrower anteriorly than posteriorly; smooth; narrow inner lamellae; L, 0.72; H, 0.38 mm (Swain, 1985).

Occurrence. Miocene, Esmeralda Formation, loc. 5, Simon Mine area, Nye County, Nevada (Swain, 1985).

Candona sp. 1, Swain

not figured

Candona sp. 1 Swain, 1986b, p. 212, not figured.

Not described in original publication.

Occurrence. Miocene, Humboldt Formation, Medicine Creek section, Winecup Ranch, Elko County, Nevada (Swain, 1986b).

Candona sp., Swain

Plate 6, figure 17

Candona sp., Swain, 1986a, p. 97, pl. 2, fig. 17.

Elongate, subtrapezoidal-subreniform, highest postmedially, dorsum nearly straight, venter concave; LV larger than RV; posterior narrower than anterior, strongly extended below; immature? shells; L, 0.52; H, 0.23 mm (Swain, 1986).

Occurrence. Pliocene, Glenns Ferry Formation, sec. 14, T. 4 S., R. 1 W., Owyhee County, Idaho (Swain, 1986a).

Candona sp., Swain

not figured

Candona sp., Swain, 1987a, p. 123, not figured.

Not described in original publication.

Occurrence. Pliocene, Salt Lake Group, loc. 13, Roselle Point, SW$^1/_4$ SE$^1/_4$ sec. 4, T. 8 N., R. 9 W., Box Elder County, Utah (Swain, 1987).

?Candona sp., indet., Peck

Plate 6, figures 18, 19

Candona sp., indet., Peck, 1941, p. 297, pl. 43, figs. 29, 30.

Suboblong, large, highest posteriorly; anterior narrower than posterior; LV larger than RV; smooth; L, 1.5; H, 0.9 mm (Peck, 1941).

Occurrence. Lower Cretaceous, Draney Formation, Freedom Quadrangle, Idaho-Wyoming (Peck, 1941).

"Candona" sp. 1, Sohn
Plate 6, figures 22, 23
"Candona" sp. 1, Sohn, 1956, p. 119, pl. 1, figs. 14-17.

Illustrated but not described in original publication. Elongate, sublanceolate-subreniform, highest submedially; ends nearly equally rounded, left slightly narrower; smooth; inner lamellae narrow; L, 0.97; H, 0.43 mm (Sohn, 1956).

Occurrence. Pliocene, Teewinot Formation, Jackson Hole, Wyoming (Sohn, 1956).

Candona? sp. 2, Sohn
Plate 6, figures 20, 21
Candona? sp. 2, Sohn, 1956, pl. 119, pl. 1, figs. 11, 12.

Figured but not described in original publication. Elongate, small, subelliptical, highest medially to anteromedially; anterior broader than posterior, smooth?; inner lamellae narrow; L, 0.68; H, 0.33 mm (Sohn, 1956).

Occurrence. Pliocene, Teewinot Formation, Jackson Hole, Wyoming (Sohn, 1956).

"Candona" sp. 3, Sohn
Plate 6, figures 24, 25
"Candona" sp. 3, Sohn, 1956, p. 119, pl. 1, figs. 22, 23.

Illustrated but not described in original publication. Elongate, subtriangular-subreniform, highest postmedially; posterior subacuminate, strongly extended below; smooth?; inner lamellae broad anteriorly; L, 1.3; H, 0.68 mm (Sohn, 1956).

Occurrence. Pliocene, Teewinot Formation, Jackson Hole, Wyoming (Sohn, 1956).

"Candona" sp. 4, Sohn
Plate 6, figure 26
"Candona" sp. 4, Sohn, 1956, pl. 119, pl. 1, fig. 13.

Illustrated but not described in original publication. Subtrapezoidal, highest postmedially, dorsum straightened, venter slightly concave; posterior bluntly pointed, strongly extended below; smooth; L, 1.2; H, 0.67 mm (Sohn, 1956).

Occurrence. Pliocene, Teewinot Formation, Jackson Hole, Wyoming (Sohn, 1956).

"Candona" sp. 5, Sohn

Plate 6, figures 27, 28

"Candona" sp. 5, Sohn, 1956, p. 119, pl. 1, figs. 29, 30.

Illustrated but not described in original publication. Elongate, subtrapezoidal, highest postmedially; dorsum nearly straight, venter slightly concave, posterior bluntly pointed, extended below; smooth; L, 1.27; H, 0.63 mm (Sohn, 1956).

Occurrence. Pliocene, Teewinot Formation, Jackson Hole, Wyoming (Sohn, 1956).

"Candona" sp. 6, Sohn

Plate 6, figures 29, 30

"Candona" sp. 6, Sohn, 1956, p. 119, pl. 1, figs. 18-21.

Illustrated but not described in original publication. Elongate, subtrapezoidal-subreniform, highest postmedially; dorsum straightened medially; venter concave; posterior subacuminate, strongly extended below; smooth; L, 1.27; H, 0.7 mm (Sohn, 1956).

Occurrence. Pliocene, Teewinot Formation, Jackson Hole, Wyoming (Sohn, 1956).

Candona sp. indet., Staplin

Plate 6, figure 31

Candona sp. indet., Staplin, 1963, p. 791, pl. 93, fig. 12.

Small, elongate, sublanceolate-subreniform, highest submedially; two dorsal sinuations; posterior narrower than anterior; smooth; L, 0.73 to 0.92; H, 0.32 to 0.40 mm. Resembles C. ohioensis but latter lacks dorsal sinuations (Staplin, 1963).

Occurrence. Pleistocene, Tazewell Stage, Bloomington slackwater silts, Fulton County, Illinois (Staplin, 1963).

?Candona sp.

Plate 6, figure 32

Elongate, subtrapezoidal, highest anteriorly, dorsum nearly straight, venter sinuous; anterior broadly rounded, posterior bluntly pointed, both ends extended below, truncate above; LV slightly larger than RV; smooth?; L, 1.13; H, 0.55 mm.

Occurrence. Lower Oligocene, Chadron Formation, Helmville, Nebraska (coll. by R. Konizeski).

Candona (Pseudocandona?) sp., Swain and Wagoner, 1985
(see Huabeinia? sp. aff. H. huidongensis Hou and Li)

Candona (Pseudocandona), n. sp.?
Plate 7, figures 13 - 15

Subtrapezoidal, highest anteromedially; dorsum straightened posterior to position of greatest height, venter slightly concave; anterior broader than posterior, both ends extended below, posterior truncate above; LV? larger than RV?; coarsely reticulate in posterior half, more weakly so in anterior half; LV hinge margin grooved; inner lamellae broadest anteriorly, vestibules present; L, 0.81; H, 0.48 mm (specimens coll. by J. E. King).

Occurrence. Pleistocene, bone bed at Boney Spring, Benton County, Missouri (J. E. King coll.).

Genus Candoniella Schneider, 1956
(Candonidae-Candoninae)
Candoniella bergstromi Swain
Plate 6, figures 33, 34; Plate 9, figures 3, 4
Candoniella sp. 2, Swain, 1985, p. 433, pl. 5, figs. 15, 16.
Candoniella bergstromi Swain, 1987c, p. 434, pl. 1, figs. 1, 2.

Small, elongate, subelliptical-subreniform, highest slightly postmedially, anterior slightly broader than posterior, but ends nearly equal; smooth; LV overlaps RV along midventer and postdorsally; L, 0.68; H, 0.34 mm (Swain, 1987c).

Occurrence. Miocene, Esmeralda Formation, Black Spring road section, 7659 feet (2332 m) S. 26° E of NW cor. Simon 1:24000 Quadrangle, Nye County, Nevada in unsurveyed area (Swain, 1987c).

Candoniella elongatolactoides (Dickinson and Swain)
Plate 3, figures 3-9; aff. 10, 11.
Candona elongatolactoides Dickinson and Swain, 1967, p. 344, pl. 37, figs. 4-9. Swain, 1987a, p. 373, pl. 4, figs. 3, 6. Swain, 1986, p. 97, pl. 2, figs. 12-14, 16, 17?. Swain, 1990, p. 150, pl. 1, figs. 10, 11, 14-16. Swain, 1995, p. 29, pl. 1, figs. 21, 22, 24; pl. 3, figs. 6-10.

Subelliptical-reniform, highest submedially, ends nearly equally rounded; posterior may be slightly truncate; smooth to weakly punctate; inner lamellae narrow; adductor scar with large dorsal spot and five smaller spots; L, 0.65; H, 0.33 mm (Dickinson and Swain, 1967) (may include immature forms as described).

Occurrence. Pliocene, Hay Ranch Formation ("Upper Humboldt") Gulf Oil Corp., No. 1 Wilkins Ranch well, Elko County, Nevada, 1244-4150 feet (379-1265 m) (Dickinson and Swain, 1967). Pliocene, Cache Valley Formation; Hoopes Creek loc. Franklin County, Idaho (Swain, 1987a). Glenns Ferry Formation, sec. 14, T. 4 S., R. 1 W., Owyhee County, Idaho (Swain, 1986a). Idaho Group, 15 miles (24 km) west of Burns, Oregon (Swain, 1990). Miocene, Horse Camp Formation, loc. 7, near Currant, Nye County, Nevada (Swain, 1987c).

Candoniella sp. 1, Swain, Becker and Dickinson

Plate 6, figures 35, 36

Candoniella sp., Swain, Becker and Dickinson, 19971, p. 484, pl. 4, fig. 10.

Candoniella sp. 1, Swain, 1985, p. 432, pl. 2, fig. 11.

Elongate, subreniform-subelliptical, highest submedially, dorsum moderately convex, venter concave, ends nearly equally rounded, posterior slightly the broader, both ends extended below; LV? larger than RV? overlapping except middorsally; shell compressed; smooth; L, 1.10; H, 0.63 mm (Swain, 1985).

Occurrence. Miocene, Esmeralda Formation, Finger Rock Wash, Mineral County, Nevada, 2 miles (3.2 km), west of Stewart Spring, about 900 feet (275 m) above base of formation; also 0.5 miles (0.8 km) south of Rawhide Ranch, Finger Rock Wash, about 1400-1500 feet (425-455 m) above base of Esmeralda Formation; also 2.85 miles (4.5 km) west of Black Spring, Nye County, Nevada, about 800 feet (244 m) above base of Esmeralda; also 7.6 miles (12.1 km) west of Finger Rock Wash, on road to Pactolus Mine, Nye County, Nevada, about 1000 feet (305 m) above base of Esmeralda; also 3.3 miles (5.3 km) east of Finger Rock Wash, Simon Mine area, about 100-150 feet (30-45 m) above base of Esmeralda. A similar form was found in the Elko Oil Shale (Eocene or Oligocene) at Elko, Nevada (Swain et al, 1971).

Candoniella? sp. aff. C suzini Schneider

Plate 9, figures 1, 2

Candoniella? sp. aff. C. suzini Schneider, Swain, 1987c, p. 435, pl. 3, figs. 9, 14a, b.

Elongate-subreniform, highest submedially, dorsum moderately convex, venter concave, anterior narrower than posterior; LV only slightly larger than RV, overlapping along middorsum; compressed; smooth; L, 1.16; H, 0.62 mm (Swain, 1987c).

Occurrence. Miocene, Horse Camp Formation, sec. 12, T. 10 N., R. 58 E., about 3.9 miles (6.24 km) east of Currant, Nye County, Nevada (Swain, 1987c).

Candoniella sp. Swain

Plate 22, figure 9

Herpetocypris? sp. aff. H. reptans (Baird), Swain, 1986b, pl. 1, fig. 19.

LV elliptical-subreniform, highest submedially, dorsum gently convex, venter concave, anterior slightly narrower than posterior, both ends extended below; LV slightly larger than RV; smooth; LV hinge margin rabbeted; inner lamellae very narrow; L, 0.62; H, 0.37 mm (Swain, 1986b).

Occurrence. Miocene, Humboldt Formation, Gulf Oil Co., No. 1 Thousand Springs well, County, Nevada, 50-55 feet (16.4-16.8 m), center SE$^1/_4$ SW$^1/_4$ sec. 8 T. 40 N., R. 66 E., Elko County, Nevada (Swain, 1986b).

Candoniella? sp. aff. C. parca Shan and Zhang, Swain

Plate 9, figures 5, 6

Candoniella: sp. aff. C. parca Shan and Zhang, in Hou et al., 1988, Swain, 1986b, p. 211, pl. 3, figs. 24, 25.

Subelliptical-subreniform, highest submedially , height about 0.6 of length, dorsum and venter only slightly convex and concave, respectively; posterior slightly broader than anterior and subtruncate medially; convexity moderate; shell blunt-ended in edge view; smooth; L, 0.88; H, 0.55 mm (Swain, 1986b).

Occurrence. Miocene, Humboldt Formation, Cobre surface section, Elko County, Nevada in lower 10 feet (3 m) of exposure; also Gulf Oil Co., No. 1 Thousand Springs well, Elko County, Nevada, 150-250 feet (46-76 m); also in conglomeratic limestone near base of formation, Medicine Creek section, Winecup Ranch, Elko County; and in lower Humboldt or Indian Well Formation (late Oligocene), Palisades section north end of Pine Creek Valley, Eureka County, Nevada (Swain, 1986b).

Candoniella pinecreekensis Swain

Plate 9, figure 7 - 9

Candoniella pinecreekensis Swain, 1986b, p. 211, pl. 4, figs. 8, 9; pl. 5, fig. 27. Swain, 1987c, p. 435, p. 4, figs. 12, 13.

Elongate subelliptical-subreniform, highest medially, height about 0.58 of length; dorsum gently convex, venter nearly straight to slightly concave; LV overlaps RV most along dorsum; smooth; LV hinge margin rabbeted to receive edge of RV; inner lamellae narrow; L, 0.83; H, 0.48 mm (Swain, 1986b).

Occurrence. Miocene, Humboldt Formation, Gulf Refining Co. No. 2 Wilkins Ranch well, Elko County, Nevada, 2945-2975 feet (898-907 m) in upper part of formation; also at

Cobre surface section, Elko County, Nevada , 204-232 feet (62071 m) above base of section; also in conglomeratic limestone near base of formation, Medicine Creek section, Winecup Ranch, Elko County; also in Humboldt or Indian Well Formation (late Oligocene), Palisade section, north end of Pine Creek Valley, Eureka County, Nevada, in lower part of exposure (Swain, 1986b); also questionably in Eocene, Twin Bridges Limestone at Twin Bridges, Elko County, Nevada (Swain, 1987c).

Candoniella whitei (Scott and Smith)
Plate 9, figures 10, 11
Plate 22, figures 16, 17
Plate 10, figures 3, 4
Erpetocypris? sp., Swain, 1949, p. 176, pl. 32, figs. 5, 18; 1956, p. 136, pl. 3, figs. 6a, b.
Candona whitei Scott and Smith, 1951, p. 324, pl. 52, figs. 1-5.
Candoniella whitei (Scott and Smith) Swain, 1964a, p. 273, pl. 41, figs. 9a-d, 10; pl. 42, fig. 1. Swain, 1987, p. 228, pl. 3, figs. 6a, b.

Elongate, subelliptical, highest medially; dorsum gently convex, venter nearly straight to slightly concave; posterior more pointed and extended below than anterior; LV over-reaches RV slightly around periphery, valves compressed; smooth; L, 0.88; H, 0.51 mm (Swain, 1964). Immature molts triangular in outline (Scott and Smith, 1951).

Occurrence. Eocene, Green River Formation, southwestern Green River Basin, Wyoming (Scott and Smith, 1951). Eocene, Green River and Wasatch (Colton) transition beds, Uinta and Picenance Creek Basins, Utah and Colorado and Green River Basin Wyoming (Swain, 1949, 1964). Paleocene, Flagstaff Limestone, near Ephraim, Utah (Swain, 1964). Paleocene or Eocene, Sheep Pass Formation, Johnson Canyon, Grant Range, Nye County, Nevada (Swain, 1987).

Genus Candonopsis Vavra, 1891
(Candonidae-Candoninae)
Candonopsis? cf. kingsleyi (Brady and Robertson)
Plate 9, figures 12, 13
Candona kingsleyi Brady and Robertson, 1870, p. 17, pl. 9, figs. 9-12.
Candonopsis kingsleyi (Brady and Robertson), Vavra, 1891, p. 54, TF 16 (1-10). Swain, 1963, p. 806, pl. 96, fig. 4, TF 3c.

Subovate-subtriangular-reniform, highest medially to postmedially; dorsum convex, venter slightly concave, anterior much narrower than posterior and extended below; smooth; compressed; inner lamellae narrow?; L, 0.76; H, 0.45 mm (immature shell?) (Swain, 1963).

Occurrence. Pleistocene, Gubik Formation, Arctic coastal plain, Alaska, Party 43, Fish Creek shot holes, shot point 1, line 1-48 at 80 feet (24 m); also Ikpikpuk shot holes, line 15-47, shot point 14 at 40-50 feet (12-15m). Widespread in Holocene of northern Europe and Siberia (Swain, 1963; Vavra, 1891).

Candonopsis sp. aff. C. shahejiensis Li and Lai

Plate 9, figure 14, 15

Candonopsis? sp. aff. C. shahejiensis Li and Lai (in Hou et al., 1988), 1978, Swain, 1985, p. 434, pl. 5, fig. 9. Swain, 1987c, p. 435, pl. 4, fig. 9.

Elongate subtriangular-subovate-subreniform; highest two-fifths of length from posterior end; dorsum convex, venter concave anteromedially; anterior narrowly rounded, extended below; LV overlaps RV along venter; smooth, moderately convex; L, 0.93; H, 0.53 mm (Swain, 1985).

Occurrence. Miocene, Esmeralda Formation, 7.6 miles (12.1 km) east of Finger Rock Wash, Mineral County, Nevada on road to Pactolus Mine, about 1000 feet (305 m) above base of lake bed sequence; also, Oligocene, Edwards Creek Tuff, T. 19 N., R. 38 E., Churchill County, Nevada near head of Edwards Creek Valley, Desatoya Range (Swain, 1987c).

Genus Carbonita Strand, 1928

(Carbonitidae)

Carbonia Jones, 1870 (invalid)

Carbonita agnes (Jones)

Plate 8, figures 50, 51

Carbonia agnes Jones, 1870, p. 218, pl. 9, figs. 6, 7.

Carbonia agnes rugulosa Jones, 1870, p. 218, pl. 9, figs. 8, 9.

Carbonia agnes subrugulosa Jones, 1870, p. 218, pl. 9, fig. 10.

Elongate subreniform-subovate, highest anteromedially; dorsum gently convex, straightened medially, venter concave; anterior narrower than posterior, extended below; RV overlaps LV along free margins; coarsely pitted to smooth or granulose; L. 0.64; H., 0.32 mm (after Cooper, 1946).

Occurrence. Pennsylvanian, Des Moinesian Series, Gimlet Cyclothem, central Illinois (Cooper, 1946).

Carbonita inflata (Jones and Kirkby)

"Carbonita humilis Group", Bless and Pollard, 1975, p. 111

(see Whipplella cunieformis Holland (part))
(see Whipplella depressa Holland (part))

Carbonita magna (Upson)
(see Whipplella carbonaria)

Carbonita orbiculata Cooper
Plate 8, figures 52, 53
Carbonita orbiculata Cooper, 1946, p. 68, pl. 8, figs. 43-45

Subcircular, highest medially, dorsum strongly convex, venter slightly convex; ends rounded broadly and nearly equally; LV overlaps RV; weakly punctate; L, 0.95; H., 0.77 mm (after Cooper, 1946).

Occurrence. Pennsylvanian, Atokan Series, Seville Limestone, western Illinois (Cooper, 1946).

Carbonita? tenuis Cooper
Plate 8, figures 54, 55
Carbonita? tenuis Cooper, 1946, p. 68, pl. 8, figs. 33-35.

Elongate subelliptical-subreniform, highest medially; dorsum gently convex, venter gently concave; posterior narrower than anterior; LV larger than RV; surface with polygonal to very elongate reticulations; L, 0.64; H., 0.31 mm (after Cooper, 1946).

Occurrence. Pennsylvanian, Virgilian Series, Cohn Cyclothem, central Illinois (Cooper, 1946).

Carbonita, n. sp., Sohn
Plate 9, figures 16-18
Carbonita, n. sp., Sohn, 1985, p. 451, TF 2.1-2.14; TF 3.10-3.18.

Internal molds ("Steinkerns") subquadrangular-subelliptical; highest postmedially; RV? slightly larger than LV?; smooth (or finely pitted on mold surface); median vertical sulcus; slightly anteromedian or submedian pitlike, aggregate muscle scar; narrow inner lamellae, widely spaced radial canals; L, 1.1; H, 0.5 mm (Sohn, 1985).

Occurrence. Upper Mississippian, Bluestone Formation, Bramwell Member, Crumpler, Wyoming County, Flattop and Bramwell Quadrangles, Mercer County, West Virginia (Sohn, 1985).

Carbonita? sp., Sohn
Plate 9, figures 19-21

Carbonita? sp., Sohn, 1985, p. 451, TF 6.18-6.24.

"Steinkerns" elongate subquadrate, highest postmedially, anterior narrower than posterior; LV? overlaps RV? dorsally and ventrally; median to anteromedian circular aggregate muscle scar; smooth; L, 1.2; H, 0.7 mm (Sohn, 1985).

Occurrence. Upper Mississippian, Bluestone Formation, Bramwell member, Flattop Quadrangle, Mercer County, West Virginia (Sohn, 1985).

Genus Caspiollina Mandelstam, 1957

(Candonidae-Candoninae)

Caspiollina microstriata Dickinson and Swain

Plate 9, figure 22

Caspiollina microstriata Dickinson and Swain, 1967, p. 341, pl. 35, fig. 24.

Elongate subreniform, highest anteromedially; dorsum broadly convex, venter concave; anterior broader than posterior; LV overlaps RV anterodorsally and midventrally; compressed; punctate and striate roughly parallel to shell margins; L, 0.98; H, 0.55 mm (Dickinson and Swain, 1967).

Occurrence. Pliocene, Hay Ranch Formation ("Upper Humboldt") roadcut, one mile south of Elko, Nevada on Highway 46 (Dickinson and Swain, 1967).

Genus Chlamydotheca Saussure, 1858

(Cyprididae-Cypridinae)

Chlamydotheca? dixieflatensis Swain

Plate 10, figures 1, 2

Chlamydotheca dixieflatensis Swain, 1987b, p. 220, pl. 1, figs. 11, 15, 16.

Subovate-subcircular, highest medially to post medially, height about three-fourths of length; dorsum convex, venter nearly straight to slightly concave; posterior more broadly rounded than anterior; RV slightly larger than LV; compressed; smooth; except for concentric lines anteriorly; exfoliated shells with polygonal shell structure pattern in an intermediate layer (part of exocuticle?); L, 1.10; H, 0.77 mm (Swain, 1987).

Occurrence. Eocene, Twin Bridges Limestone, center sec. 36, T. 32 N., R. 55 E., Twin Bridges, Elko County, Nevada (Swain, 1987).

Chlamydotheca llanoensis Pierce

Plate 9, figures 23-26

Chlamydotheca llanoensis Pierce, 1988, p. 445, TF 1.1-1.7.

Subtriangular-subtrapezoidal, highest anteromedially, height more than half of length; dorsum convex, straightened medially, venter sinuous, with anterior concavity owing to projections of flanged margin, especially in RV; anterior marginal flange broad, posterior flange narrower; smooth; adductor muscle scar with five large oblique spots and one more anteroventral; two oblique frontal spots; LV slightly larger than RV, convexity moderate; LV hinge a groove to receive edge of RV; L, 2.84; H, 1.59 mm (Pierce, 1988).

Occurrence. Late Pleistocene, Blackwater Draw Formation, Lubbock Lake archeological site, Lubbock County, Texas and Blackwater Draw site, Roosevelt County, Texas (Pierce, 1988).

Genus Clinocypris Mandelstam, 1956

(Pontocyprididae)

Clinocypris? sp., Swain

Plate 9, figure 27

Clinocypris? sp., Swain, 1986b, p. 212, pl. 4, fig. 1, pl. 5, fig. 26.

Elongate, sublanceolate-subpyriform, highest about two-fifths of length from anterior end; dorsum strongly convex, truncated posterior to point of greatest height; venter slightly concave; anterior much broader than posterior, the latter strongly extended below; LV larger than RV; compressed, weakly reticulate surface; L, 1.11; H, 0.60 mm (Swain, 1986b).

Occurrence. Miocene, Cobre, Elko County, Nevada, Humboldt Formation, 1011 feet (308 m) above base of exposure; also Miocene or late Oligocene, Humboldt or Indian Well Formation, Palisade section, north end of Pine Valley, Elko County, Nevada (Swain, 1986b).

Clinocypris? sp., Swain

Plate 10, figures 5, 6

Clinocypris? sp., Swain, 1987b, p. 227, pl. 3, figs. 9, 18.

Subpyriform, highest two-fifths of length from anterior; height about six-tenths of length; dorsum moderately convex, venter nearly straight; posterior subacuminate, strongly extended below; LV slightly larger than RV; compressed; smooth except for short, low ridge-like expansions middorsally, and a larger low expansion midventrally; L, 1.90; H, 1.10 mm (Swain, 1987c).

Occurrence. Late Cretaceous or Paleocene, Sheep Pass Formation, Sheep Pass Canyon, Egan Range, T. 10 N., R. 62 E., and 63 E., White Pine County, Nevada (Swain, 1987b).

Genus Cyclocypris Brady and Norman, 1889

(Candonidae-Cyclocypridinae)

Cyclocypris ampla Furtos

Plate 8, figures 39, 40

Cyclocypris ampla Furtos, 1933, p. 461, pl. 14, figs. 1-7. Lister, 1975, p. 11, pl. 3, figs. 1, 2, TF 9.

Subcircular-subovate, highest medially, dorsum convex, venter nearly straight; posterior slightly narrower than anterior; RV slightly larger than LV or valves subequal; smooth to sparsely punctate; inner lamellae broadest anteriorly; adductor scar median curved group of four spots, a triangular more posterior spot and an elongate more dorsal spot; two oblique frontal spots; L. 0.62; H, 0.45 mm (Lister, 1975).

Occurrence. Pleistocene, Lake Bonneville Group, Saltair core hole, SE$^{1}/_{4}$ sec. 25, T. 1 N., R. 3 W., and Sec. 28 core hole, SE$^{1}/_{4}$ SW$^{1}/_{4}$ sec. 28, T. 1 N., R. 2 W., Salt Lake County, Utah (Lister, 1975).

Cyclocypris californica Chapman

Plate 7, figures 36, 37

Cyclocypris californica Chapman, 1896, p. 96, pl. 3, figs. 4-6.

Elongate subovate, reniform, highest a little postmedially; dorsum gently convex, venter concave; posterior broader than anterior; smooth; L, 1.02 mm; H, 0.63 mm (Chapman, 1896).

Occurrence. Pliocene, Merced Formation, Wildcat Canyon, one-fourth mile south of "The Caves", east of Berkeley, California (Chapman, 1896).

Cyclocypris forbesi Sharpe

Plate 6, figure 37; Plate 10, figures 7, 8

Cyclocypris forbesi Sharpe, 1897, p. 432, pl. 41, fig. 1-7. Staplin, 1963, p. 1170, pl. 159, figs. 37-37.

Subglobular-subovate, highest medially, height about four-fifths of length; anterior narrower than posterior, dorsum strongly convex; venter straight or slightly convex; smooth or minutely pitted; color sepia brown in modern and sub-Recent specimens; RV selvage overlaps LV; LV hinge groove receives RV edge; adductor scar formed of four large and two small spots; inner lamellae broad anteriorly; L, 0.57 to 0.70; H, 0.46 to 0.53 mm (Staplin, 1963).

Occurrence. Pleistocene, Nebraskan Stage to Recent of Illinois; Mankato or late Cary Substage of western Indiana; Tazewell-Cary? Substage of Ontario; Cary? Substage of Story County, Iowa; late Kansan of Kansas and Iowa (Staplin, 1963). Recent, Illinois, Massachusetts, etc. (Furtos, 1933; Hoff, 1942).

Cyclocypris fragilis Staplin

Plate 9, figures 28, 29

Cyclocypris fragilis Staplin, 1963, p. 1180, pl. 159, fig. 14, 15

Oval, very delicate shell, highest medially, dorsum convex, venter nearly straight; anterior may be slightly broader than posterior, postdorsal slope truncated; minutely punctate, but appears smooth; scattered normal pore canals; vesibules lacking; RV selvage overlaps LV ventrally and anteriorly; LV hinge with postdorsal flange groove; RV hinge overlaps LV anteriorly; L, 0.55 to 0.60; H, 0.38 to 0.40 mm (after Staplin, 1963).

Occurrence. Pleistocene, Wisconsin Stage, Mankato or late Cary substage, 1.5 miles (2.4 km) east of Mt. Vernon, Illinois and elsewhere in Illinois, Kansas and Ohio. (Staplin, 1963).

Cyclocypris? ketneri Swain

Plate 10, figures 9-11

Cyclocypris? ketneri Swain, 1987c, p. 225, pl. 1, figs. 19a-c, 20; pl. 2, figs., 13, 18.

Small subovate, venter weakly concave; anterior slightly narrower than posterior; LV overlaps RV ventrally, RV overlaps LV dorsally; strongly convex; smooth; L, 0.42; H, 0.27 mm (Swain, 1987).

Occurrence. Eocene, Twin Bridges Limestone, sec. 36, T. 32 N., R. 55 E., Elko County, Nevada at south edge of Twin Bridges, west side of Huntingdon Creek (Swain, 1987c).

Cyclocypris laevis (O.F. Müller)

Plate 10, figures 12, 13

Cypris laevis O.F. Müller, 1785, p. 52, pl. 3, figs. 7-9.

Cypria laevis (Müller), Brady and Robertson, 1889, p. 69.

Cyclocypris laevis (Müller), G.W. Muller, 1900, p. 41, pl. 10, figs. 14-22; non Sharpe, 1908, p. 408, pl. 50, fig. 5, pl. 54, figs. 5-7; non Sars, 1925, p. 93, pl. 43, fig. 2.

?Cyclocypris laevis (Müller), Staplin, 1963, p. 1182, pl. 159, figs. 27-29.

Suboval, highest just postmedially, height nearly three-fourths of length; dorsum arched, venter nearly straight; anterior narrower than posterior; smooth; LV slightly overlaps RV except middorsally, LV hinge fits into groove in RV; inner lamellae broad anteriorly, narrow vestibules; L, 0.5 to 0.52; H, 0.36 to 0.38 mm (Staplin, 1963).

Occurrence. Pleistocene, Mankato Substage, Wisconsin Stage of Illinois; Kansan Stage of Kansas; Quaternary of Great Britain Recent, Holarctic (Staplin, 1963).

Cyclocypris ovum (Jurine)

Plate 10, figures 14, 15

Cypris laevis (part) O.F. Müller, 1776, p. 198.

Monoculus ovum (part) Jurine, 1820, p. 179, pl. 19, figs. 18, 19.

Cypris scutigera (part) and Cypris pantherina (part), Fischer, 1851, p. 162, pl. 11, figs. 3-5; p. 163, pl. 11, figs. 6-8.

Cypris serena Brady and Norman, 1889, p. 70.

Cypris pygmaea Croneberg, 1895, p. 294, pl. 7, fig. 9.

Cyclocypris laevis (Müller), Kaufmann, 1900, p. 320, p. 19, figs. 23-25, pl. 23, figs. 9-12, TF 17.

Cyclocypris laevis pygmaea (Croneberg), Ekman, 1908, p. 177.

Cyclocypris ovum (Jurine), G.W. Muller, 1912, p. 128. Staplin, 1963, p. 1181, pl. 159, figs. 20-22.

Elongate suboval, highest medially, height about two-thirds of length; dorsum arched; venter nearly straight; anterior narrower than posterior; RV slightly larger than LV; hinge flange groove in LV lies anterior to highest point; inner lamellae broad anteriorly; smooth or weakly pitted at loci of normal canals; L, 0.42 to 0.50; H, 0.29 to 0.46 mm (Staplin, 1963).

Occurrence. Pleistocene, Kansan? Stage to post-Mankato Substage, Wisconsin Stage, of Illinois; late Kansan of western Kansas; Cary? Substage of Iowa; Tazewell Stage of Ontario; late Wisconsin Stage of Ohio; Quaternary of Great Britain (Staplin, 1963). Recent of Washington, Ohio, Europe, Siberia, North Africa (authors).

Cyclocypris serena (Koch)

Plate 8, figures 41, 42

Cypris serena Koch, 1838, species 22.

Cypris ovum Jurine, Daday, 1882, p. 237.

Cypria serena (Koch), Brady and Norman, 1889, p. 79.

Cyclocypris serena (Koch), Sars, 1890, p. 55. Lister, 1975, p. 11, pl. 3, figs 3, 4 TF 10.

Subcircular, highest medially, two-thirds to three-fourths of length; dorsum convex, venter less convex in RV, concave in LV; anterior slightly narrower than posterior, the latter subtruncate above; inflated valves, lop-sided in end view; smooth; RV larger than LV; inner lamellae narrow; L, 0.63; H, 0.43 mm (Lister, 1975).

Occurrence. Pleistocene, Lake Bonneville Group, Saltair core hole SE$^1/_4$, sec. 25, R. 1 N., R. 2 W., Salt Lake County, Utah (Lister, 1975).

Cyclocypris sp. aff. C. serena (Koch)

Plate 10, figures 16, 17

Cyclocypris sp. aff. C. serena (Koch), Swain, 1987a, p. 368, pl. 1, figs. 12, 16.

Subovoid, highest submedially, dorsum more convex than venter; anterior narrower than posterior, both ends extended below; RV larger than LV, overlapping the most along dorsum and venter; strongly inflated, width exceeds height; smooth except for scattered dorsal pits; L. 0.50; H, 0.35 mm (Swain, 1987); not lop-sided end view as in C. serena, and has dorsal pitting.

Occurrence. Pliocene, Cache Valley Formation, NW$^{1}/_{4}$ SE$^{1}/_{4}$ NE$^{1}/_{4}$ sec. 10, T. 12, S., R. 40 E., Franklin County, Idaho (Swain, 1987).

Cyclocypris sharpei Furtos

Plate 10, figures 18, 19

Cyclocypris laevis Sharpe, 1908, (non O.F. Müller, 1785), p. 408, p. 50, fig. 5, pl. 54, figs. 5-7.

Cyclocypris sharpei Furtos, 1933, p. 460, pl. 14, figs. 8-14. Staplin, 1963, p. 1181, pl. 159, figs. 23-26.

Suboval, elongate, highest medially, about two-thirds of length; dorsum arched, venter slightly convex to sinuous, LV barely larger than RV, inflated, width equal height; minutely pitted, inner lamellae very narrow; L, 0.42 to 0.53; H, 0.27 to 0.34 mm (Staplin, 1963).

Occurrence. Pleistocene, Sankoty? sand, Kansan to Recent of Illinois; late Kansan Sappa Silt of Nebraska; Illinoian and Cary? of Ohio; late Wisconsin of Orleton, Ohio; Recent of New Brunswick, Ohio, Illinois, Iowa, Florida, Louisiana, Indiana, New York (Staplin, 1963); southeastern interior plains, Canada (Delorme, 1970); Minnesota (Swain, 1995).

Genus Cypria Zenker, 1854

(Candonidae-Cyclocypridinae)

Cypria anterosinuata Staplin

Plate 10, figures 20, 21

Cypria anterosinuata Staplin, 1963, p. 1174, pl. 159, figs. 8, 9.

Elongate suboval, highest anteromedially, height about 0.7 of length, dorsum arched, venter slightly concave to straight, anterior broader than posterior; anterior marginal zone compressed; surface with very fine anastomosing lines, each of which consists of minute rounded elevations, resulting in satiny sheen; LV slightly larger than RV; inflated; width about four-fifths of length; inner lamellae narrow; L, 0.45; H, 0.30 mm (Staplin, 1963).

Occurrence. Pleistocene, Mankato or late Cary Substage, Wisconsin Stage of Wabash River-Ohio River confluence area (Staplin, 1963). Recent of Minnesota (Swain, 1995).

Cypria hoopescreekensis Swain

Plate 10, figures 22-25

Cypria hoopescreekensis Swain, 1986c, p. 368, pl. 1, figures 8-11.

Ovate, highest medially, dorsum convex, venter nearly straight; posterior broader than anterior; LV slightly larger than RV; dorsally RV overlaps LV; low convexity; exterior smooth; interior weakly pitted; LV with hinge groove; inner lamellae narrow; adductor scar of four spots; two frontal spots close to adductor; L, 0.56; H, 0.37 mm (Swain, 1987).

Occurrence. Pliocene, Cache Valley Formation, NW1/4 SE1/4 NE1/4 sec. 10, T. 12 S., R. 40 E., Franklin County, Idaho (Swain, 1986c).

Cypria maculata Hoff

Plate 10, figures 26, 27

Cypria ophthalmica (part?) Turner, 1895, non C. ophthalmica (Jurine, 1820), p. 306, pl. 75, figs. 1-3, pl. 76, figs. 1-3, 5.

Cypria maculata Hoff, 1942, p. 114, pl. 7, figs. 90-95. Staplin, 1963, p. 1176, pl. 159, figs. 12, 17.

Elongate-ovate, highest postmedially, posterior broader than anterior; LV overlaps RV; smooth; dorsum convex, venter nearly straight; width exceeds half the length; RV with hinge groove medially; inner lamellae narrow; adductor scar with four large and two small subcentral spots; two frontal anteroventral spots; L, 0.45 to 0.51; H, 0.30 to 0.36 mm (Staplin, 1963).

Occurrence. Pleistocene, Tazewell to Mankato Substages, Wisconsin Stage of Illinois; Mankato or late Cary Substage of western Indiana; late Kansan Stage, Sappa Silt of Nebraska, Grand Island and Sappa Silt of western Kansas (Staplin, 1963). Late Pleistocene marl, in Bison skull, Lake Itasca, Clear water County, Minnesota (this paper). Recent of eastern United States? and Illinois (Hoff, 1942).

Cypria mediana Hoff

Plate 10, figures 28, 29

Cypria mediana Hoff, 1942, p. 110, pl. 6, figs. 82-87. Staplin, 1963, p. 1176, pl. 259, figs. 18, 19.

Elongate ovate, highest anteromedially; anterior broader than posterior; LV slightly larger than RV; dorsum convex, venter concave; inner lamellae narrow; smooth; six adductor scar spots; L, immature? 0.70; H, 0.40 mm (Staplin, 1963).

Occurrence. Pleistocene, Mankato or late Cary Substage, Wisconsin Stage, near confluence of Wabash and Ohio Rivers; Recent, southern Illinois (Staplin, 1963).

Cypria nevadensis Dickinson and Swain

Plate 10, figure 30

Cypria nevadensis Dickinson and Swain, 1967, p. 343, pl. 36, fig. 4.

Subtriangular-subovate, highest medially to anteromedially, dorsum highly arched, venter straight to slightly concave; posterior narrow, extended below; LV overlaps RV anterodorsally and midventrally; compressed; smooth; narrow inner lamellae; L, 1.15; H, 0.78 mm (Dickinson and Swain, 1967).

Occurrence. Pliocene, Hay Ranch Formation, Gulf Oil Corp., No. 1 Wilkins Ranch well, sec. 21, T. 33 N., R. 61 E., Elko County, Nevada, 1466-2557 feet (447-779 m) (Dickinson and Swain, 1967).

Cypria obesa Sharpe

Plate 10, figure 31

Cypria obesa Sharpe, 1897, p. 462, pl. 48, figs. 1-5. Staplin, 1963, p. 1177, pl. 159, figs. 16, 17.

Suboval, highest anteromedially, height equal three-fifths of length; anterior more narrowly rounded than posterior, extended below; dorsum arched, venter straight to slightly concave; LV larger than RV; compressed; smooth; inner lamellae narrow; only slight vestibule; hinge flange of LV and groove of RV twisted; adductor scar of four large and two small subcentral spots; two frontal spots; L, 0.73 to 0.32; H, 0.46 to 0.50 mm (Staplin, 1963).

Occurrence. Pleistocene, Tazewell? to Mankato or late Cary Substages, Wisconsin Stage of Illinois; Cary Substage of Story County, Iowa; Illinoian Stage of Iowa (Staplin, 1963). Recent, Illinois, Ohio, District of Columbia (Sharpe, 1897 and authors).

Cypria subangulata Chapman

Plate 7, figures 34, 35

Cypria angulata Chapman, 1896, p. 95, pl. 3, figs. 1, 2.

Subreniform, highest anteromedially, dorsum convex, angulate medially, venter concave; anterior broader than posterior; smooth; L, 0.63; H, 0.36 mm (Chapman, 1896).

Occurrence. Pliocene, Merced Formation, Wildcat Canyon, 0.25 mile south of the "Caves", loc. 589 of Chapman, near Berkeley, California (Chapman, 1896).

Cypria turneri Hoff

Plate 10, figure 32

Cypria striolata Herrick, 1887, non Brady, 1864, a synonym of C. exculpta (Fischer, 1855), p. 29, pl. 4, fig. 3.

Cypria exculpta Turner, 1894 (non C. exculpta (Fisher, 1855)). p. 13, pl. 7, figs. 2-8.
Cypria exculpta Sharpe, 1897, (non Fischer, 1855), p. 465, pl. 47, fig. 4.
Cypria elegantula Furtos, 1933 (non Liljeborg, 1853), p. 465, pl. 15, figs. 8-14.
Cypria turneri Hoff, 1942,p. 106, pl. 6, figs. 76-81. Staplin, 1963, p. 1175,pl. 159, figs. 10, 11.
Cypria ophthalmica (Jurine), Delorme, 1969, p. 263, TF 105-131.

Small, delicate, elongate-suboval, highest anteromedially, height exceeds half of length; anterior broader than posterior; dorsum gently convex, venter nearly straight to slightly concave; surface with very fine subparallel lines; posterior area semi-reticulate; inner lamellae narrow, little or no vestibule, adductor scar with four large and one small spot; one frontal spot; L, 0.52 to 0.61; H, 0.32 to 0.38 mm (Staplin, 1963).

Occurrence. Pleistocene, Illinoian Stage of Illinois; late Kansan Stage, Sappa Silt of Meade County, Kansas. Recent of Newfoundland, Delaware, Ohio, Tennessee, Illinois, southern and southeastern USA, Michigan, Wisconsin, Utah, Washington, Alaska (Staplin, 1963); Minnesota (Swain, 1995).

Cypria sp.
not figured
Cypria sp., Swain, 1987a, p. 122.

A form more elongate than C. hoopescreekensis Swain, 1987 from the Cache Valley Formation, Pliocene, Franklin County, Idaho.

Occurrence. Miocene or Pliocene, Humboldt Formation or Salt Lake Group, undifferentiated, Tanner Reservoir Dam Spillway, sec. 9, T. 11 N., R. 19 W., Box Elder County, Utah (Swain, 1987a).

Genus Cypricercus Sars, 1925
(Cyprididae - Eucypridinae)
Cypricercus elegantulus Swain
Plate 10, figures 33 - 36
Cypricercus elegantulus Swain, 1986b, p. 203, pl. 3, figs. 9, 10; p. 5, fig. 11. Swain, 1987c, p. 427, pl. 1, figs. 5, 6.

Elongate subelliptical, highest medially to anteromedially; dorsum gently convex, venter nearly straight to slightly concave; anterior somewhat broader than posterior, both are extended below; LV overlaps RV especially along venter; smooth; females? more convex in edge view; L. 1.32; H, 0.68 mm (Swain, 1987c).

Occurrence. Miocene, Humboldt Formation, near base of Medicine Creek section, T. 39 N., R. 65 E., Winecup Ranch, Elko County, Nevada, just above Triassic (Swain, 1986b). Esmeralda Formation, Black Spring Road section 7659 feet (2332 m) S 26° E. of NW corner of Simon 1:24000 Quadrangle, Nye County, Nevada, about 160 feet (49 m) above base of measured section (Swain, 1987c).

Cypricercus hunterensis Swain
Plate 11, figure 4
Cypricercus hunterensis Swain, 1986b, p. 201, pl. 1, figs. 15a-c, 20; pl. 5, fig. 30. Swain, 1987a, p. 120.

Subtriangular-subreniform, highest submedially, dorsum strongly convex, venter slightly concave; anterior slightly narrower than posterior; LV overlaps RV except postdorsally; moderately convex; smooth; adductor scar an anteromedian group of four spots; two frontal spots; L, holotype, 0.95; H, 0.45 mm (Swain, 1986b).

Occurrence. Miocene, lower part of Humboldt Formation, Gulf Refining Co., No. 1. Thousand Springs well, center SE$^1/_4$ SW$^1/_4$ sec. 8, T. 40 N., R. 66 E., Elko County, Nevada, 350-3200 feet (107-976 m); Gulf No. 2 Wilkins Ranch well, sec. 19, T. 39N., R. 69 E., Elko County, Nevada, 2945-4670 feet (898-1424 m): Medicine Creek section, T. 39 N., R. 65 E., Winecup Ranch, Elko County, Nevada, 139-199 feet (42-61 m) above base of exposure; in conglomeratic limestone and in "Bobcat" limestone near base of Medicine Creek section. Miocene or Oligocene, lower Humboldt or Indian Well Formation, Palisade section, two miles (3.2 km) south of Palisade, Eureka County, Nevada (Swain, 1986b). Humboldt or Salt Lake Group, stream course SE$^1/_4$ NW$^1/_4$ sec. 25, T., 7 N., R. 19W., Box Elder County Utah, about six miles (9.6 km) SW of Lucin (Swain, 1987a).

Cypricercus minensis Swain
Plate 10, figure 41; Plate 11, figure 1.
Cypricercus minensis Swain, 1985, p. 423, pl. 1, figs. 1, 2?, 3.

Elongate subelliptical, highest two-fifths of length from posterior end; dorsum gently convex, venter slightly concave; posterior narrower than anterior; LV overlaps RV strongly along dorsum, less strongly elsewhere; moderately convex; smooth; weak anterodorsal depression in RV; inner lamellae narrow; 18-20 radial canals anteriorly; L, holotype, 1.07; H, 0.60 mm (Swain, 1985)

Occurrence. Miocene, Esmeralda Formation, middle part, Black Spring road section, Nye County, Nevada, 2.85 miles (4.5 km) west of Black Spring, about 800 feet above base of lake bed exposure (Swain, 1985).

Cypricercus sp. aff. C. minensis (Swain)

Plate 11, figures 2, 3

Cypricercus sp. aff. minensis Swain, 1985, p. 424, pl. 1, figs. 10, 15.

Shorter and higher and greatest height is more anterior than in C. minensis (Swain, 1985).

Occurrence. Miocene, Esmeralda Formation, Black Spring road section, 2.85 miles (4.5 km) west of Black Spring, Nye County, Nevada, about 800 feet (244 m) above base of lake bed sequence (Swain, 1985).

Cypricercus mineralensis Swain

Plate 11, figures 5 - 8

Cypricercus sp. aff. affinis (Fischer), Swain, Becker, and Dickinson, 1971, p. 471, pl. 1, figs. 1a-f, 2.

Cypricercus mineralensis Swain, 1985, p. 434, pl. 1, figs. 12, 13. Swain 1987c, p. 428, pl. 3, figs. 4, 5.

Subelliptical-subreniform, highest postmedially; dorsum subtruncate anteriorly, venter gently concave; posterior narrower than anterior, both ends extended below; LV overlaps RV; moderately convex; smooth; L, holotype 1.16; H, 0.60 mm (Swain, 1985).

Occurrence. Miocene, Esmeralda Formation, Finger Rock Wash, Mineral County, Nevada, 0.5 miles (0.8 km) south of Rawhide Ranch, Granny Goose 1:24000 Quadrangle, 1400-1500 feet (425-455 m) above base of lake bed exposure; Black Spring road section 2.85 miles (4.5 km) west of Black Spring, Black Spring Quadrangle, Nye County, Nevada, about 800 feet (244 m) above base of lake beds sequence; 7.6 miles (12.1 km) east of Finger Rock Wash, on road to Pactolus Mine, Mineral County, Nevada, Goldyke 1:24000 Quadrangle, about 1000 feet (305 m) above base of lake bed sequence (Swain, 1985). Humboldt Formation, Cobre, Nevada (Swain, et al., 1971). Horse Spring Formation, Fossil Ridge, Clark County, Nevada (Swain, 1987c).

Cypricercus oblongatus Swain and Wagner

Plate 11, figures 9 - 11

Cypricercus oblongatus Swain and Wagner, 1985, p. 136, pl. 5, figs. 7, 11, 13.

Suboblong-reniform, highest medially to anteromedially; dorsum convex, venter nearly straoght to slightly concave; posterior slightly narrower than anterior, both ends extended below; LV a little larger than RV, convexity moderate, maximum width median; surface with finely bertillion markings or nearly smooth; LV hinge with terminal short ridges and intermediate groove; adductor scar of four or five spots; two frontal spots; inner lamella broadset anteriorly; L, holotype, 0.90; H, 0.53 mm (Swain and Wagner, 1985).

Occurrence. Miocene, Esmeralda Formation, Blair Junction section, Esmeralda County, Nevada, samples 1A - 2C, 4.1?, 4 - 7? (Swain and Wagner, 1985).

Cypricercus palisadensis Swain

Plate 11, figures 12, 13, 15, 16

Cypricercus palisadensis Swain, 1986b, p. 202, pl. 1, figs 11-14, pl. 2, figs 11, 12; pl. 5, figs. 21-25.

Elongate-ovate-elliptical, highest submedially; dorsum convex, venter nearly straight to slightly concave; posterior margin a little more narrowly rounded than anterior, both are extended below; LV overlaps RV along venter and ends; smooth; L, holotype, 1.07; H, 0.63 mm (Swain, 1986b).

Occurrence. Miocene or Oligocene, Humboldt Formation or Indian Well Formation, about two miles (3.2 km) south of Palisade, in northern Pine Valley, Eureka County, Nevada; also in Miocene, Humboldt Formation, Gulf Oil Co. No. 2 Wilkins Ranch well, sec. 19, T. 39 N., R. 69 E., Elko County Nevada, 4580-4610 feet (1396-1405 m) (Swain, 1986b).

Cypricercus sp. aff. palisadensis Swain

Plate 11, figure 14

Cypricercus? sp. aff. C. palisadensis Swain, 1987b, p. 22, pl. 2, fig. 19.

Elongate subelliptical, highest anteromedially, venter less convex than dorsum; posterior much narrower than anterior, strongly extended below; smooth; L, 0.63; H, 0.33 mm (Swain, 1987).

Occurrence. Eocene, Twin Bridges Limestone, west side of Huntingdon Creek, at Twin Bridges, center sec. 36, T. 32 N., R. 55 E., Elko County, Nevada (Swain, 1987b).

Cypricercus parafuscatus Dickinson and Swain

Plate 11, figures 17-19

Cypricercus parafuscatus Dickinson and Swain, 1967, p. 239, pl. 35, figs. 15-18.

Subreniform, highest medially, dorsum strongly arched, venter slightly concave; anterior narrower than posterior, extended below; LV overlaps RV anterodorsally and midventrally; moderately convex, thickest medially; smooth; inner lamellae broad anteriorly; L, holotype, 0.63; H, 0.55 mm (Dickinson and Swain, 1967).

Occurrence. Pliocene, Hay Ranch Formation ("Upper Humboldt"), Gulf Oil Co., No. 1 Wilkins Ranch well, sec. 21, T. 38 N., R. 61 E., Elko County, Nevada, 1957-4230 feet (597-1289 m) (Dickinson and Swain, 1967).

Cypricercus reticulatus Zaddach
Plate 10, figures 37 - 39
Cypris reticulatus Zaddach, 1844, p. 34.
Cypris affinis Fischer, 1851, p. 160, pl. 10, figs. 9-11.
Cypris fuscatus Sharpe, 1897 (non Jurine, 1820), p. 442, pl. 43, figs. 104.
Cypris testudinaria Sharpe, 1897, p. 444, pl. 44, figs. 1-4.
Cypris (Cypris) fuscata Sharpe, var. minor (G.W. Müller, 1900) p. 818, TF 1287 b-d.
Eucypris affinis hirsuta Furtos, 1933, p. 450, pl. 2, figs. 9-11, 16.
Cypricercus affinis (Fischer), Sars, 1925, p. 119, pl. 55, fig. 1.
Cypricercus reticulatus (Zaddach), Hoff, 1942, p. 139, pl. 1, fig. 1, pl. 7, fig. 105, pl. 8, figs. 106-109. Staplin, 1963, p. 793, pl. 94, figs. 22-25.

Subtriangular, highest anteromedially, inflated ventrally, anterior broader than posterior, both ends extended below, dorsum strongly arched, angulated anteromedially, venter slightly concave, nearly straight, "blunt selvage projection just anterior to the sinuation on the right valve" (Staplin, 1963); LV slightly overlaps RV anteriorly and ventrally; smooth; scattered tiny pits marking normal pore canals; hinge selvage flanges on LV fit into flange grooves on RV; adductor scar with four larger and two small spots; inner lamellae broad anteriorly, with inner striated chitinous zone; L, up to 1.00; H, up to 0.58 mm (Staplin, 1963).

Occurrence. Pleistocene, Illinoian Stage and Mankato or late Cary Substage, Wisconsin Stage, of Illinois and western Indiana. Holocene, holarctic region; in North America from the Arctic to Virginia and Maryland (Staplin, 1963).

Cypricercus reticulatus wrenshallensis Swian
Plate 8, figure 40

Cypricercus reticulatus wrenshallensis Swain, 1995, p. 22, pl. 2, figs. 5 - 16, 18, 19

Subovate (females?) to subpyriform (males?), highest anteromedially; dorsum convex, venter concave, posterior narrower than anterior, extended below; LV larger than RV; moderately convex; surface densely pitted to nearly smooth, in part with overprint of reticulating ridges; LV hinge rabbeted to receive RV edge, adductor scar loosely packed 6 - 9 spots, two frontal spots, eight to nine dorsal spots; inner lamellae of moderate width; normal pores may have crateriform rims; L, holotype, 0.44; H, 0.23 mm.

Occurrence. Late Pleistocene (Mankato?) clays of Lake Nemadji, Wrenshall, Carlton County, Minnesota (Swain, 1995).

Cypricercus sp. aff. C. reticulatus (Zaddach), Swain

Plate 11, figures 20, 21

Cypricercus sp. aff. C. reticulatus (Zaddach), Swain, 1986c, p. 367, pl. 1, figs. 2, 3.

Elongate-subelliptical-subreniform, highest medially to anteromedially, venter gently concave, dorsum slightly angulated medially; posterior narrower than anterior, both ends extended below; LV overlaps RV especially along venter and posterior; smooth; moderate convexity; L, 0.62; H, 0.35 mm (Swain, 1987).

Occurrence. Pliocene, Cache Valley Formation, NW$^1/_4$ SE$^1/_4$ NE$^1/_4$ sec. 10, T. 12 S., R. 40 E., Franklin County, Idaho (Swain, 1986c).

Cypricercus sacramentoensis Swain

Plate 11, figures 22-25

Cypricercus? sacramentoensis Swain, 1986b, p. 202, pl. 2, fig. 26; pl. 5, fig. 9. Swain, 1987c, pl. 427, pl. 4, fig 7, pl. 5, figs 6, 10.

Elongate, subtriangular-subreniform, highest anteromedially; dorsum convex, venter concave; posterior narrower than anterior, the latter bears narrow marginal rim in RV; LV a little large than RV, compressed, widest submedially; smooth; L, holotype, 0.87; H, 0.41 mm (Swain, 1986b).

Occurrence. Miocene or Oligocene, North Creek Formation, Sacramento Pass section, 5.85 miles (9.4 km) east of Sacramento Pass summit on north side of Highway 50, White Pine County, Nevada (Swain, 1986b). Humboldt Formation or Indian Well Formation, two miles (3.2 km) south of Palisade, Eureka County, Nevada, northern end of Pine Valley (Swain, 1987c).

Cypricercus stami Dickinson and Swain

Plate 11, figures 26, 27

Cypricercus stami Dickinson and Swain, 1967, p. 339, pl. 35, figs. 6-12. Swain, 1987a, p. 119.

Non Swain, 1986, p. 89, pl. 1, figs. 2, 3 (=C. sp. aff. C. stami).

Elongate, subelliptical-subreniform, highest anteromedially; dorsum convex, venter straight to concave; posterior narrower than anterior, both ends extended below; LV larger than RV, overlapping RV ventrally; low convexity; finely punctate; inner lamellae broad anteriorly; L, 1.0; H, 0.55 mm (Dickinson and Swain, 1967).

Occurrence. Pliocene, Hay Ranch Formation ("Upper Humboldt"), Gulf Oil Co., No. 1 Wilkins Ranch well, sec. 21, T. 38 N., R. 61 E., Elko County, Nevada, 1314-4230 feet (401-1289 m) (Dickinson and Swain, 1967). Miocene or Pliocene, Salt Lake Group, NW1/4 SW1/4 sec. 12, T. 10 S., R. 4 W., Juab County, Utah (Swain, 1987a).

Cypricercus sp. aff. C. stami Dickinson and Swain

Plate 11, figures 28, 29

Cypricercus stami Dickinson and Swain, Swain, 1986a, p. 91, pl. 1, figs 2, 3.

Differs from C. stami in being smooth rather than punctate and perhaps in having greatest height posteromedian.

Occurrence. Pliocene, Glenns Ferry Formation, sec. 14, T. 4., R. 1 W., Owyhee County, Idaho (Swain, 1986a).

Cypricercus toanoensis Swain

Plate 11, figures 30, 31

Cypricercus toanoensis Swain, 1986b, p. 200, pl. 1, figs. 16-18; pl. 2, figs. 10, 11; pl. 5, fig. 31. Swain, 1987a, p. 120; Swain, 1987c, p. 427, pl. 5, fig. 9.

Elongate subreniform; highest anteromedially; dorsum convex, truncated anteromedially, straightened medially; venter concave; posterior a little narrower than anterior, both ends extended below; LV overlaps RV except postdorsally; strongly convex, widest postmedially; smooth; LV hinge grooved to receive edge of RV edge; adductor scar with five spots; two frontal spots; inner lamellae narrow; L, holotype 1.01; H, 0.60 mm (Swain, 1986b).

Occurrence. Miocene, Humboldt Formation, Medicine Creek section, T. 39 N., R. 65 E., Winecup Ranch, Elko County, Nevada, 75-210 feet (23-64 m) above base of exposure; Gulf Refining Co., No. 1 Thousand Springs well, center SE1/4 SW1/4 sec. 8, T. 40 N., R. 66 E., Elko County, Nevada, 150-3350 feet (46-1021 m); Gulf Oil Co., No. 2 Wilkins Ranch well, Montello area, sec. 19, T. 39 N., R. 69 E., Elko county, Nevada, 2665-4290 feet (812-1308 m); Cobra surface section, Elko County, Nevada from base to 1079 feet (329 m); in "Bobcat Limestone" near base of Medicine Creek section; Miocene or Oligocene, Humboldt or Indian Well Formation, 2 miles (3.2 km) south of Palisade, Eureka County, Nevada; Miocene, Humboldt Formation, McCarthy No. 1 Rahas well, T. 35 N., R. 55 E., just south of Halleck, Elko County, Nevada; North Creek Formation, 5.85 miles (9.4 km) east of Sacramento Pass, White Pine County, Nevada; Miocene or Oligocene, Humboldt or Indian Well Formation 1.5 miles (2.4 km) northeast of Indian Well, Elko County, Nevada. Miocene or Pliocene, Salt Lake Group, undifferentiated, Rosel Point, SW1/4 SE1/4 sec. 4, T. 8 N., R. 7 W., Box Elder County, Utah (Swain, 1986b, 1987a, c, d).

Cypricercus tuberculatus (Sharpe)

Plate 11, figure 32, 33

Spirocypris tuberculatus Sharpe, 1908, p. 406, pl. 50, figs. 1, 2; pl. 54, fig. 4; pl. 55, figs, 1-6 (Spirocypris Sharpe, 1903 is a junior synonym of Strandesia Stuhlmann, 1988).

Cypricercus tuberculatus (Sharpe), Hoff, 1942, p. 133, pl. 7, figs. 103, 104. Gutentag and Benson, 1962, p. 21, pl. 1, fig. 11, TF 5-4. Staplin, 1963, p. 794, pl. 94, fig. 9. Benson, 1967, p. 228, TF 10-1, 11-4.

Subelliptical-elongate ovate, highest anteromedially, anterior broader than posterior; dorsum convex, venter slightly sinuate; LV overlaps RV anteriorly and ventrally, but overlap is variable; densely papillose, each papilla bearing normal pore and seta in living animal; tumid, width seven-ninths of length; inner surface pitted; inner lamellae broad, especially anteriorly (Staplin, 1963).

Occurrence. Pleistocene, Aftonian Stage to Tazewell Substage, Wisconsin Stage of Iowa; late Kansan Stage, Sappa Silt of Meade County, Kansas. Recent, northern Illinois and northern Indiana (Staplin, 1963). Late Kansan or early Yarmouthian Stage of Reno County, Kansas (Gutentag and Benson, 1962); Yarmouthian Stage, Dickinson County, Kansas (Benson, 1967).

Cypricercus? sp. Swain

not figured

Cypricercus? sp., Swain, 1986b, p. 203.

Subovate-subquadrate, highest medially, dorsum gently convex, venter nearly straight, slightly concave; posterior narrower than anterior; both ends extended below; LV overlaps RV most along dorsum and venter; compressed; smooth; L, 1.06; H, 0.70 mm (Swain, 1986b).

Occurrence. Miocene or Pliocene, Humboldt or Hay Ranch Formation, Gulf Oil Co. No. 2 Wilkins Ranch well, sec. 19, T. 39 N., R. 69 E., Elko County, Nevada 2665-2695 feet (813-822 m) (Swain, 1986b).

Cypricercus? sp. 1, Swain

Plate 12, figure 1

Cypricercus? sp. 1, Swain, 1987c, p. 428, pl. 4, fig. 18.

Subelliptical, highest medially, dorsum more convex than venter; anterior narrower than posterior; LV slightly larger than RV; smooth; L, 0.83; H, 0.50 mm (Swain, 1987c).

Occurrence. Oligocene, Pruett? Formation, Buck Hill volcanic series, Aqua Fria 1:62,500 Quadrangle, Brewster County, Texas. Lat. 29° 43.0' N., Long. 103° 41.3' W. (Swain, 1987c).

Cypricercus? sp., Swain
not figured
Cypricercus? sp., Swain, 1995, p. 23, pl. 2, fig. 20

Larger, more elongate than C. reticulatus wrenshallensis, and smooth rather than pitted and reticulate.

Occurrence. Late Pleistocene clays of Lake Nemadji, Wrenshall, Carlton County, Minnesota (Swain, 1995).

Genus Cypriconcha Sars, 1926

(Cyprididae-Cypridinae?)
Cypriconcha? sp., Swain
Plate 12, figure 2
Cypriconcha? sp., Swain, 1987b, p. 222, pl. 1, fig. 12.

Elongate, subelliptical-subovate, highest postmedially; dorsum gently convex, venter slightly concave to sinuous; anterior narrower than posterior, both ends extended below; smooth; compressed; LV hinge groove receives edge of RV; inner lamellae broad at both ends (generic feature): vestibules present; L, 0.77; H, 0.42 mm (Swain, 1987).

Occurrence. Eocene, Twin Bridges Limestone, center sec. 36, T. 32 N., R. 55 E., Elko County, Nevada (Swain, 1987b).

Genus Cyprideа Bosquet, 1852

(Cyprideidae)
Cypridea (Yumenia) anomala Peck
Plate 12, figures 3 - 6
Cypridea anomala Peck, 1941, pl. 299, pl. 47, figs. 18-28. Peck, 1951, pl. 311, pl. 29, figs. 12, 13.

Subpyriform, highest anteromedially, dorsum arched, angulated at position of greatest height; venter sinuous to nearly straight, concave ventroposteriorly; anterior broadly rounded, posterior narrow, strongly extended and angulated below; LV overlaps RV; ventral surface expanded, flattened; median valve surface flattened; anteroventral beak and notch absent or inconspicuous; pitted surface, with or without scattered nodes grouped terminally; inner lamellae developed terminally; L, holotype, 0.92; H, 0.66 mm(Peck, 1941).

Occurrence. Lower Cretaceous ("Lower Bear River") Formation, Lincoln County, Wyoming; Draney Formation, Freedom Quadrangle, Idaho-Wyoming; Kootenai Formation, near Twin Bridges, Dillon and Drummond, Montana (Peck, 1941); Cloverly Formation, five miles (8.04 km) north of Gardner, Montana (Peck, 1951).

Cypridea (Bisulcocypridea) arvadensis (Swain)

Plate 7, figures 18 - 22; Plate 8, figures 17, 18; Plate 12, figures 7, 8

Ilyocypris arvadensis Swain, 1949, p. 178, pl. 32, figs. 20-22. Swain, 1956, p. 136, pl. 3, fig. 16. Swain, 1964, p. 275, pl. 43, figs 8, 9a, b.

Cypridea arvadensis (Swain), Peck, 1951, p. 314, pl. 50, figs. 1-12, 16-18.

Subquadrate, highest about one-fifth of length from anterior end, dorsum and venter nearly straight, subparallel or converging slightly, posteriorly; posterior narrower than anterior and typically extended below; LV overlaps RV; anterodorsal valve surfaces each with two short sulci repeated by a small lobe; anteroventrally is a shallow groove parallel to margin, weaker on LV than on RV; surface pitted; L, 1.15; H, 0.61 mm (Swain, 1964).

Occurrence. Paleocene, Fort Union Formation, Tongue River Member, SW1/4 sec. 31, T. 9 S., R. 49 E., Powder River County, Montana; SW1/4 SE1/4 SW1/4 sec. 17, T. 9 S., R. 45 E., Powder River County, Montana; roof of Roland Coal, SE1/4 sec. 19, T. 9 S., R. 45 E., Powder River County Montana. Paleocene or Eocene, Flagstaff Limestone, one mile (1.6 km) east of Ephraim, Utah. Eocene, Colton Formation, 8 miles (12.9 km) north of Castlegate, Utah on highway 33, above a very fossiliferous mudstone; Wasatch Formation 4.5 miles (7.2 km) east of Arvada, Wyoming, NW1/4 SW1/4 SW1/4 sec. 6, T. 54 N., R. 76 W., in Montana between Arvada and Roland coals (Swain, 1949). Eocene, Colton Formation, Dripping Rock Gulch, sec. 12, T. 2 N., R. 104 W., Rio Blanco County, Colorado (Swain, 1964a).

Cypridea (Bisulcocypridea) bicostata Sohn

Plate 12, figures 9-12, 13, 14

Cypridea (Bisulcocypridea) bicostata Sohn, 1969, p. B6, pl. 2, figs. 1-10, 13, 15. Swain, 1987b, pl. 229, pl. 3, figs. 11, 13, 14.

Subquadrate, highest anteromedially, dorsum and venter nearly straight and converge toward posterior; anterior broader than posterior; LV larger than RV; bisulcate in anterior and dorsal area; intervening rounded lobe not reaching to dorsum; the more anterior sulcus shorter than median sulcus; a low ridge parallels dorsum and venter; smooth or finely pitted; scattered spinelets mostly near ends; anteroventral beak blunt, furrow extends to below midheight; L, holotype, 1.24; H, 0.82 mm (Sohn, 1969).

Occurrence. Lower? Cretaceous, Aptian? age Newark Canyon? Formation, U.S.G.S. loc. 27754, sec. 21, T. 29 N., R. 53 E., Eureka County, Nevada (Sohn, 1969); Upper Cretaceous?, lower part of Sheep Pass Formation, Egan Range, Sheep Pass Canyon section, lower member, T. 10 N., R. 62, 63 E., White Pine County, Nevada (Swain, 1987b).

Cypridea bispinosa Jones, 1878; Galeeva, 1955
(see Cypridea dequeenensis Swain and Brown)

Cypridea (Bisulcocypridea) bisulcata (Swain)
Plate 12, figures 15-18
Cypridea bisulcata Swain, 1949, p. 180, pl. 33, figs. 13-19. Swain, 1964a, p. 276, pl. 43, fig. 7; pl. 44, figs 10a-f.
Cypridea (Bisulcocypridea) bisulcata (Swain), Sohn, 1969, p. B6.

Subquadrate, highest near anterior end; dorsum nearly straight, about two-thirds of length; venter converges backward with dorsum; anterior broader than posterior; distinct anteroventral beak, hood, and notch; furrow extends nearly to midheight in some specimens; LV overlaps RV along free margins; valves relatively compressed; two short anterodorsal sulci; general surface finely and densely punctate; LV rabbeted hinge surface overlaps RV; adductor muscle scar a submedian row of three elongate spots and postjacent rounded spot; inner lamellae narrow, broadest anteriorly; L, holotype 1.00; H, 0.62 mm (Swain, 1949, 1964).

Occurrence. Paleocene or Eocene, Flagstaff Limestone, one mile (1.6 km) west of Ephraim, Utah in lower 50 feet (15 m) of formation; Eocene, Green River Formation, 100 feet (30 m) above base, Scenery Gulch, 20 miles (32 km) west of Meeker, Colorado on White River (Swain, 1949). Upper part of Colton Formation and Colton-Green River transition beds, Uinta and Piceance Creek Basins, Colorado and Utah; Knight Formation, six miles (9.6 km) east of Kemmerer, Wyoming, and interbedded Knight and Green River Formations, one mile (1.6 km) south of Piedmont, Uinta County, Wyoming (Swain, 1964a).

Cypridea brevicornis Peck
Plate 12, figures 19-21
Cypridea brevicornis Peck, 1941, p. 299, pl. 44, figs. 22-24. Peck, 1951, p. 311, pl. 49, figs. 7, 8.

Subpyriform, highest anteromedially, dorsum arched and angulated at point of greatest height; venter sinuous, gently concave; distinct anteroventral pointed beak, hood and notch; posterior much narrower than anterior, extended below; LV overlaps RV; smooth to very finely pitted; short blunt spine postmedially; may have scattered nodes posteriorly; L, holotype, 0.8; H, 0.5 mm (Peck, 1941).

Occurrence. Lower Cretaceous, Draney Formation, Freedom and Crow Creek Quadrangles, Idaho-Wyoming, Lincoln County, Wyoming (Peck, 1941, 1951).

Cypridea compta Peck
Plate 12, figures 22-26
Cypridea compta Peck, 1941, p. 300, pl. 94, figs. 25-28. Peck, 1951, p. 311, pl. 49, figs. 1-4, 14.

Subquadrate, highest anteromedially; dorsum straight, venter sinuous to nearly straight; well defined anteroventral beak and notch; posteroventral angulation of margin; posterior narrower than anterior, extended below; LV large than RV, compressed: surface reticulate network of small pits; nodes or short spines peripherally; large subcentral to anteromedian node in some specimens; L, 0.8; H, 0.5 mm (Peck, 1941, 1951).

Occurrence. Lower Cretaceous, Draney Formation, Freedom Quadrangle, Idaho-Wyoming; Lower Bear Formation, Lincoln County, Wyoming (Peck, 1941, 1951).

Cypridea dequeenensis Swain and Brown
Plate 12. figures 27-30; Plate 13, figures 30, 31.
Cypridea ventrosa var. bispinosus Vanderpool, 1928, p. 104, pl. 14, figs. 1, 2.
non Cypridea bispinosa Jones, 1878, p. 109, pl. 3, figs. 9, 10.
non Cypridea bispinosa Galeeva, 1955, p. 36, TF 1a-g; Liubimova, 1956, p. 25, pl. 5, figs. 2-4 (fide Howe and Laurencich, 1958, p. 119).
?Cypridea cf. ventrosa bispinosus (Jones), Calahan, 1939, p. 55, pl. 8, figs. 1, 2.
?Cypridea wyomingenis Jones, Peck, 1941, p. 297, pl. 43, figs. 10, 11.
Cypridea dequeenensis Swain and Brown, 1964, p. 15, pl. 4, fig. 8, pl. 5, figs. 3a-c, TF 3c, 4a. Swain and Brown, 1972, p. 13, pl. 1, fig. 18, TF 2.
Cypridea kleinbergi Galeeva, 1955, p. 31, pl. 8, figs. 2a-e (=C. bispinosa Galeeva by Howe and Laurencich).

Subtrapezoidal, highest about one-fourth of length from anterior end; hinge margin nearly straight, about five-sevenths of length; venter slightly convex, converging with dorsum posteriorly, with well defined anteroventral beak, hood and notch; posterior narrower than anterior; LV overlaps RV terminally and ventrally; dorsally RV overlaps LV; compressed; large spine lies posteromedially, another spine projects beyond dorsum also postmedially; other smaller scattered spines on surface; densely and coarsely pitted; RV hinge with long rabbet groove and terminal triangular tooth-like elevations; inner lamellae narrow, slight vestibules; L, 0.93; H, 0.60 mm (Swain and Brown, 1964).

Occurrence. Lower Cretaceous, DeQueen Limestone, at Gypsum Bluff near Murphreesboro, Pike County, Arkansas (Swain and Brown, 1964, Vanderpool, 1928). Lower Cretaceous (Barremian), Unit H, NC-DA-OT-12 well, Dare County, North Carolina, 7306-7316 feet (2227-2230 m) (Swain and Brown, 1972).

Cypridea (Cypridea) diminuta Vanderpool

Plate 12, figures 31-33

Cypridea diminutus Vanderpool, 1928, p. 103, pl. 13, figs. 7, 8.

Cypridea diminuta Vanderpool, Peck, 1941, p. 298, pl. 44, figs. 29-32; 1951, p. 312, pl. 49, fig. 6. Howe and Laurencich, 1958, p. 121, TF. Swain and Brown, 1964, p. 16, pl. 4, fig. 9; pl. 5, figs. 2a-d, TF 3b.

Cypridea (Cypridea) diminuta Vanderpool, Sohn, 1969, p. B6, pl. 1, figs. 5, 6. Swain and Brown, 1972, p. 13.

Small, subquadrate, highest one-fourth of length from anterior end; dorsum nearly straight, anterior cardinal angle more obtuse than that at posterior; venter straight to slightly convex; small anteroventral beak, hood, and notch; LV overlaps RV; compressed; densely pitted; few scattered pustules; LV outer margin with grooves along venter; LV hinge groove with terminal small depressions; inner lamellae narrow; L, 0.70; H, 0.40 (Swain and Brown, 1964).

Occurrence. Middle Cretaceous, DeQueen Formation, Arkansas (Vanderpool, 1928; Sohn, 1969); Lower Atkinson Formation, wells in Suwanee, Columbia, and Putnam Counties, Florida; beds of Eagle Ford and older Cretaceous age in wells in North Carolina (Swain and Brown, 1964, 1972).

Cypridea (Pseudocypridina) henrybelli Sohn

Plate 12, figures 38-40

Cypridea sp., Sohn, 1958, p. 123, pl. 1, figs. 13-15.

Cypridea (Pseudocypridina) henrybelli Sohn, 1979, p. 17, pl. 3, figs. 10-17, pl. 8, figs. 1-25.

Subquadrate-subovate, highest anteromedially; hinge margin nearly straight, about half of shell length; venter slightly convex; posterior narrower than anterior; anteroventral small beak, hood and notch; LV overlaps RV; each end of valve has node-like or short ridge-like elevations; finely and densely punctate surface; hinge surface indented; L, holotype, 1.16; H, 0.76 mm (Sohn, 1979).

Occurrence. Lower Cretaceous, Lakota Formation, sec. 15, T. 8 S., R. 4 E., Fall River County, South Dakota (Sohn, 1979).

Cypridea (Pseudocypridina) inornata (Peck)

Plate 7, figures 16, 17, 25 - 28. Plate 12, figures 35-37

Cypridea inornata Peck, 1941, p. 301, pl. 44, figs. 33-36; Sohn, 1958, p. 123, pl. 1, figs. 17, 18.

Pseudocypridina inornata (Peck), Peck, 1951, p. 319, pl. 48, figs. 8-11; Sohn, 1979, p. 15, pl. 4, figs. 18023.

Large, subovate-hemicircular, highest submedially, dorsum strongly convex, venter nearly straight to sinuous, with distinct anteroventral pointed beak, hood and notch; anterior narrower than posterior; LV larger than and overlaps RV most strongly along venter; smooth; L, 1.4; H, 0.9 mm (Peck, 1941).

Occurrence. Lower Cretaceous, Kootenai Formation, near Great Falls, Montana; Minnewaste Limestone, Skyline Drive east of Hot Springs, South Dakota; Lakota Formation, 3.2 miles southeast of Hot Springs, South Dakota on road to Fall River (Peck, 1941, 1951). Lakota Formation, Creek County, Wyoming and Fall River County, South Dakota (Sohn, 1958, 1979).

Cypridea (Pseudocypridina) laeli Sohn

Plate 12, figures 41-44

Cypridea (Pseudocypridina) laeli Sohn, 1979, p. 16, pl. 3, figs. 1-13, 24, 25, 32.

Ovate-semicircular, highest medially to slightly anteromedially; dorsum strongly convex, venter gently convex to sinuous; anteroventral small, beak, hood and notch; posterior narrower than anterior; LV larger than RV; moderate width; dorsal marginal ridge on each valve; ventral marginal ridge on LV; finely and densely pitted; L, 1.12; H, 0.83 mm (Sohn, 1979).

Occurrence. Lower Cretaceous, Lakota Formation, Crook County, Wyoming, U.S.G.S. coll. 31001 (Sohn, 1979).

Cypridia (Pseudocypridina) laevicula (Peck)

Plate 7, figures 25 - 28

Pseudocypridina laevicula Peck, 1951, p. 319, pl. 48, figs. 1-4.

Subovate-subquadrangular, highest anteromedially; dorsum nearly straight, venter slightly convex, converging with dorsum posteriorly; posterior narrower than anterior; anteroventral beak and notch very weakly developed; LV overlaps RV; a single row of small papillae parallels terminal margins in most specimens; L, 0.84; H, 0.53 mm (Peck, 1951).

Occurrence. Lower Cretaceous, Bear River Formation near Evanston, Wyoming; Spring Creek Gap, Uinta County, Utah; Fish Creek, Sublette County, Wyoming (Peck, 1951).

Cypridea longispina Peck 1941
(see Longispinella longispina (Peck))

Cypridea (Ulwellia) menevoides (Swain and Brown)
Plate 13, figure 8
Cypridea (Cypridea) menevoides Swain and Brown, 1972, p. 13, pl. 1, fig. 21.

Subquadrate-subtrapezoidal, highest one-fifth from anterior end; dorsum nearly straight, venter slightly convex, converging posteriorly with dorsum; distinct anteroventral beak, hood and notch; posterior narrower than anterior; RV overlaps LV on free margins.; along dorsum LV overlaps RV; valves relatively compressed; inner lamellae narrow; rabbeted hinge margin with small terminal notches in LV; finely pitted except smooth marginal zone; L, 0.98; H, 0.62 mm (Swain and Brown, 1972).

Occurrence. Lower Cretaceous, Unit H, NC-Cur-OT-12 well, Currituck County, North Carolina, 4308-4318 feet (1313-1316 m) (Swain and Brown, 1972).

Cypridea (Ulwellia) minuta Peck
Plate 34, figures 31-33
Ulwellia minuta Peck, 1951, p. 320, pl. 49, figs. 9-11.

Subquadrate, elongate, highest near anterior end; dorsum slightly concave, venter sinuous, with moderately developed beak and notch anteriorly; RV overlaps LV; dorsomedian broad depression; pitted; L, holotype, 0.73; H, 0.42 mm (Peck, 1951).

Cypridea nitidula Peck
Plate 13, figures 9-11
Cypridea nitidula Peck, 1941, p. 301, pl. 43, figs. 1-5; 1951, p. 312, pl. 49, fig. 5.

Subtrapezoidal, subquadrate, highest one-fourth to one-fifth from anterior end; dorsum and venter nearly straight, converging toward posterior; anteroventral beak short, hood and furrow also short; notch distinct; posterior narrower than anterior, angulated at ventromarginal bend; compressed; finely pitted; L, 1.02; H, 0.65 mm (Peck, 1941, 1951).

Occurrence. Lower Cretaceous, Draney Formation, Freedom Quadrangle, Montana-Idaho; Crow Creek Quadrangle, Montana-Wyoming; Lincoln County, Wyoming; "Lower Bear River" Formation, Lincoln County, Wyoming (Peck, 1941).

Cypridea (Bisulcocypridea) nyensis (Swain)
Plate 13, figures 12, 13

Cypridea nyensis Swain, 1964a, p. 278, pl. 44, figs. 9a-c.

Cypridea (Bisulcocypridea) nyensis (Swain), Sohn, 1969, p. B6; Swain, 1987b, p. 228, pl. 3, figs. 10, 12.

Elongate-subquadrate, medium size, highest near anterior end. venter with anterior notch and beak; posterior narrower than anterior; valves subequal, compressed; short dorsomedian sulcus in anterodorsal quadrant; second more anterior sulcus separated from first sulcus by a rounded lobe that does not reach dorsal margin; a second lobe lies anteroventrad of anterior sulcus; finely and densely pitted and with scattered large pustules over surface; L, 1.17; H, 0.60 mm (Swain, 1964).

Occurrence. Paleocene? or early Eocene?, Sheep Pass Formation, Johnson Canyon, Grant Range, Nye County, Nevada (Swain, 1964a, 1987b).

Cypridea obesa Peck

Plate 13, figures 14, 15

Cypridea obesa Peck, 1951, p. 318, pl. 50, figs. 19-21.

Subhemispherical, highest anteromedially; dorsum strongly arched, subumbonate, venter nearly straight except for anteroventral notch and small beak; posterior narrower than anterior, extended below; LV overlaps RV nearly equally around entire periphery; strongly convex; finely pitted, these become longitudinally arranged along venter; L, 0.92; H, 0.56 mm (Peck, 1951).

Occurrence. Lower Cretaceous, Cloverly Formation, near Lander, Wyoming, $S^1/_4$ $SE^1/_4$ sec. 5, T. 33 N., R. 100 W., and three miles (4.8 km) SW of Lander, Wyoming, halfway between roads in Squaw Creek and middle fork of Popo Agie River (Peck, 1951).

Cypridea (Cypridea) pecki Sohn

Plate 13, figures 16-18

Cypridea (Cypridea) pecki Sohn, 1969, p. B5, pl. 1, figs. 1-4, 7-9, 12-16.

Elongate-subquadrate, highest medially, ends about equally curved, but anterior is extended below; venter with weak anterior beak and notch; LV slightly overlaps RV; closely spaced minute polygonal pits of valve surfaces do not extend to edges; compressed, straight sides in dorsal view; hinge of LV a groove to receive edge of RV; L, holotype, 0.78; H, 0.45 mm (Sohn, 1969).

Occurrence. Lower Cretaceous?, Newark Canyon Formation?, center $W^1/_2$ $NW^1/_4$ sec. 21, T. 29 N., R. 53 E., Elko County, Nevada (Sohn, 1969).

Cypridea (Pseudocypridina) piedmonti (Roth)

Plate 13, figures 5, 19-22; Plate 7, figures 29, 30; Plate 8, figure 19

Pseudocypridina piedmonti Roth, 1933, p. 404, pl. 48, figs. 7a-h. Peck, 1951, p. 319, pl. 48, figs. 16-20. Sohn, 1958, p. 123, pl. 1, figs. 5-8.
Cypridea piedmonti (Roth), Harper and Sutton, 1935, p. 625, pl. 76, figs. 12-15.
Cypridea (Pseudocypridina) piedmonti Roth, Sohn, 1979, p. 15, pl. 6, figs. 1-47.
Non Cypridea (Pseudocypridina) piedmonti Roth, Swain, 1946, p. 550, pl. 82, figs. 10-12 (=Cypridea? salvadorensis Krommelbein, 1962).

Subhemicircular, highest anteromedially; dorsum strongly convex, venter sinuous, with moderately developed anteroventral beak and notch; anterior broader than posterior, both ends extended below; LV overlaps RV; along dorsum RV may extend beyond LV "yet bends downward to fit under the left valve" (Roth, 1933); hinge line depressed as a trough in dorsal view; entire surface covered by roughly polygonal, in some cases elongate, punctae; maximum L, 1.08; maximum H, 1.02 mm (Roth, 1933).

Occurrence. Lower Cretaceous, Lakota Formation, Chilson Member, sec. 28, T. 4 N., R. 6 E., Meade County, South Dakota (referred to Upper Jurassic Morrison Formation by Roth, 1933, and other authors); also at several USGS localities in Fall River and Lawrence Counties, South Dakota (Sohn, 1979).

Cypridea quadrata Peck
Plate 13, figures 23, 24
Cypridea quadrata Peck, 1951, p. 318, pl. 50, figs. 13-15.

Elongate subquadrate, highest medially, posterior of same height as anterior but with narrower curvature; LV overlaps RV; venter with anterior beak and small notch, posterior with small projection; coarsely pitted; along venter pits are in longitudinal rows; L, 0.92; H, 0.59 mm (Peck, 1951).

Occurrence. Lower Cretaceous, Cloverly Formation, S1/2 SE1/4 sec. 5, T. 33 N., R. 100 W., four miles (7.4 km) west of Lander, Fremont County, Wyoming; three miles (4.8 km) south of Lander, Wyoming, halfway between roads in Squaw Creek and middle fork of Popo Agie River; two miles (3.2 km) north of Lander, Wyoming at northern end of Lander anticline (Peck, 1951).

Cypridea (Bisulcocypridea) skeeteri Peck
Plate 13, figures 25-27
Cypridea skeeteri Peck, 1951, p. 316, pl. 49, figs. 15-26.
Cypridea (Bisulcocypridea) skeeteri Peck, Sohn, 1969, p. B6.

Elongate, subquadrate, highest about one-fourth of length from anterior end; dorsum and venter nearly straight, converging posteriorly; venter with distinct anterior beak and notch; posterior much narrower than anterior; LV overlaps RV around entire periphery; anterodorsally bisulcate; sulci short, deep, vertical, not reaching dorsal margin or midheight; intervening lobes highest anteriorly; posterior lobe may have spine near cardinal angle; terminal margins with a row of papillae or spines; general surface with or without scattered coarse nodes; densely and finely pitted; L, 0.89; H, 0.49 mm (Peck, 1951).

Occurrence. Upper Cretaceous, Bear River Formation, seven miles (11.2 km) north and nine miles (14.4. km) southeast of Evanston, Wyoming (Peck, 1951).

Cypridea tuberculata gypsumensis Vanderpool, 1928
(see Cypridea wyomingensis Jones).

Cypridea ventrosa bispinosa Vanderpool, 1928
(see Cypridea dequeenensis Swain and Brown).

Cypridea wyomingensis Jones
Plate 13, figures 1, - 3, 28, 29, 32 - 36
Cypridea tuberculata wyomingensis Jones, 1893, p. 386, pl. 15, figs. 5a, b.
Cypridea wyomingensis Jones, Peck, 1941, p. 297, pl. 43, figs. 10-17; Peck, 1951, p. 312, pl. 48, figs. 5-7. Loranger, 1951, p. 2363, pl. 2, fig. 22. Clark, 1954, p. 293, pl. 2, fig. 22. Howe and Laurencich, 1958, p. 142, TF. Swain and Brown, 1972, p. 14, pl. 1, figs. 19, 20; pl. 3, fig. 1.non Cypridea cf. wyomingensis Jones, Swartz and Swain, 1946, p. 372, pl. 52, figs. 19-22 (=C. sp. aff. C. wyomingensis Jones, Swain and Anderson, 1993, p. 104, pl. 1, figs. 5a, b.)
Cypridea tuberculata var. gypsumensis Vanderpool, 1928, p. 103, pl. 13, figs. 9-12.

Elongate, subquadrate, well-developed anteroventral hood, beak, furrow and notch, coarsely tuberculate or spinose as well as coarsely and densely pitted; shallow oblique anteroventral sulcus; L, 1.00; H, 0.62 mm (Swain and Brown, 1964).

Occurrence. Lower Cretaceous, Draney Formation, Freedom Quadrangle, Idaho-Wyoming; Kootenai Formation, near Drummond, Montana (Peck, 1941). Lower part of Bear River Formation, western Wyoming (Peck, 1951). DeQueen Formation, Gypsum Bluff, near Murphreesboro, Pike County, Arkansas (Vanderpool, 1928). Lower Cretaceous of western Canada (Loranger, 1951; Clark, 1954). Lower Cretaceous in wells in North Carolina (Swain and Brown, 1972).

Cypridea sp. aff. C. wyomingensis Jones

Plate 12, figure 34

Cypridea cf. wyomingensis Jones, Swartz and Swain, 1946, p. 372, pl. 52, figs. 19-22.

Cypridea sp. aff. C. wyomingensis Jones, Swain and Anderson, 1993, p. 104, pl. 1, figs. 5a, b.

Elongate-subquadrate, highest one-fourth of length from anterior end, dorsum and venter nearly straight, converging toward posterior; anterior broader than posterior, small anteroventral beak and notch; LV overlaps RV except dorsally where RV overlaps LV; densely pitted; about 30 scattered surface nodes; shallow anterodorsal sulcus; L, 0.95; H, 0.50 mm (Swain and Anderson, 1993).

Occurrence. Upper Jurassic Shongaloo Formation, Standard Oil No. 1 Zimmerman well, Union County, Arkansas, 7030 feet (2143 m).

Cypridea? sp., Sohn

Plate 13, figures 37, 38

Cypridea? sp., Sohn, 1958, p. 123, pl. 1, figs. 3-15, 16, not described.

Subquadrate, relatively short and high; straight dorsum and venter converge slightly toward posterior; highest one-fourth from anterior end; small anteroventral beak and notch; LV slightly overlaps RV ventrally; termino-median nodes on each valve; smooth; L, approximately 1.17; H, 0.75 mm (from Sohn's illustrations).

Occurrence. Lower? Cretaceous, Fall River Formation, SW1/4 sec. 15, T. 8 S., R. 4 E., Fall River County, South Dakota. A variant having very weak posteromedian node occurs in NW1/4 SW1/4 sec. 15, T. 8 S., R. 4 E., (Sohn, 1958).

Cypridea (C.) sp., Swain and Brown

Plate 13, figure 39

Cypridea (C.) sp., Swain and Brown, 1972, p. 14, pl. 1, figs. 22-25.

Elongate, subquadrate-subtrapezoidal, highest one-third of length from anterior end; dorsum moderately convex, venter nearly straight; anterior broadly curved, angulated at venter, posterior narrow, extended below, also angulated at venter; LV slightly overlaps RV; densely and finely pitted; anteroventral beak very short; inner lamellae moderate in width, narrow anterior vestibule; few short radial canals anteriorly; hinge margin with rabbet grooves on each valve; adductor scar a median group of five spots; L, 0.53; H, 0.42 mm (imperfect shell) (Swain and Brown, 1972).

Occurrence. Upper Jurassic or (probably) Lower Cretaceous, Unit H, well NC-DA-OT-12, Dare County, North Carolina, 7146-7156 feet (2178-2181 m) (Swain and Brown, 1972).

"Cypridea" sp. 1, Sohn

Plate 13, figures 40, 41

"Cypridea" sp. 1, Sohn, 1979, p. 18, pl. 7, figs. 8-12.

Subtrapezoidal, highest one-fourth of length from anterior end; dorsum sinuous, owing to projection at position of greatest height; venter nearly straight, with very small anterior beak and notch; LV large than RV; posterior narrower than anterior; both extremities form abrupt obtuse angles with venter; densely and coarsely pitted; scattered larger pustules; L, 0.8; H, 0.55 mm (Sohn, 1979).

Occurrence. Lower Cretaceous, Unit 2 of Chilson Member of Lakota Formation, U.S.G.S. coll. 26098, Fall River County, South Dakota (Sohn, 1979).

Cypridea sp., Betty Kellett Nadeau, unpublished

Plate 13, figure 42

Subpentagonal, highest about one-third of length from anterior end; dorsum angulated at position of greatest height; venter nearly straight, forming an acute angle with posterior margin; tiny anteroventral beak; smooth; LV larger than RV; dimensions unknown (Nadeau photograph, 1945).

Occurrence. Upper Jurassic or Lower Cretaceous, Morrison? Formation, Canyon City, Colorado? (Nadeau, 1945).

Cypridea (Pseudocypridina?) n. sp., Sohn

Plate 7, figures 23, 24

Pseudocypridina?, n. sp., Sohn, 1958, p. 123, pl. 1, figs. 9-12 (not described).

Subhemispherical-subquadrate, highest medially to anteromedially; dorsum convex, venter sinuous with small anterior beak and notch; anterior equal to or broader than posterior; LV slightly overlaps RV; dorsal border of each valve and ventral border of LV bear a marginal ridge; smooth?; L, 1.26; H, 0.86 mm (from Sohn's illustrations, 1958).

Occurrence. Lower Cretaceous, Lakota Formation, NW$^{1}/_{4}$ sec. 29, T. 67 N., R. 65 W., Crook County, Wyoming (Sohn, 1958).

Cypridea sp., Betty Kellett Nadeau, unpublished

Plate 13, figure 43

Elongate, subquadrate, highest about one-third of length from anterior end; dorsum and venter nearly straight and converge posteriorly; LV larger than RV; rough weathered surface;

venter with anterior beak and notch, the former broken; posterior narrower than anterior; dimensions unknown (Nadeau, photograph, 1945).

Occurrence. Upper Jurassic or Lower Cretaceous, Morrison? Formation, Canyon City, Colorado?

Cypridacea incertae sedis 1, Betty Kellett Nadeau, unpublished
Plate 13, figures 44, 45; Plate 14, figure 1

Subtriangular-subovate, highest about one-fourth of length from anterior end; dorsum strongly convex, venter nearly straight; posterior narrowly rounded, strongly extended below; LV slightly larger than RV; moderately convex; smooth; dimensions unknown; resembles immature Cypridea obesa Peck, 1951; plate 12, fig. 1 may represent different taxon from plate 11, figs. 44, 45.

Occurrence. Upper Jurassic or Lower Cretaceous, Morrison Formation?, Canyon City, Colorado?

Cypridacea incertae sedis 2, Betty Kellett Nadeau, unpublished
Plate 14, figure 4

Subtrapezoidal, highest medially to postmedially; dorsum gently convex, venter nearly straight; posterior narrowly rounded, strongly extended below, truncate above; LV larger than RV, strong ventral overlap; smooth; dimension unknown.

Occurrence. Upper Jurassic or Lower Cretaceous, Morrison Formation?, Canyon City, Colorado?

Cypridacea incertae sedis 3, Betty Kellett Nadeau, unpublished
Plate 14, figures 5, 6

Subtrapezoidal, highest anteromedially; dorsum, nearly straight, for about half of shell length; venter nearly, slightly sinuous; posterior bluntly pointed, extended below, truncate above; LV larger than RV; smooth; dimensions unknown; may be same as figs. 4 and 5 but dorsum straighter.

Occurrence. Upper Jurassic or Lower Cretaceous, Morrison Formation?, Canyon City, Colorado?

Cypridacea incertae sedis 4, Betty Kellett Nadeau, unpublished
Plate 14, figure 7

Elongate-subreniform, highest anteromedially; anterior broad, posterior narrow; dorsum moderately convex, venter slightly concave; smooth, dimensions unknown.

Occurrence. Upper Jurassic or Lower Cretaceous, Morrison Formation?, Canyon City, Colorado?

Cypridacea incertae sedis 5, Betty Kellett Nadeau, unpublished
Plate 14, figure 8

Subtrapezoidal-subreniform, highest slightly anteromedially; anterior broadly rounded; posterior bluntly pointed, strongly extended below, truncate above; LV larger than RV; dorsum straight medially, venter concave; smooth; dimensions unknown.

Occurrence. Upper Jurassic or Lower Cretaceous, Morrison Formation?, Canyon City, Colorado?

Cypridacea incertae sedis 6, Betty Kellett Nadeau, unpublished
Plate 14, figures 9, 10

Elongate, subovate, highest anteromedially; dorsum slightly convex, venter nearly straight; posterior narrower than anterior and extended below; LV slightly larger than RV; smooth; dimensions unknown.

Occurrence. Upper Jurassic or Lower Cretaceous, Morrison Formation?, Canyon City, Colorado?

Cypridacea incertae sedis 7, Betty Kellett Nadeau, unpublished
Plate 14, figures 11, 12

Elongate-subovate-subpyriform, highest anteromedially to medially; dorsum truncate posteriorly in adult forms; venter nearly straight; posterior narrower than anterior; LV? larger than RV?; smooth; dimensions unknown.

Occurrence. Upper Jurassic or Lower Cretaceous, Morrison Formation?, Canyon City, Colorado?

Cypridacea incertae sedis 8, Betty Kellett Nadeau, unpublished
Plate 14, figure 13

Subovate, highest anteromedially; height about 0.65 of length; dorsum moderately convex, venter nearly straight; anterior broader than posterior; LV larger than RV; smooth; dimensions unknown.

Occurrence. Upper Jurassic or Lower Cretaceous, Morrison Formation?, Canyon City, Colorado?

Cypridacea incertae sedis 9, Betty Kellett Nadeau, unpublished

Plate 14, figure 14

Subcircular, highest postmedially, dorsum more convex than venter; posterior narrower than anterior; finely striated; dimensions unknown.

Occurrence. Upper Jurassic or Lower Cretaceous, Morrison Formation?, Canyon City, Colorado?

Cypridacea incertae sedis 10, Betty Kellett Nadeau, unpublished

Plate 14, figure 15

Subtrapezoidal-subelliptical, highest submedially; dorsum with median straightened section, about half of shell length; venter nearly straight to slightly concave; anterior more broadly and uniformly rounded than posterior; smooth; LV slightly overlaps RV, especially midventrally; dimensions unknown.

Occurrence. Upper Jurassic or Lower Cretaceous, Morrison Formation?, Canyon City, Colorado?

Ostracoda incertae sedis, Betty Kellett Nadeau, unpublished

Plate 14, figure 16

Subpentagonal, highest medially, dorsum more convex than venter; posterior very narrow and extended below; bilobate thlipsurid depression of posterior third, otherwise smooth; dimensions unknown.

Occurrence. Upper Jurassic or Lower Cretaceous, Morrison Formation?, Canyon City, Colorado?

Ostracoda incertae sedis, Betty Kellett Nadeau, unpublished

Plate 14, figure 17

Elongate, subrectangular, highest near anterior end; dorsum and venter long and straight, converging slightly toward posterior; posteroventral rear-projecting thick spine; smooth; dimensions unknown.

Occurrence. Upper Jurassic or Lower Cretaceous, Morrison Formation?, Canyon City, Colorado?

Cyprididae incertae sedis, Benson, 1967

Plate 14, figures 2, 3

Cyprid ostracode, unknown, Benson, 1967, p. 222, TF 6-8, 10-2.

Elongate subtriangular-subpyriform, highest anteromedially; dorsum strongly convex, venter slightly concave; anterior broad, posterior narrow, extended below; bertillion surface ornamentation, with narrow ridges, furrows and pits; adductor muscle scar consists of four oblique main spots; two oblique frontal spots; inner lamellae narrow; L, 1.42; H, 0.81 mm (Benson, 1967); resembles Dogelinella Schneider, 1957.

Occurrence. Pleistocene, Yarmouth Stage, Dickinson County, Kansas, USNM 649609, loc. P (after Benson, 1967).

Genus Cyprideis Jones, 1856

(Cytherideidae - Cytherideinae)

Cyprideis littoralis Brady

Plate 14, figures 18-20

Cyprideis littoralis Brady, 1870, p. 125; Gutentag and Benson, 1962, p. 47, pl. 2, figs. 4-7.

Elongate subelliptical-subovate; male larger and more elongate than female; highest submedially; dorsum slightly convex, venter slightly sinuous; posterior narrower than anterior; anterodorsal shallow sulcus; finely pitted; LV slightly larger than RV; radial canals numerous, no vestibules; RV hingement with anterior crenulate tooth, inter-terminal groove and posterior non-crenulate tooth; L, female 0.99 to 1.03, H, 0.63 mm (Gutentag and Benson, 1962).

Occurrence. Pliocene, Laverne Formation and Pleistocene sink deposits, Meade County, Kansas. Holocene, Europe and North America, mostly in brackish water (Gutentag and Benson, 1962).

Cyprideis maladensis Swain

Plate 14, figures 21-24

Cyprideis maladensis Swain, 1987a, p. 126, pl. 2, figs. 10-13, 14, 17, 18.

Subquadrate-subpyriform, dimorphic, highest about two-fifths of length from anterior; dorsum nearly straight, venter slightly sinuous; anterior broader than posterior; weakly bisulcate, weakly pitted; shell surface swirled in appearance near sulci; terminal flattened borders; sieve plates vary from oval to triangular; L, 0.97; H, 0.53 mm (Swain, 1987a).

Occurrence. Pliocene, Salt Lake Group, secs. 24-26, T. 16 S., R. 36 E., along strike, five to seven miles (8-11 km) southwest of Malad City, Oneida County, Idaho (Swain, 1987a).

Cyprideis salebrosa van den Bold

Plate 8, figures 43 - 45; Plate 14, figures 25, 25a

Cyprideis salebrosa van den Bold, 1963, p. 377, pl. 7, figs. 9a-d, pl. 11, figs. 1a-c. Benson, 1967, p. 234, TF 12-1, 8. Lister, 1975, p. 22, pl. 2, figs. 11-14, TF 24.

Elongate quadrangular-elliptical, highest anteromedially; male more elongate than female; dorsum gently convex, venter slightly sinuous; anterior much broader than posterior, the latter extended below and crenulate; LV larger than RV, hinge holomerodont; radial canals numerous, no vestibule; L, female, holotype, 0.99; H, 0.58 mm (van den Bold, 1963).

Occurrence. Late Pleistocene, $NE^1/_4$ $NW^1/_4$ sec. 9, T. 5 N., R. 28 E., Beaver County, Oklahoma (Benson, 1967); Lake Bonneville Group, Saltair core hole $SE^1/_4$ sec. 25, T. 1 N., R. 3 W., also Sec. 28 core hole, $SE^1/_4$ $SW^1/_4$ sec. 28, T. 1 N., R. 2 W., Salt Lake County, Utah (Lister, 1975). Pliocene, Talpara Formation, Caparo Clay Member, Todd's Road, northern Basin, Trinidad (van den Bold, 1963).

Cyprideis sp., Benson
Plate 15, figures 1-3
Cyprideis sp., Benson, 1969, p. 109, pl. 21, fig. 4; pl. 22, figs. 4-6.

Elongate-subreniform, highest anteromedially; dorsum gently convex, venter nearly straight; LV larger than RV; anteromedian sulcus; pitted; inner lamellae narrow, no vestibule; L, 0.75; H, 0.43 mm (Benson, 1969).

Occurrence. Early Pleistocene, Rita Blanca lake deposits, near Channing, Hartley County, Texas (Benson, 1969).

Cyprideis? sp., Swain
Plate 15, figure 4
Cyprideis? sp., Swain, 1986c, p. 378, pl. 4, fig. 7.

Elongate-elliptical, highest submedially; dorsum truncated medially; venter slightly concave; LV larger than RV; two weak sulci dorsomedially and anterodorsally; double pitlike depression medially; L, 0.70; H, 0.37 mm (Swain, 1986).

Occurrence. Pliocene, Cache Valley Formation, $NW^1/_4$ $SE^1/_4$ $NE^1/_4$ sec. 10, T. 12 S., R. 40 E., Franklin County, Idaho (Swain, 1986c).

Cyprideis, n. sp., Swain
Plate 15, figure 5
Cyprideis, n. sp., Swain, 1986b, p. 215, pl. 4, fig. 19.

Ovate-elliptical, highest anteromedially, dorsum gently convex, venter nearly straight; anterior broader than posterior; LV larger than RV; dorsomedian long vertical sulcus; anteromedian node-like elevation; L, 1.1; H, 0.63 mm (Swain, 1986).

Occurrence. Miocene, Humboldt Formation, Gulf Oil Co. No. 1. Thousand Springs well, Elko County, Nevada, 1670-1680 feet (509-512 m) (Swain, 1986b).

Genus Cypridopsella Kaufmann, 1900

(Cypridopsidae-Cypridopsinae)

Cypridopsella? esmeraldensis Swain and Wagoner

Plate 15, figures 8-10, 11

Cypridopsella? esmeraldensis Swain and Wagoner, 1985, p. 136, pl. 3, figs. 1-4. Swain, 1985, p. 431, pl. 2, figs. 7, 8, 14.

Subtrapezoidal to subreniform, highest postmedially; dorsum slightly truncate medially; venter nearly straight to slightly concave; posterior narrower than anterior, both ends extended below; LV larger than RV, over-reaching it the most dorsally and ventrally; dimorphic, females broader posteriorly than males; smooth; L, holotype male, 0.98; H, 0.58 mm; LV hinge grooved, weakly crenulate posteriorly; inner lamellae broad terminally, with vestibules; radial canals short, numerous. (Swain and Wagoner, 1985; Swain, 1985).

Occurrence. Miocene, Esmeralda Formation, Blair Junction section, Esmeralda County, Nevada (Swain and Wagoner, 1985); Black Spring road section, 2.85 miles (4.5 km) west of Black Spring, Nye County, Nevada, about 800 feet (244 m) above base of lake beds (Swain, 1985).

Genus Cypridopsis Brady, 1867

(Cypridopsidae-Cypridopsinae)

Cypridopsis dickinsoniana Tasch

Plate 15, figures 12, 13

Cypridopsis dickinsoniana Tasch, 1963, p. 1250, pl. 174, figs. 1-4.

Ovate, tumid, highest postmedially?, dorsum convex, venter straight to slightly concave; anterior narrower than posterior; valves subequal but RV overlaps LV posteriorly; minutely pitted; L, 0.74; H, 0.44 mm (Tasch, 1963).

Occurrence. Permian, Wellington Formation, Elmo VIII B, bed 10, NE$^1/_4$ NW$^1/_4$ sec. 21, T. 15 S., R. 2 E., Dickinson County, Kansas (Tasch, 1963).

Cypridopsis fabulina (Jones and Kirkby)

Plate 15, figures 14-16

Cythere fabulina Jones and Kirkby, 1867, p. 217.

Carbonia fabulina Jones, 1870, p. 218; Jones and Kirkby, 1879, p. 2, figs. 1-10.

Cypridopsis fabulina (Jones and Kirkby), Scott and Summerson, 1943, p. 668, pl. 1, figs. 6-11, pl. 2, fig. 5.

Subovate-reniform, highest medially, ends equally rounded; RV larger than LV; overlap most along venter; pitted to finely reticulate; round muscle scar area anteromedian in position; L, 0.80; H, 0.55 mm (after Scott and Summerson, 1943).

Occurrence. Lower Pennsylvanian, Briceville, Jellico, Scott and Hanna Formations, between Coalfield and Christmas, Morgan County, Tennessee (Briceville); just north of Molus, Harlan County, Kentucky (Hance) on U.S. Highway 25 W.; five miles (8 km) NE of Lafollette, Campbell County, Tennessee (Briceville); one mile (1.6 km) west of U.S. highway 25 on coal mine road, Campbell County, Tennessee (Jellico); northeast of Newcomb, Williamsburg Quadrangle, Campbell County, Tennessee (Scott and Summerson, 1943). Also in Carboniferous of British Isles (Jones and Kirkby, 1867).

Cypridopsis ovata (Holland)
Plate 15, figures 17, 18, 21, 22
Whipplella ovata Holland, 1934, p. 348, pl. 25, figs. 4a-c.
Cypridopsis ovata (Holland), Scott, 1944, p. 144, pl. 24, figs. 15-17.

Subpyriform-subreniform, highest slightly anteromedially; anterior more narrowly rounded than posterior and strongly extended below; RV larger than LV; smooth; adductor muscle scar anteromedian, with three or more spots; L, 0.78; H. 0.41 mm (Holland, 1934; Scott, 1944).

Occurrence. Permian, upper Dunkard Series, upper Greene Group, Nineveh Formation, eight miles (13 km) northwest of Waynesburg, Pennsylvania (Holland, 1934).

Cypridopsis parvula (Holland)
Plate 15, figures 23, 24
Whipplella parvula Holland, 1934, p. 347, pl. 25, figs. 6a-c.
Cypridopsis parvula (Holland), Scott, 1944, p. 144, pl. 24, figs. 12-14. Tasch, 1963, p. 1249, pl. 174, fig. 18.

Small, subovate-subtriangular, highest medially, dorsum strongly convex, anterodorsal slope truncate, venter nearly straight; RV overlaps LV; smooth; adductor scar circular, anteromedian; L, 0.59; H, 0.35 mm (Holland, 1934; Scott, 1944).

Occurrence. Permian, upper Dunkard Series, upper Greene Group, Nineveh Formation; Limestone Hill, southernmost Wood County, northwestern West Virginia; also at Nineveh, 8 miles (13 km) northwest of Waynesburg, Greene County, Pennsylvania (Holland, 1934); Wellington Formation, Marion 1A bed 6, NE$^1/_4$ NE$^1/_4$ sec. 10, T. 17 S., R. 2 E., Marion County, Kansas (Tasch, 1963).

Cypridopsis? pliocaenica Chapman
Plate 7, figures 40, 41
Cypridopsis pliocaenica Chapman, 1896, p. 98, pl. 3, figs. 16-18.

Subovate, highest anteromedially, anterior broader than posterior, dorsum convex, venter nearly straight, smooth?; L, 0.98 mm; H, 0.58 mm; compressed (atypical of Cypridopsis) (Chapman, 1896).

Occurrence. Pliocene, Merced Formation, San Pablo Valley, in hills east of Berkeley, California (Chapman, 1896).

Cypridopsis saltlakensis Swain

Plate 15, figures 25, 26

Cypridopsis saltlakensis Swain, 1947, p. 519, pl. 76, figs. 17-20; Swain, 1987, p. 121.

Subovate, highest medially; dorsum moderately convex, venter nearly straight; RV overlaps LV along venter; most convex in posterior third; smooth; hinge of RV with groove to receive edge of LV; inner lamellae, vestibules narrow; L. 0.65; H, 0.46 mm (Swain, 1947).

Occurrence. Pliocene, Salt Lake Formation, sec. 15, T. 13 N., R. 2 W., Cache County, Utah (Swain, 1947). Miocene or Pliocene, Humboldt or Salt Lake Group, undifferentiated, on hillside NW$^{1}/_{4}$ SE$^{1}/_{4}$ NE$^{1}/_{4}$ sec. 33, T. 7 S., R. 19 W., Box Elder County, Utah (Swain, 1987).

Cypridopsis vidua (O.F. Müller)

Plate 8, figures 46, 47; Plate 15, figures 27-29

Cypris vidua O.F. Müller 1776, p. 199.

Cypridopsis vidua (Müller), Brady, 1868, p. 117. Gutentag and Benson, 1962, p. 26, pl. 1, fig. 10, TF 6. Staplin, 1963, p. 1183, pl. 159, figs. 30-36. Benson, 1967, p. 230, TF 11-2, 11-3. Lister, 1975, p. 15, pl. 3, figs. 7, 8 , TF 14. Cameron and Lundin, 1977, p. 349, pl. 1, fig. 20.

Subtriangular-subovoid, highest submedially, height about two-thirds of length; dorsum strongly convex to angulate; venter nearly straight to slightly convex; LV slightly larger than RV; pitted, anterior RV margin tuberculate; inner lamellae broadest anteriorly, vestibules present; hinge groove in RV, in part with mid-hinge minute denticles; adductor scar of four large and two smaller spots; other scars present; L, female, 0.58 to 0.73; H, 0.37 to 0.46 mm (Staplin, 1963).

Occurrence. Pleistocene, Illinoian to Wisconsin Stages, Odee Formation, Meade County, Kansas; late Kansan to early Yarmouthian Stages, Reno County, and Dickinson County, Kansas, below Pearlette Ash (Gutentag and Benson, 1962). Nebraskan Stage to Recent of Illinois; Mankato or Cary Substage, Wisconsin Stage of Indiana; Cary Substage of Story County, Iowa; Tazewell Substage of Ontario; late? Wisconsin Loveland Loess? of Ohio; Sappa Silt (late Kansan) of Nebraska, Kansas and Iowa (Staplin, 1963). Lake Bonneville Group, Salt Lake County, Utah (Lister, 1975). Late Pleistocene Lake Cochise deposits,

Willcox Playa, Cochise County, Arizona (Cameron and Lundin, 1977). Recent of Holarctic and Neotropical regions (authors).

Cypridopsis? sp., Swain

Plate 15, figure 30

Cypridopsis? sp. Swain, 1986b, p. 209, pl. 3, fig. 7.

Subovate, highest medially, anterior more narrowly rounded than posterior; LV larger than RV; tumid; smooth; L, 0.60; H, 0.37 mm (Swain, 1986b).

Occurrence. Miocene, Humboldt Formation, Gulf Refining Co., No. 2 Wilkins Ranch well, 2635-2665 feet (803-813 m) to 4270-4300 feet (1302-1311 m), Elko County, Nevada, sec. 21, T. 38 N., R. 61 E. (after Swain, 1986b).

Cypridopsis? sp. immature

Plate 15, figure 31

LV subovate, highest anteromedially, dorsum straightened posterior to greatest height; anterodorsal slope also truncated; anterior narrower than posterior; pitted, inner lamellae broad anteriorly, with vestibule; L, 0.62; H, 0.37 mm.

Occurrence. Pleistocene, bone bed at Boney Spring, Benton County, Missouri (J.E. King, coll., unpublished).

Cypridopsis sp., Lister, 1975

(see Paracypridopsis sp., Lister)

Genus Cyprinotus Brady, 1886

(Cyprididae-Cyprinotinae)

Cheikella Sohn and Morris, 1963 (see Sohn and Swain, 1999, J. Paleontology, in press).

Cyprinotus sp. aff. C. cingalensis Brady

Plate 15, figures 32, 33

Cyprinotus sp. aff. C. cingalensis Brady, Swain, 1986b, p. 205, pl. 5, figs. 19, 20.

Subtriangular, highest medially; dorsum strongly convex, venter nearly straight; posterior narrower than anterior, both ends extended below; convexity moderate; LV overlaps RV on free margins; RV has strong hump or ridge middorsally; smooth; L, 1.30; H, 0.77 mm (Swain, 1986b).

Occurrence. Miocene, Humboldt Formation, Cobre section, Cobre, Elko County, Nevada, about 1011 feet (308 m) above base of exposure (Swain, 1986b).

Cyprinotus glaucus Furtos

Plate 15, figures 34, 35.

Cyprinotus glaucus Furtos, 1933, p. 444, pl. 5, figs. 9-16. Staplin, 1963, p. 792, pl. 94, figs. 30, 31. Lister, 1975, p. 13, pl. 2, figs. 104, TF 12. Cameron and Lundin, 1977, p. 349, pl. 1, fig. 19.

Subtriangular-subreniform, highest anteromedially, height exceeds half of length; anterior slightly broader than posterior; smooth; LV slightly larger than RV; minute tubercles on ventro-anterior margin of RV; inner lamellae broad anteriorly; LV hinge flange fits RV hinge groove; adductor scar of four large and one small spot; two frontal scars; L, 1.10; H, 0.55 mm. Pleistocene forms with more rounded dorsal margin of LV than modern specimens (Staplin, 1963).

Occurrence. Pleistocene, Mankato Substage, Wisconsin Stage, Lake Chicago, Illinois (Staplin, 1963). Lake Bonneville Group, Saltair core hole SE1/4 sec. 25, T. 1 N., R. 3 W.; sec. 28 core hole, SE1/4 SW1/4 sec. 28, T. 1 N., R. 2 W., Salt Lake County, Utah (Lister, 1975). Late Pleistocene Lake Cochise deposits, Willcox playa, Cochise County, Arizona (Cameron and Lundin, 1977). Recent, Lake Erie (Furtos, 1933).

Cyprinotus? sp. aff. C. reniformis Hou et al, 1978 (Swain, 1985)

(see Cyprinotus sp. aff. yuchengensis Hou and Li)

Cyprinotus? sp. aff. C. yuchengensis Hou and Li

Plate 16, figures 1, 2

Cyprinotus? sp. aff. C. reniformis Hou and Li, 1988 (Hou et al, 1978), Swain, 1985, p. 423, pl. 5, figs. 17, 18.

Elongate subreniform, highest postmedially, dorsum convex, straightened medially, venter slightly concave; posterior broader than anterior; LV larger than RV, overlapping RV dorsally and ventrally; moderately convex; smooth; L, 0.93; H, 0.53 mm (Swain, 1985).

Occurrence. Miocene, Esmeralda Formation , 7.6 miles (12.1 km) east of Finger Rock Wash on road to Pactolus Mine, Goldyke 1:24000 Quadrangle, Mineral County, Nevada; about 1000 feet (305 m) above base of lake bed sequence (after Swain, 1985).

Cyprinotus sp., Staplin

Plate 15, figure 36

Cyprinotus sp., Staplin, 1963, pl. 94, figs. 26, 27, undescribed.

Subtriangular, highest medially, dorsum strongly convex, subumbonate, venter nearly straight; posterior more narrowly rounded than anterior; both ends extended below; L, 1.1; H, 0.65 mm (from Staplin photos, 1963).

Occurrence. Pleistocene, late Kansan to early Yarmouthian Stage, Sappa Silt; SE$^{1}/_{4}$ sec. 15, SW$^{1}/_{4}$ sec. 14, T. 98 N., R. 48 W., Lyon County, Iowa (Staplin, 1963).

Cyprinotus sp. 1, Sohn

Plate 16, figures 6, 7

Cyprinotus sp. 1, Sohn, 1956,p. 119, pl. 1, figs. 31-34, undescribed.

Elongate subtriangular-subpyriform, highest anteromedially, anterior broader than posterior; dorsum arched, venter nearly straight; LV overlaps RV; anterior area compressed, flangelike; L, 1.36; H, 0.70 mm (from Sohn photos, 1956).

Occurrence. Pliocene, Teewinot Formation, Jackson Hole, Wyoming (Sohn, 1956).

Cyprinotus sp., Benson

Plate 16, figures 5, 5a

Cyprinotus sp., Benson, 1967, p. 222, TF 6-12, 10-8.

Subovate-subreniform, highest medially, dorsum strongly arched, venter slightly concave; posterior much narrower than anterior and extended below; finely punctate; adductor muscle scar of four main oblique spots; two frontal spots; many small dorsal spots; L, 1.1; H, 0.55 mm (Benson, 1967).

Occurrence. Pleistocene, Yarmouth Stage, Dickinson County, Kansas (Benson, 1967).

Cyprinotus? sp. 1, Swain

Plate 16, figures 3, 4

Cyprinotus? sp. 1, Swain, 1986b, p. 205, pl. 3, figs. 1, 2.

Subtriangular, highest anteromedially; dorsum strongly convex, venter nearly straight; posterior narrower than anterior, extended below; LV larger than RV, overlapping RV strongly anterodorsally; convexity moderate; smooth; L, 0.93; H, 0.60 mm (Swain, 1986b).

Occurrence. Miocene, Humboldt Formation, Gulf Oil Co., No. 1 Thousand Springs well, 1250-1260 feet (381-384 m), Elko County, Nevada (Swain, 1986b).

Cyprinotus? sp. 2, Swain

not figured

Cyprinotus? sp. 2, Swain, 1986b, p. 206, not figured.

Subtrapezoidal, highest anteromedially; dorsum strongly convex, straightened postmedially; anterior narrower than posterior, possibly with anteroventral spines; both ends strongly extended below; LV larger than RV, overlapping around entire periphery; compressed; smooth; L, 0.77; H, 0.48 mm (Swain, 1986b).

Occurrence. Miocene, upper part of Humboldt Formation, Gulf Refining Co., No. 2, Wilkins Ranch well, 2945-2975 feet (898-907 m), Elko County, Nevada (Swain, 1986b).

Genus Cypris O.F. Müller, 1776

(Cyprididae-Cypridinae)

Cypris candida O.F. Müller, 1776

(see Candona candida (Müller)

Cypris osmondi Dickinson and Swain

Plate 16, figure 8

Cypris osmondi Dickinson and Swain, 1967, p. 339, pl. 35, fig. 1.

Subtriangular-ovoid, highest medially, dorsum highly arched, venter very slightly concave; posterior narrower than anterior; LV larger than RV, with slight anterodorsal and midventral overlap; strongly convex; smooth, inner lamellae broad; L, 0.93; H, 0.60 mm (Dickinson and Swain, 1967).

Occurrence. Pliocene, Hay Ranch Formation (=Upper Humboldt), Gulf Oil Corp. No. 1 Wilkins Ranch well, Elko County, Nevada, 1224-2074 feet (373-632 m) (Dickinson and Swain, 1967).

Cypris pagei (Swain), Swain, 1956

(see Pseudoeucypris pagei (Swain)

Cypris potamides Dickinson and Swain

Plate 16, figures 9-12

Cypris potamides Dickinson and Swain, 1967, p. 340, pl. 35, figs. 2-5; pl. 36, figs. 8, 12. Swain, 1987a, p. 120, pl. 2, figs. 16, 19, 20.

Subtriangular, highest medially, dorsum moderately arched, venter straight or slightly concave, posterior narrower than anterior; LV larger than RV with slight dorsal overlap of RV; moderately convex; smooth; inner lamellae broader anteriorly than posteriorly; L, 0.95; H, 0.58 mm (Dickinson and Swain, 1967).

Occurrence. Pliocene, Hay Ranch Formation (=Upper Humboldt), Gulf Oil Corp. No. 1 Wilkins Ranch well, Elko County, Nevada, 1344-2677 feet (410-816 m) (Dickinson and Swain, 1967). Salt Lake Group, Cache Valley Formation, secs. 4 to 26, T. 16 S., R. 36 E., five to seven miles along strike southwest of Malad City, Oneida County, Idaho (Swain, 1987a).

Cypris procera Chapman

Plate 7, figures 38, 39

Cypris procera Chapman, 1986, p. 96, pl. 3, figs. 7-9.

Elongate subelliptical subreniform, highest slightly anteromedially; dorsum weakly convex, venter concave; anterior broader than posterior; smooth; L, 1.27 mm; H, 0.71 mm (Chapman, 1896).

Occurrence. Pliocene, Merced Formation, Wildcat Canyon and Bollinger Canyon, in hills east of Berkeley, California (Chapman, 1896).

Cypris purbeckensis Forbes?, Jones

Plate 7, figures 10, 11; Plate 8, figures 1, 2; Plate 16, figures 13, 14

Cypris purbeckensis Jones, 1885, p. 347, pl. 9, figs. 1-6. Jones, 1893, p. 386, pl. 15, figs. 15a, b. Peck, 1941, p. 301, pl. 43, figs. 31-33.

?Cypris purbeckensis Jones, 1886, p. 147, pl. 4, figs. 5a-c.

Subreniform-ovate, highest medially; dorsum moderately convex, venter slightly concave; anterior broader than posterior; LV larger than RV, overlapping it except at hinge; smooth; L, 1.1; H, 0.6 mm (Peck, 1941).

Occurrence. Lower Cretaceous, Draney Limestone, Freedom Quadrangle, Idaho-Wyoming (Peck, 1941). Lower Cretaceous? Bear River Formation (-Lower Bear River), Wyoming (Jones, 1893). Upper Jurassic? Morrison Formation, Colorado (Jones, 1886).

Genus Cyprois Zenker, 1854

(Cyprididae-Cyproidinae)

Cyprois ephraimensis Swain

Plate 16, figures 15 - 18

Cyprois cf. C. marginata (Strauss), Swain, 1949, p. 177, pl. 32, figs. 8-10, 14, 15. Swain, 1956, p. 135, pl. 3, figs. 4a, b.

Cyprois ephraimensis Swain, 1964a, p. 271, pl. 42, figs. 2a, b; pl. 44, fig. 4.

Subovate, highest medially to anteromedially; dorsum convex, venter nearly straight; anterior broader than posterior; LV slightly larger than RV, overreaching RV dorsally and terminally; compressed; smooth, inner lamellae narrow, slight vestibule anteriorly; L, 1.01; H, 0.71 mm (Swain, 1964).

Occurrence. Eocene, upper part of Colton Formation, in Colton-Green River transition beds, and in lower part of Green River Formation, Uinta and Piceance Creek Basins, Utah and Colorado; upper Wasatch or Knight Formation and Wasatch-Green River transition beds at Fossil and 17 miles (27 km) north of Kemmerer, Wyoming; also in middle Green River Formation east of Piedmont, Wyoming (Swain, 1964). Paleocene or Eocene, Flagstaff Limestone, near Ephraim, Utah (Swain, 1949).

Cyprois cf. marginata (Strauss), Swain, 1949
(see Cyprois ephraimensis Swain)

Cyprois whiteflatensis Swain
Plate 16, figures 19-21
Cyprois whiteflatensis Swain, 1987b, p. 226, pl. 1, figs. 4, 9, 10, 18a, b.

Subtriangular, highest medially, height about 0.7 of length; dorsum strongly convex, subumbonate, venter nearly straight to slightly concave; anterior broader than posterior; both ends extended below; LV larger than RV; compressed; faintly reticulate surface terminally; internal surface pitted terminally; inner lamellae broadest anteriorly, with vestibules; adductor muscle scar of five spots; two frontal spots; L, holotype, 1.50; H, 0.97 mm (Swain, 1987).

Occurrence. Eocene, Twin Bridges Limestone, west side of Huntingdon Creek, sec. 36, T. 32 N., R. 55 E., Elko County, Nevada, at south end of Twin Bridges (Swain, 1987b).

Cyprois sp. 1
Plate 16, figure 22

Subovate-subreniform, highest medially, dorsum moderately convex, venter slightly concave; posterior narrower than anterior and extended below; LV larger than RV; compressed; L, 1.54; H, 1.02 mm .

Occurrence. Lower Oligocene, Chadron Formation, Helmville, Nebraska (R. Konizeski, coll., unpublished).

Cyprois sp. 2
Plate 16, figure 23
R. Konizeski, unpublished.

Ovate-subpyriform, highest anteromedially; dorsum moderately convex, venter slightly sinuous; posterior narrow, extended below; LV larger than RV; anterior broad, slightly extended above; smooth; L, 1.38; H, 0.81 mm.

Occurrence. Lower Oligocene, Chadron Formation, Helmville, Nebraska (R. Konizeski, coll.).

Genus Cythere Müller, 1785

(Cytheridae)
Cythere monticula Jones, 1893)
(see Looneyella monticula (Jones))

Cytherideidae? sp. indet.
not figured
Cytherideidae? sp. indet., Swain, 1986a.

Subelliptical-subquadrate, straight hinge margin; hinge of LV with weak terminal sockets and an interterminal furrow; vertical row of three? adductor spots; two frontal spots; broad anterior inner lamellae, wide vestibule; very weak anteromedian sulcus; L, 0.67; H, 0.39 mm (Swain, 1986a).

Occurrence. Pliocene, Glenns Ferry Formation, sec. 14, T. 4 S., R. 1 W., Owyhee County, Idaho (Swain, 1986a).

Genus Cytheridea Bosquet, 1852
(Cytherideidae - Cytherideinae)

"Cytheridea" atlantosaurica Jones
Plate 8, figures 5, 6
Cytheridea atlantosaurica Jones, 1886, p. 147, pl. 4, figs. 7a-c

Oblong, highest postmedially, dorsum and venter nearly straight but converge posteriorly; anteromiddorsal vertical depression; weak marginal rim; smooth, L, 0.8; H, 0.4 mm; said to resemble smooth variety of Cytheridea "[Cyprideis]" torosa (after Jones, 1996).

Occurrence. Upper Jurassic, Morrison Formation, "Atlantosaurus" beds, Canyon City, Colorado (Jones, 1996).

"Cytheridea" tenuis Jones
Plate 8, figures 9, 10
Cytheridea tenuis Jones, 1893, p. 390, pl. 15, figs. 7a, b

Elongate, sublanceolate, highest anteromedially; dorsum convex, venter weakly concave; anterior broader than posterior, both are rounded; the posterior subtruncate; sparsely tuberculate; L, 0.6; H, 0.25 mm (after Jones, 1893).

Occurrence. Lower Cretaceous, Lower Bear River Formation, Cokeville, Wyoming (Jones, 1893).

"Cytheridea" truncata Jones
Plate 8, figures 7, 8
Cytheridea truncata Jones, 1893, p. 390, pl. 15, figs. 4a, b

Subtrapizoidal, highest anteromedially, dorsum and venter straight, converging posteriorly; posterior obliquely truncated above, anterior also truncate; coarsely pitted surface; L, 0.6; H, 0.25 mm (after Jones, 1893). May be Trapezoidella.

Occurrence. Lower Cretaceous, Lower Bear River Formation?, Cokeville, Wyoming (Jones, 1893).

Genus "Cytherideis" Jones, 1856
(=Cushmanidea Blake, 1933, part)
Cytherideis aequalis Jones
Cytherideis aequalis Jones, 1893, p. 300, pl. 15, figs. 11a, b
Plate 8, figures 11, 12

Elongate elliptical reniform, highest medially; dorsum convex, venter concave; ends equally rounded; smooth; orientation uncertain; L, 0.6; H, 0.25 mm (after Jones, 1893).

Occurrence. Lower Cretaceous, lower Bear River Formation, Cokeville, Wyoming (Jones, 1893).

"Cytherideis" impressa Jones
Plate 8, figures 13, 14
Cytherideis impressa Jones, 1893, p. 391, pl. 15, figs. 12a, b

Very elongate reniform to elliptical, highest medially; dorsum convex, venter concave; midventral depression; ends nearly equal; smooth; L, 0.6; H, 0.25 mm (after Jones, 1893). Orientation uncertain.

Occurrence. Lower Cretaceous, Lower Bear River Formation, Cokeville, Wyoming (Jones, 1893).

"Cytherideis" marshii Jones
Plate 8, figures 3, 4

Cytherideis marshii Jones, 1886, p. 147, pl. 4, figs. 6a - c

Subpyriform, highest anteriorly; dorsum straight, venter slightly concave; anterior much broader than posterior, truncate; middorsal depression?; median pit?; marginal flattened rim; smooth; L, 0.86; H, 0.47 mm (after Jones, 1886).

Occurrence. Upper Jurassic, Morrison Formation, Atlantosaurus beds, Canyon City, Colorado (Jones, 1886).

Genus Cytheridella Daday, 1905

(Limnocytheridae-Metacypridinae)

Cytheridella alpinensis Swain

Plate 16, figures 24-27

Cytheridella alpinensis Swain, 1987c, p. 437, pl. 5, figs. 17-22.

Dimorphic, females elongate-subquadrate; males short, subquadrate-subreniform; dorsum nearly straight, about two thirds of length in females; slightly convex and five-eighths of length in males; anterior slightly broader than posterior; LV slightly larger than RV; strongly inflated in females, less so in males; dorso-anteromedian sulcus extends to midheight; ridgelike ventromedian shell expansion; densely pitted with coarser overprint of narrow reticulating ridges; ends compressed in males; anterior compressed in females; L, holotype female 0.68; H, 0.35 mm; L, male 0.57; H, 0.35 mm (Swain, 1987c).

Occurrence. Oligocene, Pruett? Formation, Buck Hill volcanic series, Alpine Quadrangle, Brewster County, Texas (Swain, 1987c).

Cytheridella sp. aff. C. ilosvayi Daday

Plate 16, figure 28

Cytheridella sp. aff. C. ilosvayi Daday, 1905, Swain, 1987c, p. 438, pl. 5, fig. 23.

Elongate subquadrate, highest one-third of length from anterior end; dorsum nearly straight, about half of shell length, venter slightly convex, about two-thirds of length; posterior narrower than anterior, both ends extended below; LV only slightly larger than RV; width exceeds height; anteroventral marginal zone compressed; smooth; L, 0.53; H, 0.27 mm (Swain, 1987).

Occurrence. Oligocene, Pruett? Formation, Buck Hill volcanic series, Alpine Quadrangle, Brewster County, Texas (Swain, 1987c).

Cytheridella? todiltensis (Swain)

Plate 7, figures 31-33

Metacypris todiltensis Swain, 1946, p. 553, pl. 84, figs. 20-22.

Cytheridella todiltensis (Swain), Kietzke, 1992, p. 173.

Elongate, oblong-elliptical, highest postmedially; dorsum nearly straight to slightly concave, venter slightly convex; ends nearly equal; valve size nearly equal; vertical narrow sulcus, constricted at midheight, lies two-fifths of length from anterior; narrow terminal rims; finely pitted; L, holotype, 0.65; H, 0.30 mm (Swain, 1946).

Occurrence. Upper Jurassic, Morrison Formation, Todilto Limestone Member, about six miles (9.6 km) north of Thoreau, New Mexico on State Highway 56 (Swain, 1946).

Cytheridella? sp., immature
Plate 17, figure 7

Very small, subpyriform, highest anteromedially, dorsum convex, venter slightly concave but nearly straight; anterior very broad, posterior narrow, extended below; smooth border and narrow ridges anteriorly; three small spines posteriorly; densely pitted, with overprint of coarse reticulating ridges; L, 0.24; H, 0.15 mm.

Occurrence. Pleistocene, bone bed at Boney Spring, Benton County, Missouri (J.E. King, coll, unpublished.). If correctly identified, it suggests subtropical environment.

Genus Cytherissa Sars, 1925

(Cytherideidae-Cytherideinae)
Cytherissa lacustris (Sars, 1863)
Plate 16, figures 29, 30
Cythere lacustris Sars, 1863, p. 30.
Cyprideis torosa Jones, 1856 (part), p. 21, pl. 2, figs. 1a-i.
Acanthopus resistans Vernet, 1878, p. 506, pl. 17, figs. 1-13.
Cytheridea lacustris (Sars), Brady, 1868, p. 26, figs. 18-21; pl. 40, fig. 2.
Cytherissa lacustris (Sars), Sars, 1925, p. 153, pl. 70. Staplin, 1963, p. 1202, pl. 159, figs. 52-54.
Cytherissa simplissima Swain, 1963, Delorme, 1970, p. 1257.

Subpyriform, highest about one-third of length from anterior end; dorsum straight, about two-thirds of length, venter slightly concave medially, converging with dorsum to rear; anterior broader than posterior, extended below; anteromedian sulcus; several protuberances may be present; pitted; terminal hinge sockets and interminal groove in LV: inner lamellae narrow; adductor muscle scar a vertical submedian row of four spots; L, female, 0.89 to 0.93; H, 0.53 to 0.59 mm (Staplin, 1963).

Occurrence. Pleistocene of Lake Michigan cores; Aftonian Stage of Illinois. Holocene, typically in deep cold lakes of northern Hemisphere (authors).

Cytherissa simplissima Swain
Plate 17, figures 2-5, 6
Cytherissa simplissima Swain, 1963, p. 811, pl. 95, figs. 4a-c; pl. 96, figs. 15a-g, TF 6c, d.
Cytherissa lacustris (Sars), Delorme, 1970, p. 1257.

Subovate-subquadrate, highest about one-third of length from anterior end; dorsum straight, two-thirds of shell length; anterior cardinal angle more obtuse than posterior cardinal angle; venter slightly concave medially; anterior broader than posterior; shallow dorsomedian sulcus; nodes flank sulcus; weakly pitted; RV hinge with terminal elongate teeth and interminal groove; adductor muscle scar a vertical row of four spots; two frontal spots; narrow inner lamellae, widely spaced radial canals; L, holotype, 0.91; H, 0.56 mm (Swain, 1963).

Occurrence. Pleistocene, Gubik Formation, non-marine facies, Arctic coastal plain, northern Alaska, Ikpikpuk area shot holes, line 6, shot point 8, at 30-40 feet (9-12 m) (Swain, 1963).

Genus Cytheromorpha Hirschmann, 1909
(Cytheridae-Cytherinae?)
Cytheromorpha? sp., Swain
Plate 17, figure 8
Cytheromorpha? sp., Swain, 1985, p. 435, pl. 4, figs. 21, 22.

Elongate, subquadrate, highest one-fifth of length from anterior end; dorsum straight, venter slightly concave; posterior narrower than anterior; subequal valves; compressed; small median node and adjacent pit; pustulose surface; anterior coarse marginal spines; RV hinge with small anterior socket postjacent tooth, interterminal bar, posterior tooth and socket; inner lamellae broad anteriorly; L, 0.57; H, 0.23 mm (Swain, 1985).

Occurrence. Miocene, Esmeralda Formation, 3.3 miles (5.3 km) east of Finger Rock Wash in Simon Mine area, Mineral County, Nevada (Swain, 1985).

Genus Darwinula Brady and Norman, 1889
(Darwinulidae)
Argilloecia (part) + Polycheles Brady and Robertson, 1870, p. 10, 25.
Darwinella Brady and Robertson, 1872, p. 50

Darwinula aurea (Brady and Robertson, 1870)

(see Darwinula stevensoni (Brady and Robertson))

Darwinula? hollandi Scott

Plate 17, figures 11, 12

Darwinula hollandi Scott, 1944, p. 146, pl. 24, figs. 6-8.

Elongate sublanceolate, highest in posterior third, dorsum convex, venter straight to concave; posterior broader than anterior, the latter narrow and extended below; LV overlaps RV along free margins; thickest postmedially; smooth; L, holotype, 0.42; H, 0.20 mm (Scott, 1944) (generic assignment uncertain).

Occurrence. Permian, Dunkard Group, Nineveh Limestone, Nineveh, Pennsylvania (Scott, 1944).

Darwinula leguminella (Forbes)

Plate 8, figures 15, 16

Darwinella leguminella Forbes, Jones, 1885, p. 346, pl. 8, figs. 30, 31

Darwinula leguminella (Forbes), Jones, 1886, p. 147, pl. 4, figs. 4a-c

Elongate subpyriform, heighest postmedially; dorsum convex, venter nearly straight; anterior narrower than posterior; valve size uncertain, LV? shown as the larger; L, 0.93; H, 0.4 mm (after Jones, 1886).

Occurrence. Upper Jurassic, Morrison Formation, "Atlantosaurus beds", Canyon City, Colorado (Jones, 1886).

Darwinula sp. aff. D. liulingchuanensis Zhong

Plate 17, figure 42

Darwinula sp., Kietzke, 1987, p. 123, TF 2-ln

Darwinula sp. aff. D liulingchuanensis Zhong, 1964, Lucas and Kietzke 1993, p. 59, TF 3-6.

Elongate elliptical, highest postmedially; dorsum slightly convex, venter nearly straight to weakly concave; anterior a little narrower than posterior; LV larger than RV and overlaps it except dorsally?; smooth; valves inflated, oval in end and view; L, 0.66; H, 0.36 mm (after Lucas and Kietzke, 1993).

Occurrence. Upper Triassic, Chinle Group, Petrified Forest Formation, Painted Desert Member, SW1/4 SE1/4 sec. 4, T. 20 N., R. 24 E., Petrified Forest National Park, Arizona (Lucas and Kietzke, 1993).

Darwinula rogersii (Jones, 1862)

Plate 17, figure 13

Candona? rogersii Jones, 1862, p. 124, pl. 5, figs. 20-22.

Darwinula rogersii (Jones), Swain and Brown, 1972, p. 16, pl. 1, figs. 6, 7, TF 3.

Subelliptical-sublanceolate, highest medially to postmedially; dorsum slightly more convex than venter; anterior narrower than posterior; relatively compressed; smooth; L, 0.85; H, 0.39 mm (Swain and Brown, 1972).

Occurrence. Upper Triassic, Newark Supergroup, Cumnock Formation, 0.5 miles (0.8 km) west of Gulf, Chatham County, North Carolina (Swain and Brown, 1972); Keuper (Upper Triassic) shales near Richmond, Virginia and Deep River, North Carolina (Jones, 1862).

Darwinula sarytirmensis Sharapova

Plate 16, figures 31, 32

Darwinula sarytirmensis Sharapova in Mandelstam, 1947, p. 254, pl. 2, fig. 8. Kietzke and Lucas, 1994, p. 27, TF 2E-L.

Elongate ovate, highest one-fourth of length from posterior end; dorsum gently convex, venter slightly concave; anterior narrower than posterior; LV larger than RV; smooth; shell composed of an inner columnar crystalline layer, four zones of granular crystals in endocuticle and an outer unlayered exocuticle of calcite; adductor muscle scar medially located, comprising a rosette of up to 11 spots; L, 1.2; H, 0.55 mm (Kietzke and Lucas, 1994).

Occurrence. Lower Jurassic, Kayenta Formation, SW$^{1}/_{4}$ sec. 35, T. 27 N., R. 12 E., Coconina County, Arizona (Kietzke and Lucas, 1995). Also in Jurassic non-marine rocks of Congress of Industrial States (Sharapova, 1947) and in China (Hao, et al., 1983; Meizhen, 1984; Xu, 1988)

Darwinula simplus (Roth)

Plate 17, figures 14, 15

Paracypris simplus Roth, 1933, p. 403, pl. 43, figs. 3a-d.

Elongate elliptical, highest postmedially; dorsum slightly convex, venter sinuous; anterior narrower than posterior; LV larger than and overlaps RV according to Roth; smooth; L, 1.08; H, 0.48 mm (Roth, 1933).

Occurrence. Upper Jurassic, Morrison Formation, sec. 28, T. 4 N., R. 6 E., Lawrence County, South Dakota (Roth, 1933). According to Sohn (11958, 1977, 1979) this horizon and locality are in the Lower Cretaceous Lakota Formation.

Darwinula stevensoni (Brady and Robertson)

Plate 17, figures 9, 10, 16-20

Argilloecia? aurea Brady and Robertson, 1870, p. 16, pl. 8, figs. 4, 5.

Polycheles stevensoni Brady and Robertson, 1870, p. 25, pl. 7, figs. 1-7; pl. 10, figs. 4-14.
Darwinella stevensoni (Brady and Robertson), Brady and Robertson, 1872, p. 50 (non Müller, 1865).
Darwinula stevensoni (Brady and Robertson), Brady and Norman, 1889, p. 122, pl. 10, figs. 7- 13; pl. 13, figs. 1-9; pl. 23, fig. 5.
Darwinula improvisa Turner, 1895, p. 336, pl. 81, figs. 1-3, 13.
Darwinula aurea (Brady and Robertson), G.W. Müller, 1912, p. 240; Staplin, 1963, p. 1191, pl. 151, figs. 50-51.
Darwinula cf. stevensoni (Brady and Robertson), Swain, 1949, p. 180, pl. 32, figs. 16, 17. Lister, 1975, p. 30, pl. 3, figs. 15, TF 34. Swain, 1986a, p. 375, pl. 4, figs. 14a, b; Swain, 1987c, p. 436, pl. 2, figs. 3, 4, 8.

Elongate subelliptical, highest postmedially; height:length ratio about 1:2.4; dorsum slightly convex, venter sinuous; anterior narrower than posterior; RV larger than LV and overlaps it, but in some specimens overlap is slight; strongly convex, subcylindrical in end view; smooth; RV hinge margin bears a furrow for reception of edge of LV: calcified inner lamellae weak to absent; adductor muscle scar an anteromedian rosette of nine wedge-shaped spots; in some specimens two "frontal" spots lie anterad of adductor group; L, 0.79; H, 0.34 mm; males poorly known (Swain, 1986).

Occurrence. Paleocene, Fort Union Formation, Tongue River Member, NW$^1/_4$ sec. 31, T. 9 S., R. 49 E., Powder River County, Montana (Swain, 1949). Miocene Esmeralda Formation, Black Spring and Cedar Springs surface sections, Nye County, Nevada; Horse Camp Formation, near Currant, Nye County, Nevada (Swain, 1987c). Pliocene Cache Valley Formation, Hoopes Creek loc. Franklin County, Idaho (Swain, 1986a). Pleistocene, Lake Bonneville Group, Saltair core hole SE$^1/_4$ sec. 25, T. 1 N., R. 3 W., and Section 28 core hole, SE$^1/_4$ SW$^1/_4$ sec. 28, T. 1 N., R. 2 W., Salt Lake County, Utah (Lister, 1975). Mankato Substage, Wisconsin Stage and pre-Recent of Illinois (Staplin, 1963). Recent, cosmopolitan (authors).

Darwinula subquadrata Swain and Brown
Plate 17, figures 21, 22
Darwinula subquadrata Swain and Brown, 1972, pl. 16, figs. 1-5; pl. 2, figs. 2, 4, TF 4.

Subelliptical to subquadrate, highest medially; dorsum straight, about half of shell length; venter nearly straight and subparallel to dorsum; anterior narrower than posterior and truncate above; compressed; adductor muscle scar an anteromedian circular spot with longitudinal internal markings; smooth exterior, microreticulate interior; L, 0.83; H, 0.33 mm (Swain and Brown, 1972).

Occurrence. Upper Triassic, Newark Supergroup, Cumnock Formation, 0.5 miles (0.8 km) west of Gulf, Chatham County, North Carolina; Upper Triassic shales near Carpenter, Wake County, North Carolina (Swain and Brown, 1972).

Darwinula sp., Sohn
Plate 17, figures 23-25
Darwinula sp., Sohn, 1977a, p. 135, TF r-x.

Elongate, sublanceolate, highest about one-fourth of length from posterior end; anterior much narrower than posterior; LV? overlaps RV?; smooth; L, 0.66; H, 0.29 mm (Sohn, 1977).

Occurrence. Pennsylvanian, Monongahela Formation, Monongalia County, West Virginia, USNM 16827 (Sohn, 1977a).

Darwinula sp., Swain and Wagoner
not figured
Darwinula sp., Wagoner, 1974, unpublished M.S. thesis, Humboldt State University, California. Swain and Wagoner, 1985, p. 143.

Poorly preserved specimens probably representing this genus; not described.

Occurrence. Miocene, Esmeralda Formation, Blair Junction section, sample 4-7, representing nearly the highest ostracode occurrence in the section, Esmeralda County, Nevada (Swain and Wagoner, 1985).

Darwinula sp., Swain
Plate 17, figure 26
Darwinula sp., Swain, 1986b, p. 213, pl. 2, fig. 14.

RV elongate, sublanceolate, highest postmedially; dorsum gently convex; venter nearly straight; anterior narrower than posterior; compressed, smooth; L, RV, 0.80; H, 0.30 mm (Swain, 1986b).

Occurrence. Miocene, Humboldt Formation, Gulf Refining Co., No. 2 Wilkins Ranch well, sec. 19, T. 39 N., R. 69 E., Elko County, Nevada, 4910-4940 feet (1497-1506 m); also section at Cobre, Elko County, Nevada 340-360 feet (104-110 m) above base (Swain, 1986b).

Darwinula sp., Swain
Plate 17, figure 27
Darwinula sp., Swain, 1987b, p. 230, pl. 3, fig. 8.

Very elongate, subelliptical, highest about one-third of length from posterior end; dorsum less convex than venter; RV overlaps LV around most of periphery; strongly convex;

smooth; adductor muscle scar an anteromedian rosette of about 15 elongate spots in two opposing rows; L, 0.69; H, 0.25 mm (Swain, 1987b).

Occurrence. Paleocene or Eocene, Sheep Pass Formation, Johnson Canyon loc., Grant Range, Nye County, Nevada (Swain, 1987b).

Darwinula sp., Swain
Plate 17, figures 28, 29
Darwinula sp., Swain, 1987c, p. 436,pl. 5, figs. 15, 16.

Elongate, subelliptical, highest postmedially; dorsum and venter gently convex; anterior narrower than posterior; RV larger than LV, moderately convex; smooth; L, 0.55; H, 0.23 mm (Swain, 1987c).

Occurrence. Oligocene, Pruett ? Formation, Buck Hill volcanic series, Alpine Quadrangle, Brewster County, Texas (Swain, 1987c).

Darwinula, sp. 1, Sohn
Plate 17, figures 30-33
Darwinula sp. 1, Sohn, 1985, p. 457, TF 6.1-6.4, undescribed internal mold.

Internal mold (Steinkern) elongate subelliptical, highest near posterior end; posterior broader than anterior, the latter truncate above; LV larger than RV; adductor muscle scar and anteromedian rosette of six or more elongate spots; L, 0.6; H, 0.3 mm (Sohn, 1985; described from author's photographs).

Occurrence. Upper Mississippian, Bluestone Formation, Bramwell Member, SW$^1/_4$ Flat Top $7^1/_2$ minute Quadrangle, Mercer County, West Virginia (Sohn, 1985).

Darwinula sp. Sohn
not illustrated or described
Darwinula sp. Sohn, 1964

Occurrence. Upper Triassic, Shublik Formation, northern Alaska (Sohn, 1964).

Darwinula sp. 2, Sohn
Plate 17, figure 34
Darwinula sp. 2., Sohn, 1985, p. 457, TF 6.5-6.8, 6.14-6.15 (undescribed internal mold).

Internal mold elongate, sublanceolate, highest near posterior end; dorsum gently convex, venter nearly straight; anterior narrowly rounded, extended below; LV larger than RV; adductor scar an anteromedian irregular spot with indistinct internal elements; L, 1.10; H, 0.50 mm (Sohn, 1985).

Occurrence. Upper Mississippian, Bluestone Formation, Bramwell Member, SW$^1/_4$ Flat Top 7$^1/_2$ minute Quadrangle, Mercer County, West Virginia (Sohn, 1985).

Darwinula sp. 3, Sohn

Plate 17, figure 35

Darwinula sp. 3, Sohn, 1985, p. 459, TF 6.12-6.13 (undescribed internal mold).

Internal mold elongate subelliptical, highest postmedially; anterior narrower than posterior; LV? larger than RV?; adductor muscle scar a submedian spot having indistinct rounded internal elements; L, 1.24; H, 0.60 mm (Sohn, 1985).

Occurrence. Upper Mississippian, Bluestone Formation, Bramwell Member, SW$^1/_4$ Flat Top 7$^1/_2$ minute Quadrangle, Mercer County, West Virginia (Sohn, 1985).

Darwinula sp. 1, Swain and Brown

Plate 17, figures 36 - 37

Darwinula sp. 1, Swain and Brown, 1972, p. 17, pl. 1, figs. 8, pl. 2, fig. 5, TF 5.

Elongate sublanceolate, highest postmedially; dorsum gently convex, venter slightly sinuous; anterior narrower than posterior; moderately convex; smooth; adductor muscle scar a submedian group of about six spots in a circular pattern; L, 0.90; H, 0.32 mm (Swain and Brown, 1972).

Occurrence. Upper Triassic, Cumnock Formation, 0.5 miles (0.8 km) west of Gulf, Chatham County, North Carolina; undifferentiated late Triassic rocks exposed near Carpenter, Wake County, North Carolina (Swain and Brown, 1972).

Darwinula, sp. 1, Swain and Brown

Plate 17, figure 38

Darwinula?, sp. 3, Swain and Brown

Plate 17, figure 39

Darwinula? sp. 3, Swain and Brown, 1972,p. 18, pl. 1, figs. 10, TF 7.

Elongate-ovate, highest postmedially; anterior much narrower than posterior; compressed (compacted?); smooth; L, 0.65; H, 0.33 mm (Swain and Brown, 1972).

Occurrence. Upper Triassic, Cumnock Formation, Bethany Church, 0.5 miles (0.8 km) west of Gulf, Chatham County, North Carolina (Swain and Brown, 1972).

Darwinula? sp. 4, Swain and Brown

Plate 17, figure 40

Darwinula? sp. 4, Swain and Brown, 1972, p. 18, pl. 1, fig. 11, TF 8.

Subtriangular-sublanceolate, highest anteromedially; dorsum more convex than venter; posterior narrower than anterior; compressed (compacted?); smooth; adductor muscle scar slightly anteromedian; L, 0.65; H, 0.35 mm (Swain and Brown, 1972).

Occurrence. Upper Triassic, Cumnock Formation, Bethany Church, 0.5 miles (0.8 km) west of Gulf, Chatham County, North Carolina (Swain and Brown, 1972).

Darwinula sp., Sohn

Plate 18, figure 1

Darwinula sp., Sohn, 1976, p. 305, TF 1-7. (undescribed except for muscle scars).

Internal mold sublanceolate, highest near posterior end; dorsum and venter nearly straight, converging toward anterior; anterior much narrower than posterior, bluntly pointed and extended below, truncate above; adductor muscle scar an anteromedian rosette of eight (LV) or nine (RV) spots (Sohn, 1976) (described from author's illustrations).

Occurrence. Upper Pennsylvanian, Monogahela Formation, just below Little Waynesburg Coal, Pursglove, Monongalia County, West Virginia (Sohn, 1976).

Darwinula? sp.

Plate 18, figures 2, 3

Darwinula? sp., Betty Kellet Nadeau, 1945, unpublished.

Elongate subelliptical-sublanceolate, highest postmedially; dorsum slightly convex, venter straight to slightly concave; anterior narrowly rounded, extended below; smooth; magnification unknown.

Occurrence. Upper Jurassic or Lower Cretaceous, Morrison Formation?, near Canyon City, Colorado? (Nadeau, 1945, unpublished).

Darwinula sp. A, Kietzke

Plate 16, figure 34

Darwinula sp. A., Kietzke, 1989, p. 185, TF 4C-D, 6A.

Elongate oval, highest postmedially; dorsum slightly convex, venter nearly straight, RV overlaps LV weakly; anterior much narrower than posterior; smooth; L, 0.61 to 0.76; H, 0.3 to 0.4 mm (Kietzke, 1989).

Occurrence. Lower Triassic, Moenkopi Formation, NW$^1/_4$ NE$^1/_4$ NW$^1/_4$ sec. 36, T. 7 N., R. 4 W., Chipola County, New Mexico, at White Ridge; upper part of basal unit (Kietzke, 1989).

Darwinula sp. B, Kietzke

Plate 16, figure 33

Darwinula sp. B., Kietzke, 1989, p. 185, TF 4E.

Elongate subovate, highest medially, anterior narrower than posterior; RV overlaps LV ventrally; dorsum more convex than venter; smooth; L, 0.63 to 0.6; H, 0.33 to 0.42 mm (Kietzke, 1989).

Occurrence. Lower Triassic, Moenkopi Formation, Unit 4, NW1/4 NE1/4 NW1/4 sec. 36, R. 7 N., R. 4 W., Cibola County, New Mexico, at White Ridge (Kietzke, 1989).

Darwinula sp., Kietzke

Plate 17, figure 41

Darwinula sp., Kietzke, 1987

Subelliptical, highest posteriorly; dorsum and ventor nearly straight, converging toward anterior; anterior narrower than posterior; RV larger than LV?; smooth; L, 0.5; H, 0.25 mm (after Kietzke, 1987).

Occurrence. Upper Triassic, Chinle Formation, upper shale, Bull Canyon, NW1/4 ces. 28, T. 9N., R. 26 E., Guadalupe County, New Mexico (Kietzke, 1987).

Genus Darwinuloides Mandelstam, 1956

(Darwinuloidacea - Darwinuloididae)

Darwinuloides? magna (Jiang), (Kietzke and Lucas)

Plate 16, figures 36, 37

Darwinula magna Jiang in Xu, 1983, pl. 8, figs. 5, 6. Kietzke and Lucas, 1995, TF 3A-E.

Darwinula cf. sarytirmensis Sharapova, Xu, 1988, pl. 2, fig. 16 (fide Kietzke and Lucas, 1995).

Subovoid, highest postmedially; dorsum strongly convex, venter nearly straight to slightly convex; anterior narrower than posterior, extended below and merging abruptly with venter; smooth; LV larger than RV, overlapping ventrally and posteriorly; L, 1.23; H, 0.7 mm (Kietzke and Lucas, 1995). (strongly resembles Darwinuloides Mandelstam, 1956).

Occurrence. Lower Jurassic, Kayenta Formation SW1/4 sec. 35, T. 27 N., R. 12 E., Coconino County, Arizona (Kietzke and Lucas, 1995).

Darwinuloides? sp. (Kietzke)

Plate 17, figure 41

Darwinula sp. Kietzke, 1987, p. 123, TF 25-N.

Elongate subelliptical, highest postmedially; dorsum slightly convex; venter straight to very slightly concave; anterior narrower than posterior; LV overlaps RV; valves inflated postmedially; smooth to finely punctate; adductor muscle scar located at midpoint; L, 0.36 to 0.98; H, 0.19 to 0.55 mm (Kietzke, 1987).

Occurrence. Upper Triassic, Chinle Formation, upper shale, Bull Canyon, NW$^1/_4$ sec. 28, T. 9 N., R. 26 E., Guadalupe County, New Mexico; also Redonda Member, Shark Tooth Hill, SW$^1/_4$ NE$^1/_4$ sec. 28, T. 98 N., R. 34 E., Quay County, New Mexico; also Sloan Canyon Formation, NW$^1/_4$ sec. 12, T. 35 N., R. 35 E., Union County, New Mexico (Kietzke, 1987).

Darwinuloides sp. B, Kietzke

Plate 16, figure 35

Darwinuloides sp. B., Kietzke, 1989, p. 189, TF 4B.

Elongate subtriangular, highest postmedially; dorsum convex, ventor nearly straight; posterior broader than anterior; RV overlaps LV along venter; widest postmedially; smooth ?; L, 0.75 to 0.9; H, 0.33 to 0.43 mm (after Kietzke, 1989).

Occurrence. Lower Triassic, Moenkopi Formation, NW1/4 NE1/4 NW1/4 sec. 36, T. 7 N., R. 4 W., Cibola County, New Mexico at White Ridge (Kietzke, 1989).

Darwinuloides sp. A, Kietzke

Plate 17, figure 1

Darwinuloides sp. A., Kietzke, 1989, p. 187, TF 4A.

Small subtriangular, highest postmedially; dorsum strongly convex, venter nearly straight, anterior much narrower than posterior; RV overlaps LV along venter and mid dorsum, widest medially; adductor muscle area ventromedian; smooth?; L, 0.64 to 0.69; H, 0.39 to 0.44 mm (after Kietzke, 1989).

Occurrence. Lower Triassic, Moenkopi Formation, NW1/4 NE1/4 NW1/4 sec. 36, T. 7 N., R. 4 W., Cibola County, New Mexico at White Ridge (Kietzke, 1989).

Genus Disopontocypris Mandelstam, 1956

(Cyprididae-Disopontocypridinae)

Disopontocypris? hendersoni Swain

Plate 18, figures 4, 5

Disopontocypris? hendersoni Swain, 1987c, p. 43, pl. 1, figs. 3, 4.

Elongate subelliptical-subreniform, highest postmedially; dorsum gently convex, venter concave; posterior slightly narrower than anterior; LV slightly larger than RV; anterior eighth of shell compressed; postventral marginal zone compressed; smooth; L, 0.97; H, 0.50 mm (Swain, 1987c).

Occurrence. Miocene, Esmeralda Formation, Black Spring road measured section, 7659 feet (2332 m) S. 26 E. of NW corner of Simon 1:24000 Quadrangle in unsurveyed area, Nye County, Nevada, 330 feet (100 m) above base of section (Swain, 1987c).

Genus Dogelinella Schneider in Mandelstam et al, 1957

(Cyprididae-Cyprinotinae)

Dogelinella? coaldalensis Swain and Wagoner

Plate 18, figures 6 - 8

Dogelinella? coaldalensis Swain and Wagoner, 1985, p. 134, pl. 3, figs. 9-12. Swain, 1990, p. 149, pl. 2, fig. 15.

Small, elongate, subovate to sublanceolate, highest about one-third of length from anterior end; dorsum convex, venter slightly concave; posterior narrower than anterior, both ends extended below; LV overlaps RV; compressed at both ends; valves have low convexity; ornamentation a meshwork of fine ridges, reticulate except ventrally where concentric ridges have only a few crossbars; valve edges with knife edge rims; rabbet groove in LV received hinge margin of RV; adductor muscle scar a median group of 4 to 5 spots; two more anteroventral, frontal spots; inner lamellae narrow, radial canals few; normal pores lie in part on raised tubercles; L, 0.50; H, 0.30 mm (Swain and Wagoner, 1985).

Occurrence. Miocene, Esmeralda Formation, Blair Junction section, Esmeralda County, Nevada, samples 1C2, 10 (Swain and Wagoner, 1985). Truckee Formation, 2 miles (3.2 km) west-southwest of Hazen, Churchill County, Nevada (Swain, 1990).

Dogelinella? sp., Swain

Platc 18, figures 9, 10

Dogelinella? sp., Swain, 1987c, p. 249, pl. 4, figs. 10, 11.

Subreniform, subquadrate, highest about one-third of length from anterior end; dorsum straightened medially, venter concave; posterior narrower than anterior, both ends extended below, truncate above; LV overlaps RV on free margins; moderately convex, flattened and depressed medially; smooth; L, 0.43; H, 0.36 mm (Swain, 1987c).

Occurrence. Oligocene or Miocene, North Creek Formation, secs. 3 and 4, T. 14 N., R. 69 E., White Pine County, Nevada, north of U.S. Highway 50, east of Sacramento Pass (Swain, 1987c).

Genus Dongyingia Hou et al, 1978

(Cyprididae-Cypridinae?)

Dongyingia? lariversi Swain

Plate 18, figures 11 - 15

Dongyingia? lariversi Swain, 1985, p. 429, pl. 2, figs. 9, 10; pl. 5, figs. 4, 5. Swain, 1987c, p. 430, pl. 1, figs. 13-16.

Subquadrate to subelliptical-subreniform, highest anteromedially; dorsum straightened medially, venter nearly straight to concave; LV strongly overlaps RV; strong posterodorsal node, two strong ventral nodes; densely pitted; lophodont hinge, RV with terminal projecting flanges, interterminal depressed hinge edge; inner lamellae moderate in width ventroterminally; narrow vestibules terminally; L, 0.84; H, 0.52 mm (Swain, 1985, 1987c).

Occurrence. Miocene, Esmeralda Formation, Black Spring road section, 2.85 miles (4.5 km) west of Black Spring, Nye County, Nevada, about 800 feet (244 m) above base of lake bed sequence (Swain, 1985). Warrior Mine measured section, Nye County, Nevada 110 feet (33 m) and 400 feet (122 m) above base. Black Spring measured section, Nye County, Nevada, 340 feet (104 m) above base (Swain, 1987c).

Genus Dryelba Sohn, 1982

(Dryelbidae)

Dryelba? angularis (Peck)

Plate 27, figures 16-21

Metacypris angularis Peck, 1941, p. 302, pl. 44, figs. 1-14.

Oblong, highest medially, dorsum concave to nearly straight, venter slightly convex to nearly straight; anterior broader than, or nearly equal to, posterior, both ends truncate above; bisulcate in anterodorsal sulcus; anterior marginal zone flattened, nodose; posterior surface with three or more nodes, or nodes may be lacking; surface densely pitted; dimorphic, females with broad flattened striated ventral surface; males narrower; L, 1.0; H; 0.6 mm (Peck, 1941).

Occurrence. Lower Cretaceous, Peterson Limestone, Freedom Quadrangle, Idaho-Wyoming, and in Lincoln County, Wyoming; Kootenai Formation near Great Falls, Whitehall and Drummond, Montana (Peck, 1941).

Dryelba? persulcata (Peck)

Plate 8, figures 20 - 24

Metacypris persulcata Peck, 1941, p. 302, pl. 44, figs. 15-21.

Dryelba? persulcata (Peck), Sohn, 1982, p. 313.

Medium size; suboblong to subovate, highest at or posterior to anterocardinal angle; ends about equally broad, the anterior subtruncate above and below; dorsum nearly straight to sinuous; venter slightly convex; ventral surface flattened; median sulcus extends to below midheight; anterodorsal rounded node, anterodorsal short sulcus; anterior border compressed; surface densely and coarsely pitted; smooth marginal rims ventrally and dorsally; valve contact incised dorsally; L, 1.0; H, 0.61 mm (Peck, 1941). Generic position questionable because it is not pustulose, a generic characteristic.

Occurrence. Lower Cretaceous, "Lower Bear River", T. 5 S., R. 45 E., 1.7 miles (2.7 km) east of mouth of South Fork of Tincup Creek, Freedom Quadrangle, Idaho-Wyoming; also 0.25 mile (0.64 km) east of Smith's Fork Ranger Station on Smith's Fork of Bear River, Lincoln County, Wyoming (Peck, 1941).

Dryelba pustulosa Sohn

Plate 18, figures 16-19

Dryelba pustulosa Sohn, 1982, p. 314, pl. 3, figs .1-13; pl. 4, figs. 13-30.

Subquadrate, highest postmedially, dorsum and venter nearly straight and converge slightly toward anterior; anterior narrower than posterior, truncate above; large pustulose node anterad of submedian sulcus; four or more nodes posterior to sulcus except subcentral node near venter may be subdivided; surface with polygonal reticules; L, 0.73 to 1.4; H, 0.4 to 0.8 mm (Sohn, 1982).

Occurrence. Upper Jurassic, Morrison Formation, SW$^{1}/_{4}$ sec. 7, T. 53 N., R. 60 W., Crook County, Wyoming; borehole sec. 28, T. 6 S., R. 2 E., Custer County, South Dakota (Sohn, 1982).

Genus Elkocythereis Dickinson and Swain, 1967

(Limnocytheridae-Limnocytherinae)

Elkocythereis bramletti Dickinson and Swain

Plate 18, figures 20-23

Elkocythereis bramletti Dickinson and Swain, 1967, p. 346, pl. 38, figs. 5-14; pl. 39, figs. 1-6, TF 6. Swain et al., 1971, p. 473.

Females subquadrate, highest in anterior fourth; males elongate subquadrate, maintaining approximately equal height throughout most of length; dorsum nearly straight, venter nearly straight to sinuous; anterior broader than posterior, both ends may be extended below; valves nearly equal in size; width moderate; surface bears a vertical median sulcus that encloses three connecting pits and dies out near midheight; a sinuous longitudinal groove lies dorsomedially; alae weak or absent; marginal ridges vary from weak to strong; general surface reticulate; LV hinge a denticulate bar truncated terminally by rounded sockets; inner lamellae broad anteriorly;

vestibule weak or absent; adductor muscle scar a median vertical row of four spots; L, female 0.82; H, 0.48 mm (Dickinson and Swain, 1967).

Occurrence. Pliocene, Hay Ranch Formation, Gulf Oil Corp. No. 1 Wilkins Ranch well, sec. 21, T. 38 N., R. 61 E., Elko County, Nevada, 957 to 5441 feet (292-1658 m); also roadcut on pass about one mile south of Elko, Nevada on State Highway 46, sec. 19, T. 34 N., R. 55 E.,; also roadcut on State Highway 8A, 20 miles north of Austin, Lander County, Nevada (Dickinson and Swain, 1967).

Elkocythereis sp. aff. E. bramletti Dickinson and Swain

Plate 18, figure 38

Elkocythereis sp. aff. E. bramletti Dickinson and Swain, Swain, 1985, p. 435, pl. 5, fig. 3.

Elongate subquadrate-subreniform, highest anteromedially; dorsum nearly straight, venter slightly concave; posterior narrower than anterior; valves subequal, compressed; terminal margins in part coarsely spinose; median large rounded node; other scattered nodes; densely reticulate surface; RV hinge bears closely spaced crenulate ridges and terminal crenulate teeth; L, 0.82; H, 0.43 mm (Swain, 1985).

Occurrence. Miocene, Esmeralda Formation, 3.3 miles (5.3 km) east of Finger Rock Wash on road to Simon Mine, Stewart Spring 1:24000 Quadrangle, Mineral County, Nevada, in lower 100-150 feet (30-45 m) of lake bed section (Swain, 1985).

Elkocythereis minutidontis Dickinson and Swain

Plate 18, figures 24-28

Elkocythereis minutidontis Dickinson and Swain, 1967, p. 348, pl. 37, figs. 25-34; pl. 38, fig. 15, TF 7. Swain, 1986b, p. 214, pl. 1, fig. 5.

Subquadrate (female) to elongate-subquadrate (male), highest near anterior end; dorsum nearly straight to slightly convex, venter concave medially; anterior broader than posterior; valves subequal, relatively compressed; median sulcus consists of interconnecting pits and extends to within one-third of height from venter; two anteromedian pits; weak terminal marginal ridges; venter weakly but variably alate; general surface reticulate; LV hinge a long bar with terminal small elongate tooth depressions; inner lamellae of moderate width anteriorly, narrower posteriorly; adductor muscle scar a vertical row of four small spots; L, female 0.75; H, 0.45 mm (Dickinson and Swain, 1967).

Occurrence. Pliocene, Hay Ranch Formation, Gulf Oil Corp., No. 1 Wilkins Ranch well, sec. 21, T. 38 N., R. 61 E., Elko County, Nevada, 957 to 6860 feet (596-2091 m); also outcrop about 2 miles (3.2 km) north of Twin Bridges, sec. 24, T. 32 N., R. 55 E., Elko County, Nevada; also roadcut about one mile south of Elko, Nevada on State Highway 46, sec.

19, T. 34 N., R. 55 E.; also roadcut on State Highway 8A, 20 miles (32 km) north of Austin, Lander County, Nevada (Dickinson and Swain, 1967). Pliocene or Miocene, Hay Ranch or Humboldt Formation, Gulf Oil Corp. No. 2, Wilkins Ranch well, Montello area, sec. 19, T. 39 N., R. 69 E., Elko County, Nevada, 2635-2665 feet (803-877 m) (Swain, 1986b).

Elkocythereis postdorsoalata Swain

Plate 18, figures 29-33

Elkocythereis postdorsoalata Swain, 1986a, p. 99, pl. 3, figs. 22, 23; pl. 4, figs. 3-9, 12, 16; pl. 5, figs. 1, 18. Swain, 1987a, p. 377, pl. 4, fig. 12.

Elliptical-subovate to subquadrate, highest anteriorly; dorsum nearly straight to concave; posterior narrower than anterior, the latter extended below; LV a little larger than RV, moderately convex; anteromedian short sulcus, anterodorsal oblique sulcus, intervening low, lobelike elevation not reaching dorsal margin; larger elevated knob lies posterior to midlength; postventrally a moundlike node or low ala at posterior end of flattened midventer; variably reticulate surface; LV hinge with terminal shallow sockets and intervening furrow; inner lamellae narrow, small vestibules terminally; radial canals numerous; adductor muscle scar an anteromedian vertical row of four spots; two frontal spots; L, holotype 0.95; H, 0.44 mm (Swain, 1986).

Occurrence. Pliocene, Glenns Ferry Formation, sec. 14, T. 4 S., R. 1 W., Owyhee County, Idaho (Swain, 1986a). Cache Valley Formation, Hoopes Creek loc. Franklin County, Idaho (Swain, 1987a).

Genus Eocypridina Kesling and Ploch, 1960

(Cylindroleberidacea-Cylindroleberididae-Cycloasteropinae)

Eocypridina? aciculata (Scott and Summerson)

Plate 15, figure 6, 6a

Cypridina aciculata Scott and Summerson, 1943, p. 674, p. 2, fig. 1.3.

Eocypridina aciculata (Scott and Summerson), Sohn, 1977, p. 132.

Large, subovate to subcircular, highest medially; dorsum more convex than venter; anterior with typical cypridinoid beak and notch; entire surface with complex set of acicular radii, about nine sets of radii on each valve; three trapezia of radii arranged obliquely from an anteromedian point ("epigenetic" structures by Sohn, 1977); L, 5.3; H, 3.2 mm (Scott and Summerson, 1943) (may be brackish-marine).

Occurrence. Lower Pennsylvanian, Hance Formation, along U.S. highway 119 just north of Molus, Harland County, Kentucky (Scott and Summerson, 1943).

Eocypridina radiata (Jones, Kirkby and Brady)

Plate 15, figure 7

Cypridina radiata Jones, Kirkby and Brady, 1874, p. 14, pl. 5, figs. 6a-f. Scott and Summerson, 1943, p. 673, pl. 2, fig. 2.

Eocypridina radiata (Jones). Kesling and Ploch, 1960, p. 284.

Large, subovate, highest medially; dorsum and venter about equally convex; deep cypridinoid notch and strong beak; surface with sets of acicular radii, some form hexagons, some trapezia, others circular, 15-22 aciculae in each set; 50-60 sets on each valve ("epigenetic" structures by Sohn, 1977); L, 5.0; H, 4.0 mm (Scott and Summerson, 1943).

Occurrence. Lower Pennsylvanian, Hance Formation, along US. highway 119, just north of Molus, Harlan County, Kentucky (Scott and Summerson, 1943).

Genus Erpetocypris Brady and Norman, 1889

(see Herpetocypris Brady and Norman, 1890)

Erpetocypris lata Chapman, 1896

(see Herpetocypris lata (Chapman))

Erpetocypris merriamiana Chapman, 1896

(see Herpetocypris merriamiana (Chapman))

Erpetocypris?., Swain, 1949

(see Candoniella whitei (Scott and Smith))

Genus Eucypris Vavra, 1891

(Cyprididae-Eucypridinae)

Eucypris bobcatensis Swain

Plate 18, figures 34-36

Eucypris bobcatensis Swain, 1986b, p. 198, pl. 4, figs. 13, 14; pl. 5, figs. 1, 2.

Subtriangular, highest two-fifths from anterior end; dorsum moderately convex, venter slightly concave; anterior broader than posterior, both ends extended below; LV overlaps RV; moderately convex; densely reticulate; L, holotype, 0.98; H, 0.61 mm (Swain, 1986b).

Occurrence. Miocene, lower part of Humboldt Formation, "Bobcat limestone", named for bobcat den at one of the exposures, 161-199 feet (49-61 m) above base of Medicine Creek section, Winecup Ranch, Elko County, Nevada (Swain, 1986b).

Eucypris sp. aff. E. bobcatensis Swain

Plate 18, figures 37a, b

Eucypris sp. aff. E. bobcatensis Swain, Swain, 1987b, p. 221, pl. 2, figs. 12a, b.

Elongate subpyriform, highest submedially; dorsum moderately convex, venter nearly straight to slightly concave; posterior narrower than anterior, strongly extended below; LV larger than RV; smooth; moderate convexity of valves; adductor muscle scar with four spots; two frontal spots; several dorsal muscle spots; L, 1.16; H, 0.63 mm (Swain, 1987).

Occurrence. Eocene, Twin Bridges Limestone, sec. 36, T. 32 N., R. 55 E., east side of Huntingdon Creek, Elko County, Nevada (Swain, 1987).

Eucypris fingerrockensis Swain

Plate 19, figures 1-5

Eucypris sp. aff. E. ornatoides Dickinson and Swain, Swain, Becker and Dickinson, 1971, pl. 3, fig. 6.

Eucypris fingerrockensis Swain, 1985, p. 420, pl. 1, figs. 12, 14, 16; pl. 5, fig. 6. Swain, 1987c, p. 425, pl. 2, figs. 13, 14. Swain, 1990, p. 148, pl. 2, fig. 17.

Subovate-subpyriform, highest anteromedially; dorsum strongly convex, venter nearly straight to slightly sinuous; posterior broader than anterior, both ends extended below; LV slightly larger than RV, compressed; smooth; LV hinge margin furrowed to receive edge of RV; inner lamellae narrow; terminal narrow vestibules; L, 1.07; H, 0.68 mm (Swain, 1985).

Occurrence. Miocene, Esmeralda Formation, Finger Rock Wash, 0.5 mile (0.8 km) south of Rawhide Ranch, Granny Goose 1:24000 Quadrangle, Mineral County, Nevada (Swain, 1985). Horse Camp Formation 0.5 mile (0.8 km) southeast of U.S. Highway 6 at Currant Creek School site, Nye County, Nevada (Swain, 1987c). Esmeralda Formation, Black Spring Road section, 2.85 miles (4.5 km) west of Black Spring, Nye County, Nevada, about 800 feet (244 m) above base of section. Humboldt Formation, lower part, Gulf Refining Co. No. 1 Thousand Springs well, Elko County, Nevada, 1690-1700 feet (515-518 m) (Swain, 1987). Truckee Formation 2 miles (3.2 km) WSW of Hazen, Churchill County, Nevada (Swain, 1990).

Eucypris meadensis Gutentag and Benson

Plate 19, figures 6-8

Eucypris meadensis Gutentag and Benson, 1962, p. 23, pl. 2, figs. 8, 9, TF 5. Benson, 1967, p. 235, TF 13-9, 14-6.

Subtriangular, highest medially to anteromedially, dorsum arched, venter concave medially; anterior broader than posterior; LV overlaps RV; punctate; inner lamellae broad terminally, vestibules wide; radial canals numerous; LV hinge formed of a flange that fits into rabbet groove in RV; adductor muscle scar has five spots that form a 45° angle with tangent to ventral margin; L, 1.2; H, 0.68 mm (Gutentag and Benson, 1962).

Occurrence. Recent spring, Meade County, Kansas State Park, NW1/4 SW1/4 sec. 14, T. 33 S., R. 29 W. (Gutentag and Benson, 1962). Included here for comparison with Pleistocene freshwater ostracodes from Kansas.

Eucypris microreticulata Swain

Plate 19, figures 9-13

Eucypris ornatoides microreticulata Swain, 1986b, pl. 1, figs. 9, 10; pl. 2, figs. 1, 2; pl. 3, fig. 18.

Eucypris microreticulata Swain, 1986b, p. 198, pl. 5, fig. 10.

Eucypris cf. microreticulata Swain, 1987, p. 425, pl. 5, figs. 3-5, 7, 8. Swain, 1990, p. 148, pl. 1, figs. 6, 7, 9.

Dimorphic, subovate-subtriangular to subquadrate, highest anteromedially; dorsum moderately convex, venter nearly straight to slightly concave; posterior more narrowly rounded than anterior in presumed males, more broadly rounded in females; LV overlaps RV; moderately convex, widest medially in males, postmedially in females, weakly reticulate; L, 0.96; H, 0.53 mm (Swain, 1986b).

Occurrence. Miocene, Humboldt Formation, Gulf Refining Co., No. 1 Thousand Springs well, center SE1/4 SW1/4 sec. 8, T. 40 N., R. 66 E., Elko County, Nevada, 1670-3100 feet (509-945 m): Gulf Oil Co., No. 2 Wilkins Ranch well, sec. 19, T. 39 N., R. 69 E., Elko County, Nevada, 4360-4940 feet (1329-1506 m); Cobre surface section, Elko County, Nevada, 0-10 feet (0-3 m) above base of section; Medicine Creek surface section, T. 39 N., R. 65 E., east of Black Mountain, Winecup Ranch SW 1:24000 Quadrangle, Elko County, Nevada, 161-837 feet (49-255 m) above base of exposure (Swain, 1986b). Horse Spring Formation, Fossil Ridge, Gass Peak 1:62500 Quadrangle, approximately NW1/4 sec. 31 (unsurveyed area), T. 17 S., R. 16 E., Clark County, Nevada (Swain, 1987). Lower Pliocene, Beartrap Quadrangle, Los Angeles County, California on State Highway 99 (Swain, 1990).

Eucypris ornatoides Dickinson and Swain

Plate 19, figures 14-18

Eucypris ornatoides Dickinson and Swain, 1967, p. 340, pl. 36, figs. 5-7, 9-11. Swain, 1986b, p. 197, pl. 1, figs. 1-4, 6-8; pl. 5, figs. 4, 5, 8.

Eucypris cf. ornatoides Dickinson and Swain, Swain 1987a, p. 120.

Subtrapezoidal-subreniform, highest anteriorly; dorsum convex, straightened posterior to greatest height; venter straight to slightly concave; posterior narrower than anterior, both ends extended below; LV overlaps RV anterodorsally and midventrally; smooth; L, 1.18; H, 0.70 mm (Dickinson and Swain, 1967).

Occurrence. Pliocene, Hay Ranch Formation (=Upper Humboldt), Gulf Oil Corp. No. 1 Wilkins Ranch well, sec. 21, T. 38 N., R. 61 E., Elko County, Nevada, 1435-3910 feet (437-1192 m). Miocene, Humboldt Formation, Gulf Ref. Co. No. 1 Thousand Springs well, center SE$^1/_4$ SW$^1/_4$ sec. 8, T. 40 N., R. 66 E., Elko County, Nevada, 250-3150 feet (76-960 m); Gulf Ref. Co. No. 2 Wilkins Ranch well, sec. 19, T. 39 N., R. 69 E., Elko County, Nevada, 0-1022 feet (311 m) above base of exposure; Medicine Creek surface section, T. 39 N., R. 65 E., Winecup Ranch, Elko County, Nevada, from 0-1711 feet (522 m) above base of exposure (Swain, 1986b). Miocene or Pliocene, Humboldt Formation or Salt Lake Group, Tanner Reservoir dam spillway, sec. 9 T. 7 S., R. 19 W., Box Elder County, Utah (Swain, 1987a).

Eucypris ornatoides elongata Swain
Plate 19, figures 19, 20
Eucypris ornatoides elongata Swain, 1986b, p. 197, pl. 3, figs. 11, 12.

Elongate, sublanceolate-subpyriform, highest anteromedially; venter nearly straight to concave; posterior narrower than anterior, both ends extended below; LV overlaps RV; strongly convex; smooth; L, 0.97; H, 0.50 mm (Swain, 1986b).

Occurrence. Miocene, Humboldt Formation, middle? part, Gulf Refining Co., No. 1 Thousand Springs well, center SE$^1/_4$ SW$^1/_4$ sec. 8, T. 40 N., R. 66 E., Elko County, Nevada, 600-2150 feet (183-655 m); Gulf No. 2 Wilkins Ranch well, sec. 19, T. 39 N., R. 69 E., Elko County, Nevada, 4270-4300 feet (1302-1311 m); Cobre surface section, Elko County, Nevada, 272-288 feet (83 88 m) above base of exposure (Swain, 1986b).

Eucypris ornatoides microreticulata Swain, 1986b
(see Eucypris microreticulata Swain, 1986b)

Eucypris punctatula Dickinson and Swain
Plate 19, figures 21, 22
Eucypris punctatula Dickinson and Swain, 1967, p. 341, pl. 36, figs. 1, 2.

Elongate subreniform, highest medially; dorsum moderately convex, venter concave; posterior narrower than anterior, extended below; valves subequal, convexity moderate; densely punctate surface; inner lamellae narrow, vestibules terminally; L, 1.18; H, 0.63 mm (Dickinson and Swain, 1967).

Occurrence. Pliocene, Hay Ranch Formation (=Upper Humboldt), roadcut one mile south of Elko on State Highway 46 (Dickinson and Swain, 1967).

Eucypris serrata (Müller)
Plate 17, figure 45
Cypris clavata Baird, 1837, p. 137.
Cypris lienenklausi serrata Müller, 1900, p. 70.
Eucypris serrata (Müller), Alm, 1915, p. 158. Lister, 1975, p. 14, pl. 2, fig. 15, TF 13.

Large, elongate subtriangular-subpyriform, highest anteromedially, about half of length; dorsum convex, venter nearly straight to concave; posterior much narrower than anterior, extended below; smooth, wavy surface near anterior margin; adductor muscle scar of four to six spots; two frontal scars; inner lamellae broad anteriorly, with vestibule; radial canals numerous; L, 1.91; H. 0.99 mm (Lister, 1975).

Occurrence. Pleistocene, Lake Bonneville Group, Saltair core hole, SE1/4 sec. 25, T. 1 N., R. 3 W., Salt Lake County, Utah (Lister, 1975). Holocene of Europe.

Eucypris trapeziformis Staplin
Plate 19, figure 23
Eucypris trapeziformis Staplin, 1963, p. 792, pl. 94, fig. 28.

Elongate subtrapezoidal, highest anteromedially, equal to half of length; anterior broader than posterior; dorsum straightened postmedially, venter concave; smooth; inner lamellae broad at both ends, more so anteriorly; vestibules deep; radial canals numerous; LV overlaps RV; hinge of LV overlaps RV; L, 0.94; H, 0.47 mm (Staplin, 1963).

Occurrence. Pleistocene, Kansan Stage, Arenzville Quadrangle, Cass County, Illinois, NW1/4 SE1/4 NW1/4 sec. 27, T. 19, N., R. 11 W. (Staplin, 1963).

Eucypris? sp., Swain
Plate 19, figure 24
Eucypris? sp., Swain, 1986a, p. 89, pl. 1, fig. 1.

Small, subovate, highest one-third of length from anterior end, venter nearly straight to slightly convex; posterior more narrowly rounded than anterior, extended below; LV larger than RV, overlapping it dorsally; compressed, smooth; L, 0.39; H, 0.27 mm (Swain, 1986a).

Occurrence. Pliocene, Glenns Ferry Formation, sec. 14, T. 4 S., R. 1 W., Owyhee County, Idaho (Swain, 1986a).

Eucypris sp. 1, Swain

Plate 19, figures 25, 26

Eucypris sp. 1, Swain, 1986b, p. 199, pl. 2, figs. 3, 4.

Elongate, subovate, highest medially; dorsum convex, venter nearly straight; posterior a little narrower than anterior, both ends extended below; LV overlaps RV especially anterodorsally; strongly convex; smooth; L, 1.08; H, 0.63 mm (Swain, 1986b).

Occurrence. Miocene, Humboldt Formation, Gulf Oil Co. No. 1 Thousand Springs well, center SE$^{1}/_{4}$ SW$^{1}/_{4}$ sec. 8, T. 40 N., R. 66 E., Elko County, Nevada, 2550-2600 feet (777-793 m) (Swain, 1986b).

Eucypris sp. 2, Swain

Plate 19, figures 27, 28

Eucypris sp. 2, Swain, 1986b, p. 199, pl. 2, figs. 5, 6.

Elongate, subovate-subreniform, highest submedially, dorsum arched, venter concave; posterior more narrowly rounded than anterior, both ends extended below; anterior with weak marginal rim and flattened border; LV overlaps RV except postdorsally; strongly convex; smooth; L, 0.98; H, 0.55 mm (Swain, 1986b).

Occurrence. Miocene, Humboldt Formation, Medicine Creek surface section, east of Black Mountain, T. 39 N., R. 65 E., Winecup Ranch, Elko County, Nevada, 199-200 feet (61m) above base of exposure (Swain, 1986b).

Eucypris sp. 3, Swain

not figured

Eucypris sp. 3, Swain, 1986b, p. 199.

Small elongate-subovate, highest medially, dorsum moderately convex, venter nearly straight, ends rounded, extended below; posterior the narrower; LV slightly larger than RV; compressed, smooth, L, 0.57; H, 0.47 mm (Swain, 1986b).

Occurrence. Miocene, Humboldt Formation, Gulf Refining Co., No. 2 Wilkins Ranch well, sec. 19, T. 39 N., R. 69 E., Elko County, Nevada, 4240-4270 feet (1292-1302 m) (Swain, 1986b).

Eucypris sp. 4, Swain

Plate 19, figure 29

Eucypris sp. 4, Swain, 1986b, p. 200, pl. 2, fig. 19.

Partly exfoliated shell, elongate subquadrate, highest medially, arched dorsum, slightly concave venter; anterior broader than posterior; valves subequal, convexity low; coarsely punctate; L, 0.92; H, 0.53 mm (Swain, 1986b).

Occurrence. Oligocene, Indian Well Formation or Miocene Humboldt Formation, Palisade surface section, north end of Pine Valley, Elko County, Nevada (Swain, 1986b).

Genus Fusocandona Hou and Shan, 1978

(Candonidae-Candoninae)

Fusocandona cachevalleyensis Swain

Plate 19, figure 30; Plate 20, figures 1-4

Fusocandona cachevalleyensis Swain, 1987a, p. 374, pl. 2, figs. 1-4. Swain, 1990, p. 151, pl. 1, figs. 21, 22.

Elongate, subelliptical-subquadrate to subtrapezoidal, highest just anterior to midlength; dorsum gently convex, venter nearly straight to slightly concave; LV terminal margins with ventral pointed projections; anterior broader than posterior in some specimens; LV overlaps RV except dorsally; LV hinge margin grooved to receive edge of RV; inner lamellae broadest anteriorly; vestibules terminally; LV with free margin selvage groove; adductor muscle scar with six spots; two frontal spots; about seven dorsal muscle scars; L, holotype, 1.20; H, 0.67 mm (Swain, 1987a).

Occurrence. Pliocene, Cache Valley Formation, Hoopes Creek locality, NW1/4 NE1/4 sec. 10, T. 12 S., R. 40 E., Franklin County, Idaho (Swain, 1987a). Idaho Group, NE1/4 SW1/4 sec. 9, T. 22 S., R. 46 E., Malheur County, Oregon (Swain, 1990).

Fusocandona exotica Swain

Plate 20, figures 5, 6

Fusocandona exotica Swain, 1987a, p. 375, pl. 2, figs. 5a-c, 6, 9.

Elongate subreniform, highest postmedially; dorsum moderately convex, somewhat sinuous, venter concave; anterior rounded above, acuminately extended below; posterior sinuous, with strongly extended ventroterminal acuminate extension of LV below; LV overlaps RV strongly except along dorsum; moderately convex; smooth except for pattern of fine reticulate hash marks; L, 1.20; H, 0.67 mm (Swain, 1980).

Occurrence. Pliocene, Cache Valley Formation, NW1/4 SE1/4 NE1/4 sec. 10, T. 12 S., R. 40 E., Franklin County, Idaho (Swain, 1987a).

Genus Fusocandonopsis Swain, 1986

(Candonidae-Candoninae)

Fusocandonopsis owyheensis Swain

Plate 20, figures 7-11

Fusocandonopsis owyheensis Swain, 1986a, p. 98, pl. 2, figs. 19-21; pl. 3, figs. 1-9.

Elongate subquadrate-subtrapezoidal, highest anteromedially; dorsum nearly straight, a little more than half of shell length; venter concave to nearly straight and sinuous; anterior rounded except for protruding spines, extended below; posterior narrowly rounded strongly extended below; anteroventral forward-projecting hollow spines are relatively shorter than ventroposterior backward-projecting hollow spines; LV overlaps RV; especially midventrally; female shells much wider than male shells; surface with densely spaced pustules, plus ridges and nodes, but some specimens nearly lack nodes, there may also be an oblique anterodorsal ridge, a mid-dorsal oblique ridge, a postdorsal ridge or series of nodes and a postmedian longitudinal ridge; hinge of LV a rabbet groove to receive hinge edge of RV; inner lamellae narrow but vestibules occur terminally; adductor muscle scar an anteromedian group of five spots; two? frontal spots; L, holotype RV 2.04; H, 0.88 mm (Swain, 1986a).

Occurrence. Pliocene, Glenns Ferry Formation, sec. 14, T. 4 S., R. 1 W., Owyhee County, Idaho (Swain, 1986a).

Family Geisinidae Sohn, 1961

Geisinidae gen. indet., sp. 1, Sohn

Plate 20, figures 12, 13

Geisinidae gen. indet., sp. 1, Sohn, 1985, p. 457, TF 6.9 to 6.11.

Internal mold subquadrate, highest postmedially, dorsum sinuous, venter gently convex; anterior narrower than posterior; posterior half with terminal and ventral ridges, representing location of vestibules; anteromedian shallow sulcus reaching to one-third of height from dorsum; possibly a second more anterior short oblique sulcus; L, 0.8; H, 0.5 mm (Sohn, 1985). Differs from marine Jonesina in presence of vestibules.

Occurrence. Late Mississippian, Bluestone Formation, Bramwell Member, SW$^{1}/_{4}$ Flattop 7$^{1}/_{2}$ minute Quadrangle, Mercer County, West Virginia, USGS 12982-PC (Sohn, 1985).

Geisinidae gen. indet., sp. 2, Sohn

Plate 20, figures 14, 15

Geisinidae gen. indet., sp. 2, Sohn, 1985, p. 457, TF 6.16, 6.17.

Internal mold elongate, subrhomboidal, highest medially; dorsum nearly straight; anterior narrow, posterior extended above; anteromedian sulcus extends to about two-fifths of height from dorsum; anterojacent lobe; possible trace of anterior vestibules on lobe; L, 0.92; H, 0.50 mm (Sohn, 1985).

Occurrence. Upper Mississippian, Bluestone Formation, Bramwell Member, SW$^{1}/_{4}$ Flattop 7$^{1}/_{2}$ minute Quadrangle, Mercer County, West Virginia, USGS coll. 12984-PC (Sohn, 1985).

Genus Gerdalia Belousova, 1961

(Darwinulidae)

Gerdalia cf. triassica Belousova

Plate 16, figures 38

?Gerdalia triassica Belousova 1961. Lucas, 1991, TF 6D

Gerdalia sp. cf. G. triassica Belousova, Lucas and Kietzke, 1993, p. 64 TF 7

Very elongate-elliptical, highest medially, dorsum slightly more convex than venter; ends narrowly rounded, the posterior more pointed; valve size and overlap uncetain; smooth; L, 1.1; H, 0.46 mm (Lucas and kietzke, 1993).

Occurrence. Upper Triassic, Chinle Group, Petrified Forest Formation, Painted Desert Member, red mudstone just below "Painted Desert Sandstone 3" of Billingsley, NW1/4 NE1/4 NE1/4 sec. 33. T. 20 N., R. 24 E., Petrified Forest National Park, Arizona (Lucas and Kietzke, 1993); also San Pedro Arroyo Formation, Ojo Huelos Member, south central New Mexico (Kietzke, 1989a; Lucas, 1991).

Gerdalia sp., Kietzke

Plate 17, figures 43, 44

Gerdalia sp., Kietzke, 1989, p. 186, TF 4F, G, 6C.

Very elongate, sublanceolate, highest postmedially; anterior narrow than posterior; RV overlaps LV on venter and posterior; inflated in dorsal view; smooth; adductor muscle scar at midpoint; L, 0.96 to 0.9; H, 0.28 to 0.38 mm (Kietzke, 1989).

Occurrence. Triassic, Meonkopi Formation, basal part, White Ridge, NW1/4 NE1/4 NW1/4 sec. 36, T. 7N., R. 4 W., Cibola County, New Mexico (Kietzke, 1989)

Genus Gutschickia, Scott, 1944

(Carbonitidae)

Gutschickia convexa Tasch

Plate 20, figure 16

Gutschickia convexa Tasch, 1963, p. 1248, pl. 174, fig. 17.

Large, ovate, dorsum arched, venter nearly straight; RV overlaps LV along venter; strongly inflated; pitted; adductor muscle scar rounded, anteromedian; L, 1.74; H, 1.04 mm (Tasch, 1963).

Occurrence. Permian, Wellington Formation, NE$^1/_4$ NE$^1/_4$ sec. 10, T. 17 S. R. 2 E., Marion County, Kansas (Tasch, 1963).

Gutschickia deltoidea (Holland)

Plate 20, figures 17-20

Whipplella deltoidea Holland, 1934, p. 345, pl. 25, figs. 1a-c.

Gutschickia deltoidea (Holland), Scott, 1944, p. 146, pl. 23, figs. 1-4. Sohn, 1977a, p. 137.

Large, subovate, highest medially, dorsum strongly convex, venter nearly straight, height three-fourths of length; RV overlaps LV along venter; LV overreaches RV along dorsum; tumid in edge view, densely pitted; circular adductor muscle scar anteroventral to midpoint; narrow flanges terminally; L, 1.48; H, 1.16 mm (Holland, 1934; Scott, 1944).

Occurrence. Permian, upper part of Dunkard Series, Nineveh Formation, at Nineveh, eight miles (12.8 km) northwest of Waynesburg, Greene County, Pennsylvania (Holland, 1934).

Gutschickia ninevehensis (Holland)

Plate 20, figures 21-26

Whipplella ninevehensis Holland, 1934, p. 347, pl. 25, figs. 3a-c.

Gutschickia ninevehensis (Holland), Scott, 1944, p. 146, pl. 23, figs. 5-8. Sohn, 1977, p. 135, TF 3a-e (The latter as Gutschickia sp. but may not be conspecific).

Ovate, highest medially, dorsum strongly convex, venter nearly striaght to slightly convex; ends more or less equally rounded, extended below; posterior a little narrower than anterior; LV overlaps RV ventrally; dorsally LV overreaches RV but does not overlap it; finely and densely pitted; adductor muscle scar a median group of about 17 spots (referred to by Sohn, 1977 as a "healdiid" scar); L, 1.07; H, 0.67 mm (Holland, 1934; Sohn, 1977).

Occurrence. Permian, upper part of Dunkard Series, Upper Greene Group, Nineveh Formation, about 8 miles (12.8 km) northwest of Waynesburg, Greene County, Pennsylvania (Holland, 1934).

Gutschickia ninevehensis (Holland), Cooper, 1946
(see Gutschickia sp., Sohn, 1977)

Gutschickia wellingtonensis Tasch
Plate 20, figures 27, 28
Gutschickia wellingtonensis Tasch, 1963, p. 1247, pl. 174, figs. 5-8.

Large, subovate, highest medially, dorsum strongly convex, venter nearly straight to slightly convex; ends rounded, extended below; RV overlaps LV along venter; LV overreaches but does not overlap RV dorsally; surface minutely pitted; adductor muscle scar circular, anteriorad of midpoint; L, 1.52; H, 1.11 mm (Tasch, 1963).

Occurrence. Permian, Wellington Formation, NE$^1/_4$ NE$^1/_4$ sec. 10, T. 17 S., R. 2 E., Marion County, Kansas (Tasch, 1963).

Gutschickia sp., Sohn
Plate 20, figures 29-31
Gutschickia sp., Sohn, 1977, p. 135, TF 2c-e.

Internal mold subquadrate, highest medially, dorsum slightly convex, venter slightly sinuous, but both are nearly straight; anterior narrower than posterior, truncate above; adductor muscle scar a submedian rounded area of about 15 spots; L, 1.12; H, 0.71 mm (Sohn, 1977).

Occurrence. Pennsylvanian, Monongahela Formation, USGS coll. 12900-PC, Monongalia County, West Virginia (Sohn, 1977).

Gutschickia sp., Sohn
Plate 20, figures 32, 33
Gutschickia ninevehensis (Holland), Cooper, 1946; Sohn 1977 (may not be conspecific)
Gutschickia sp., Sohn, 1977, p. 140, TF 3f, g, not described.

Subtrapezoidal, highest medially, dorsum straight, about half of shell length, venter straight, about two-thirds of length; ends rounded extended below, the posterior narrower than anterior; compressed narrow terminal flanges; finely and densely pitted; LV extends beyond RV dorsally; L, 1.05; H, 0.77 mm (from Sohn's illustration, 1977).

Occurrence. Pennsylvanian, Cohn cyclothem, Clark County, Illinois (Cooper, 1946; Sohn, 1977).

Gutschickia sp., Sohn
Plate 20, figure 34; Plate 21, figures 1, 2

Gutschickia sp., Sohn, 1977, p. 140, TF 3h-n, not described.

Subquadrate, highest postmedially; dorsum slightly convex, nearly straight, venter nearly straight, converging with dorsum anteriorly; LV overlaps RV posteriorly and ventrally; smooth; adductor muscle scar area rounded, submedian; L, 1.03; H, 0.78 mm (from Sohn's illustrations).

Occurrence. Pennsylvanian, Monongahela Formation, USGS coll. 12899-PC, Monongalia County, West Virginia (Sohn, 1977).

Gutschickia spp., Sohn
Plate 21, figures 3-6
Gutschickia spp., Sohn, 1985, p. 453, TF 3.1 to 3.9, 4.1 to 4.9, 5.1 to 5.4, not described.

The forms illustrated by Sohn vary from subovate to subelliptical, highest medially; dorsum arched, venter nearly straight to concave; anterior a little narrower than posterior; LV larger than RV, overlap variable; dorsally LV overreaches RV; smooth; adductor muscle scar a median rounded group of multiple spots; L, 0.7 to 0.8; H, 0.45 to 0.55 mm (Sohn, 1985).

Occurrence. Upper Mississippian, Bluestone Formation, Bramwell Member, Crumpler Quadrangle, Wyoming County, West Virginia; Flattop Quadrangle, Mercer County, West Virginia, USGS coll. 12981-12983-PC (Sohn, 1985).

Gutschickia levis Cooper
Plate 19, figures 31, 32
Gutschickia levis Cooper, 1946, p. 68, pl. 8, figs. 15, 16.

Ovate, highest postmedially, dorsum more convex than venter; anterior narrower than posterior; LV overlaps RV except ventrally where RV overlaps LV; tumid carapace; smooth; L, 0.64; H., 0.44 mm (after Cooper, 1946).

Occurrence. Pennsylvanian, Des Moinesian Series, Summum Cyclothem, central Illinois (Cooper, 1946).

Gutschickia ovata Cooper
Plate 19, figures 33, 34
Gutschickia ovata Cooper, 1946, p. 68, pl. 8, figs. 17-20.

Large, subovate, very tumid, highest medially; dorsum gently convex, venter nearly straight; anterior a little narrower than posterior, LV overlaps RV; venter flattened; surface reticulate; L., 1.16; H., 0.73 mm (after Cooper, 1946).

Occurrence. Pennsylvanian, Des Moinesian Series, Summum Cyclothem, central Illinois (Cooper, 1946).

Gutschickia subangulata (Jones and Kirkby)

Plate 20, figures 35, 36

Carbonia fabulina subangulata Jones and Kirkby, 1879, p. 31, pl. 2, figs. 20-24.

Gutschickia subangulata (Jones and Kirkby), Cooper, 1946, p. 69, pl. 8, figs. 26-32.

Large, short, subtrapezoidal, highest anteromedially; dorsum nearly straight, short, venter longer and nearly straight; both ends extended below, truncate above; LV overlaps RV except ventrally where RV overlaps LV; smooth to granulose; L., 1.36; H., 0.87 mm (after Cooper, 1946).

Occurrence. Pennsylvanian, Virgilian Series, Newton Limestone, southern Illinois (Cooper, 1946).

Genus Hastifaba Cooper, 1946

(Kloedenellidae)

Hastifaba pervulgata Cooper

Plate 21, figures 27, 34

Hastifaba pervulgata Cooper, 1946, p. 111, pl. 18, figs 33-35.

Subhemispherical, highest postmedially; dorsum sinuous, anterior cardinal angle more obtuse than posterior angle; venter convex; anterior broader than posterior; anterodorsal short sulcus curves slightly toward anterior; postmedian dorsal inflation, incised hinge; smooth; L., 0.97; H., 0.64 mm (after Cooper, 1946).

Occurrence. Pennsylvanian, Des Moinesian Series, Jamestown Limestone, central Illinois (Cooper, 1946).

Hastifaba robusta Cooper

Plate 22, figures 26, 27

Hastifaba robusta Cooper, 1946, p. 112, pl. 18, figs. 22-26.

Subovate, highest postmedially; dorsum sinuous, anterior cardinal angle more obtuse than posterior angle; venter convex; anterior narrower than posterior; marginal narrow rims; anterodorsal short sulcus; incised hinge; inflated post dorsal surface; smooth; L., 1.05; H., 0.64 mm (after Cooper, 1946); mid-anterior margin with tiny spine.

Occurrence. Pennsylvanian, Des Moinesian Series, Summum Cyclothem, central Illinois (Cooper, 1946).

Hastifaba spinosa Cooper

Plate 22, figures 38, 39

Hastifaba spinosa Cooper, 1946, p. 111, pl. 18, figs. 27-32

Large, subquadrate, highest anteromedially; anterior cardinal more obtuse than posterior angle; dorsum long and straight, venter concave medially; ends subequal, the posterior subtruncate below; marginal rims; anterodorsal short sulcus; dorsomedian hornlike projections; mid-anterior margin with tiny spine; smooth; L., 1.08; H., 0.66 mm (after Cooper, 1946). .

Occurrence. Pennsylvanian, Virgilian Series, Cohn Cyclothem, central Illinois (Cooper, 1946).

Genus Hemicyprinotus Schneider, 1957

(Cyprididae-Cyprinotinae)

Hemicyprinotus ionensis Swain

Plate 21, figures 7-9, 12

Hemicyprinotus ionensis Swain, 1985, p. 420, pl. 1, figs. 7, 8, 11; pl. 4, figs. 3, 4; pl. 5, figs. 10-14. Swain, 1990, p. 149, pl. 2, fig. 14.

Subquadrate-subreniform, highest one-third of length from anterior end; dorsum gently convex, venter gently concave; posterior narrower than anterior, both ends extended below; LV overlaps RV ventrally and anterodorsally; low convexity; weakly and densely pitted; anteromedian adductor muscle scar a group of four spots; two frontal scars; inner lamellae narrow, vestibules terminally; L, holotype, 0.75; H, 0.49 mm (Swain, 1985).

Occurrence. Miocene, Esmeralda Formation, Black Spring road section, 2.85 miles (4.5 km) west of Black Spring, Nye County, Nevada (Swain, 1985). Truckee Formation, 2 miles (3.2 km) WSW of Hazen, Churchill County, Nevada (Swain, 1990).

Hemicyprinotus? sp., aff. H. ionensis Swain

Plate 21, figures 10, 11

Hemicyprinotus? sp. aff. H. ionensis Swain, Swain, 1987c, p. 426, pl. 4, figs. 6, 7; Pl. 5, fig. 1.

Subquadrate-subelliptical, highest medially, dorsum nearly straight, venter slightly concave; ends broadly rounded, anterior truncated above; LV larger than RV, moderately convex; smooth?; L, 0.87; H, 0.53 mm, lateral surface depressed medially (Swain, 1987c).

Occurrence. Miocene, Horse Spring Formation, Fossil Ridge, Gass Peak Quadrangle; approximately NW1/4 sec. 31 (unsurveyed), T. 17 S., R. 16 E., Clark County, Nevada (Swain, 1987c).

Hemicyprinotus krafti Dickinson and Swain

Plate 21, figure 13

Hemicyprinotus krafti Dickinson and Swain, 1967, p. 341, pl. 36, fig. 3.

Elongate, subovate, highest anteromedially, dorsum angulated at point of greatest height, venter nearly straight; anterior narrower than posterior; LV a little larger than RV; moderately convex; smooth; L, 1.0; H, 0.6 mm (Dickinson and Swain, 1967).

Occurrence. Pliocene, Hay Ranch Formation, Gulf Oil Corp., No. 1 Wilkins Ranch well, sec. 21, T. 38 N., R. 61 E., Elko County, Nevada 1526-3768 feet (465-1149 m) (Dickinson and Swain, 1967).

Hemicyprinotus oneidensis Swain

Plate 21, figures 14-16

Hemicyprinotus oneidensis Swain, 1987a, p. 121, pl. 1, figs. 1-11.

Elongate, subovate, highest anteromedially, males more elongated and less convex than females; dorsum convex, venter nearly straight; anterior narrower than posterior; LV overlaps RV; moderately convex; smooth; L, holotype, 1.06; H, 0.67 mm (Swain, 1987a).

Occurrence. Pliocene, Salt Lake Group, secs. 4-26, T. 16 S. R. 36 W., along strike, five to seven miles (8-11.3 km) southwest of Malad City, Oneida County, Idaho (Swain, 1987a).

Hemicyprinotus reniformis Swain

Plate 21, figures 17, 18

Hemicyprinotus reniformis Swain, 1964a, p. 268, pl. 41, figs. 4, 5, 8.

Elongate, subovate-reniform, highest one-fourth of length from posterior end, venter concave; anterior narrower than posterior; LV larger than RV; low convexity; finely pitted; L, holotype, 0.78; H, 0.48 mm (Swain, 1964a).

Occurrence. Eocene, Colton-Green River transition beds, Cypridea bisulcata Zone, on north side of White River, eight miles (12.9 km) below Rangley, Colorado (Swain, 1964a).

Hemicyprinotus? simplex Swain

Plate 21, figures 19, 20

Hemicyprinotus? simplex Swain, 1986b, p. 206, pl. 2,figs. 17, 18.

Elongate, subelliptical-subquadrate-subreniform, highest anteromedially, venter concave, anterior slightly broader than posterior, both ends extended below; LV slightly larger than RV, convexity moderate; smooth; adductor muscle scar an anteromedian group of 4 to 5 spots; L, holotype, 0.72; H, 0.38 mm (Swain, 1986b).

Occurrence. Miocene, Humboldt Formation, in lower 232 feet (71 m), section at Cobre, Elko County, Nevada (Swain, 1986b).

Hemicyprinotus? subreniformis (Swain and Wagoner), Swain, 1985
(see Ussuriocypris? subreniformis Swain and Wagoner).

Hemicyprinotus watsonensis Swain
Plate 21, figures 25-26
Candona whitei Scott and Smith (part), 1951, p. 344, pl. 52, figs. 6, 11.
Hemicyprinotus watsonensis Swain, 1964a, p. 267, pl. 41, figs. 1a-i, 3a-d.

Small to medium size, subelliptical; highest at or posterior to median; dorsum gently convex, venter nearly straight to slightly concave; anterior narrower than posterior, subtruncate above; LV larger than RV; female dimorphs more convex than males; very finely and weakly pitted or reticulate; rabbeted hinge, RV fits over LV, adductor muscle scar a median group of four or five spots; two frontal spots; L, holotype, 0.83; H, 0.50 mm (Swain, 1964a).

Occurrence. Eocene, Green River-Colton transition beds, 130 feet (39.6 m) above base, east end of Raven Ridge, Rio Blanco County, Colorado (figured specimens); the most widespread and abundant early Tertiary ostracode specimens in northeastern Utah and northwestern Colorado; ranges from Colton through Green River and into Uinta and Bridger Formations in Uinta, Piceance Creek and southern Green River Basins, Colorado, Utah and Wyoming (Swain, 1964a).

Hemicyprinotus sp. aff. H. watsonensis Swain
Plate 21, figures 30, 31
Hemicyprinotus sp. aff. H. watsonensis Swain, Swain, 1987b, p. 224, pl. 2, figs. 1, 2.

Elongate, subquadrate-subreniform, highest postmedially; dorsum convex, venter concave; anterior narrower than posterior; LV overlaps RV except middorsally; moderate convexity; smooth; L, 0.85; H, 0.64 mm (Swain, 1987b).

Occurrence. Eocene, Twin Bridges Limestone, near center sec. 36, T. 32 N., R. 55 E., Elko County, Nevada, on west side of Huntingdon Creek (Swain, 1987b).

Hemicyprinotus watsonensis emaciata Swain
Plate 21, figures 28, 29
Heterocypris, n. sp. and var. Swain, 1956, p. 135, pl. 3, figs. 2a-c.
Hemicyprinotus watsonensis emaciata Swain, 1964a, p. 268, pl. 41, figs. 2a-c.

Lateral outline, overlap and ornamentation as in H. watsonensis; shell relatively more compressed than typical of species; L, LV, 0.88; H, 0.53, width, 0.23 mm (Swain, 1964).

Occurrence. Eocene, upper part of Colton-Green River transition beds, Dripping Rock Gulch section, east end of Raven Ridge, Rio Blanco County, Colorado (Swain, 1964).

Hemicyprinotus? sp. Swain
Plate 21, figures 32, 33
Hemicyprinotus? sp., Swain, 1986b, p. 206, pl. 2, figs. 7, 8.

A form similar to H?. simplex Swain, 1986b in general outline and smooth surface, but is less elongate; adductor muscle scar a group of four obliquely oriented spots; two frontal spots; L, 0.90; H, 0.57 mm (Swain, 1986b).

Occurrence. Miocene, Humboldt Formation, Gulf Oil Co., No. 1 Thousand Springs well, center, SE$^{1}/_{4}$ SW$^{1}/_{4}$ sec. 8, T. 40 N., R. 66 E., Elko County, Nevada, 5 to 2125 feet (1.5 to 648 m); Gulf, No. 2 Wilkins Ranch well, sec. 19, T., 39 N., R. 69 E., Elko County, Nevada, 4490-4520 feet (1369-1378 m); Cobre surface section, Cobre, Nevada, 288 to 360 feet (88-110 m) above base (Swain, 1986b).

Hemicyprinotus? sp., Swain
Plate 22, figures 1, 2
Hemicyprinotus? sp., Swain, 1985, p. 421, pl. 4, figs. 1, 2.

Elongate, subovate to subelliptical, highest medially; dorsum convex, venter nearly straight; anterior slightly narrower than posterior and extended below; LV overlaps RV especially ventrally and anterodorsally; moderately convex; smooth; weak concentric posterior markings; L, 0.95; H, 0.60 mm (Swain, 1985).

Occurrence. Miocene, Esmeralda Formation, Black Spring Road section, 2.85 miles (4.5 km) west of Black Spring, Nye County, Nevada, about 80 feet (244 m) above base of lake beds sequence (Swain, 1985).

Genus Herpetocypris Brady and Norman, 1889, Sars, 1890
(Cyprididae-Herpetocypridinae)
Erpetocypris Brady and Norman, 1889

Herpetocypris andersoni Swain
Plate 22, figures 3, 4
Herpetocypris andersoni Swain, 1986a, p. 91, pl. 1, figs. 4, 5?; pl. 2, fig. 15.

Elongate, subelliptical-reniform, highest one-fourth of length from anterior end; dorsum straightened medially, venter concave; anterior narrower than posterior, especially in females; both ends extended below; LV overlaps RV along midventer and dorsally cardinal slopes; compressed, males more so than females; smooth; hinge of LV a rabbet groove to receive RV edge; inner lamellae moderately wide terminally; narrow vestibules; adductor muscle scar a median group of five spots; two frontal spots; L, holotype, 0.74; H, 0.34 mm (Swain, 1986a).

Occurrence. Pliocene, Glenns Ferry Formation, sec. 14, T. 4 S., R. 1 W., Owyhee County, Idaho (Swain, 1986a).

Herpetocypris franklinensis Swain

Plate 22, figures 5-8

Herpetocypris franklinensis Swain, 1987a, p. 367, pl. 1, figs. 4-7. Swain, 1987c, p. 120 not figured.

Elongate, subelliptical-reniform, highest anteromedially; dorsum straightened medially; venter concave; posterior narrower than anterior, both ends extended below; LV overlaps RV except middorsally; compressed; smooth exteriorly, interior pitted terminally in addition to openings of normal pores; hinge of LV grooved to receive edge of RV and straightened midportion of LV hinge is depressed; terminal LV hinge areas form elongate socket-furrows into which fit flange of RV; adductor muscle scar a submedian group of six spots; two frontal spots; inner lamellae narrow, with terminal vestibules; L, holotype 1.20; H, 0.53 mm (Swain, 1987a).

Occurrence. Pliocene, Cache Valley Formation, NW1/4 SE1/4 NE1/4 sec. 10, T., 12 S., R. 40 E., Franklin County, Idaho (Swain, 1987a). Miocene or Pliocene, Humboldt Formation or Salt Lake Group, Tanner Reservoir overflow canal, sec. 9, T., 11 N., R. 19 W., Box Elder County, Utah (Swain, 1987c).

Herpetocypris lata (Chapman)

Plate 23, figures 5, 6

Erpetocypris lata Chapman, 1896, p. 98, pl. 3, figs. 13-15.

Elliptical-subreniform, highest slightly anteromedially; dorsum convex, venter slightly concave, posterior narrower than anterior; relatively compressed, widest postmedially; smooth; L, 2.12; H, 1.02 mm (after Chapman, 1896).

Occurrence. Pliocene, Merced Formation, Bollinger Canyon, east of Berkeley, California (Chapman, 1896).

Herpetocypris merriamiana (Chapman)

Plate 23, figures 7, 8

Erpetocypris merriamiana Chapman, 1896, p. 97, pl. 3, figs. 10-12.

Elongate reniform, highest medially, anterior slightly narrower than posterior, dorsum convex, venter concave; smooth; L, 1.49 mm; H, 0.85 mm (Chapman, 1896).

Occurrence. Pliocene, Merced Formation, Wildcat Canyon in hills east of Berkeley, California (Chapman, 1896).

Herpetocypris stenoformis Dickinson and Swain

Plate 22, figure 10-12

Herpetocypris stenoformis Dickinson and Swain, 1967, p. 342, pl. 36, figs. 13, 14. Swain and Wagoner, 1985, p. 137, pl. 5, fig. 8. Swain, 1990, p. 150, pl. 1, fig. 20.

Elongate-reniform, highest medially; dorsum a low broad arch, venter slightly concave; ends rounded; LV overlaps RV middorsally; smooth; inner lamellae broadest anteriorly; L, holotype 1.0; H, 0.48 mm (Dickinson and Swain, 1967) (closely similar to Candoniella).

Occurrence. Pliocene, Hay Ranch (=Upper Humboldt) Formation, Gulf Oil Corp., No. 1, Wilkins Ranch well, sec. 21, T. 38 N., R. 61 E., Elko County, Nevada (Dickinson and Swain, 1967). Miocene, Esmeralda Formation, sample 1C2, Blair Junction surface section, Esmeralda County, Nevada (Swain and Wagoner, 1985). Pliocene, Idaho Group, undifferentiated, just below a lava flow, 15 miles (24 m) west of Burns, Oregon on U.S. Highway 10 (Swain, 1990).

Herpetocypris? winfreyi Swain

Plate 22, figures 13-15

Herpetocypris winfreyi Swain, 1987b, p. 225, pl .1, fig. 13; pl. 3, figs. 1, 2.

Elongate, subelliptical-subreniform, highest postmedially; dorsum gently convex, venter slightly concave; anterior narrower than posterior; LV slightly larger than RV; compressed; smooth; LV hinge furrowed to receive RV edge; adductor muscle scar a submedian group of four or five spots; one or two frontal spots; widely spaced normal pores have depressed inner openings; L, holotype, 0.70; H, 0.28 mm (Swain, 1987b).

Occurrence. Paleocene or Eocene, Sheep Pass Formation, T. 8 N., R. 59 W., Grants Range, Johnson Canyon, Nye County, Nevada; Eocene, Twin Bridges Formation, sec. 33, T. 32 N., R. 55 E., Elko County, Nevada (Swain, 1987b).

Herpetocypris sp., Benson

Plate 22, figure 18

Herpetocypris sp., Benson, 1967, p. 235, TF 13-4, not described.

Elongate, subtrapezoidal-subreniform, highest postmedially; dorsum slightly convex, venter concave; posterior narrower than anterior, both ends extended below; smooth?; inner lamellae of moderate width; adductor muscle scar an anteromedian group of four? spots; two frontal spots; L, 0.98; H, 0.40 mm (from Benson, 1967).

Occurrence. Pliocene, Laverne Formation, USNM 649613, Beaver County, Oklahoma (Benson, 1967).

Herpetocypris sp., Swain

Plate 22, figure 19

Herpetocypris sp., Swain, 1986b, p. 204, pl. 1, fig. 19; pl. 4, fig. 12.

Elongate, subelliptical, highest medially; dorsum gently convex, truncated anteriorly, venter nearly straight to slightly concave; ends nearly equal; LV overlaps RV; compressed; smooth; adductor muscle scar an anteromedian group of four oblique spots; two frontal spots; L, 0.63; H, 0.32 mm (Swain, 1986b).

Occurrence. Miocene, Humboldt Formation, Gulf Refining Co., No. 2 Wilkins Ranch well, 4270-4300 feet (1301-1311 m); Gulf No. 1 Thousand Springs well, SE1/4 SW1/4 sec. 8, T. 40 N., R. 66 E., Elko COunty, Nevada, 50 - 55 feet (15 - 17 m) (Swain, 1986b).

Genus Heterocypris Claus, 1893

(Cyprididae-Cyprinotinae)

Heterocypris blairensis Swain and Wagoner

Plate 22, figures 20-27

Heterocypris blairensis Swain and Wagoner, 1985, p. 134, pl. 1, fig. 5; pl. 2, figs. 1-16. Swain, 1985, p. 425, pl. 2, figs. 1, 2. Swain, 1986b, p. 206, pl. 3, figs. 5, 6.

Elongate, subovate to subreniform, highest medially; dorsum gently convex, subtruncate postmedially, venter nearly straight to concave; anterior narrower than posterior, both ends extended below; LV overlaps RV most along dorsal slopes and midventrally; moderately convex; surface with fine reticulating bertillion ridges; shallow anterodorsal depression; LV hinge weakly rabbeted below to receive RV edge; adductor muscle scar an anteromedian group of four spots; two frontal spots; inner lamellae broadest anteriorly; small vestibules; about 25 short radial canals anteriorly; L, holotype, 0.87; H, 0.53 mm (Swain and Wagoner, 1985).

Occurrence. Miocene, Esmeralda Formation, Blair Junction section, Esmeralda County, Nevada, samples 1B to 1Q (Swain and Wagoner, 1985). Esmeralda Formation, Finger Rock Wash, 2 miles (3.2 km) west of Stewart Spring, Stewart Spring 1:24000 Quadrangle, about 900 feet (275 m) above base of exposed lake beds sequence (Swain, 1985). Humboldt Formation,

Gulf Oil Co., No. 1 Thousand Springs well, center SE1/4 SW1/4 sec. 8, T. 40 N., R. 66 E., Elko County, Nevada, 1420-1440 feet (433-439 m) (Swain, 1986b).

Heterocypris humboldtensis Dickinson and Swain

Plate 22, figures 28, 29

Heterocypris humboldtensis Dickinson and Swain, 1967, p. 342, pl. 35, figs. 13-19.

Subelliptical-subreniform, highest medially, height about three-fifths of length; dorsum moderately convex, venter straight to slightly concave; anterior broader than posterior, the latter extended below; LV overlaps RV anterodorsally and ventrally; smooth; moderately convex; L, holotype, 0.88; H, 0.55 mm (Dickinson and Swain, 1967).

Occurrence. Pliocene, Hay Ranch Formation (=Upper Humboldt), Gulf Oil Corp., No. 1 Wilkins Ranch well, sec. 21, T. 38 N., R. 61 E., Elko County, Nevada, 1405-4180 feet (428-1274 m) (Dickinson and Swain, 1967).

Heterocypris riflensis Swain

Plate 22, figures 30, 31

Heterocypris? sp., Swain, 1956, p. 134, pl. 3, figs. 3a, b.

Heterocypris riflensis Swain, 1964,p. 268, pl. 44, figs. 1a, b.

Elongate subovoid, highest postmedially; dorsum moderately convex, venter slightly concave medially, but nearly straight; anterior narrower than posterior, both ends extended below; LV overlaps RV except along middorsal margin; moderately convex; low median swelling, otherwise smooth; L, holotype, 0.93; H, 0.56 mm (Swain, 1964b).

Occurrence. Eocene, Green River Formation, lower part, or Colton Formation, upper part; truck road leading to Anvil Point Demonstration Plant of U.S. Bureau of Mines, Moffat County, Colorado (Swain, 1964).

Heterocypris whiteriverensis Swain

Plate 22, figure 32

Heterocypris whiteriverensis Swain, 1964, p. 269, pl. 42, fig. 13.

Elongate-subovate, subreniform, highest medially; dorsum moderately convex, venter slightly concave; posterior broader than anterior, extended below; valves subequal, weakly convex; finely pitted; adductor muscle scar a group of four or five spots; L, holotype, 0.92; H, 0.53 mm (Swain, 1964).

Occurrence. Eocene, Colton-Green River transition beds, north side of White River, 8 miles (12.9 km) below (west) of Rangely, Colorado (Swain, 1964).

Heterocypris? wilkinsensis Dickinson and Swain

Plate 22, figure 33

Heterocypris? wilkinsensis Dickinson and Swain, 1967, p. 342, pl. 35, figs. 21-23.

Subtriangular-subreniform, highest postmedially; dorsum strongly arched, truncated anteromedially, venter slightly concave; anterior narrower than posterior; LV overlaps RV anterodorsally and midventrally; moderately convex, compressed anteriorly; smooth?; inner lamellae broad anteriorly; L, holotype, 1.05; H, 0.63 mm (Dickinson and Swain, 1967).

Occurrence. Pliocene, Hay Ranch Formation (=Upper Humboldt), Gulf Oil Corp., No. 1 Wilkins Ranch well, sec. 21, T. 38 N., R. 61 E., Elko County, Nevada, 1435-2043 feet (437-623 m) (Dickinson and Swain, 1967).

Genus Hilboldtina Scott and Summerson, 1943

(Cypridacea, Family uncertain)

Hilboldtina multiplicata Scott and Summerson

Plate 22, figures 34-37

Hilboldtina multiplicata Scott and Summerson, 1943, p. 670, pl. 1, figs. 21-25.

Ovate-oblong, highest postmedially, dorsum more convex than venter; anterior narrower than posterior; RV overlaps LV on free margins; LV overreaches RV along hinge; finely striated longitudinally in middle two thirds, branching and anastomosing?; adductor muscle scar anteromedian and circular; L, 1.07; H, 0.60 mm (Scott and Summerson, 1943).

Occurrence. Lower Pennsylvanian, Kanawha Group, Scott Formation, on WPA (Work Projects Administration) road northwest of Newcomb, Campbell County, Tennessee, 0.25 mile (0.4 km) SE of between Burnt Pone and Hatfield Branches, Williamsburg Quadrangle (Scott and Summerson, 1943).

Genus Huabeinia Li and Lai, 1978, Hou et al, 1988

(Cyprididae-Huabeininae)

Huabeinia? sp. aff. H. huidongensis Hou and Li

Plate 23, figures 1-3, 9, 10

Huabeinia? sp. aff. H. huidongensis Hou and Li, 1978, Swain, 1985, p. 429, pl. 2, fig. 3; Swain, 1987c, p. 429,pl. 2, figs. 1, 2.

Elongate, subquadrate-subtrapezoidal; highest medially; dorsum nearly straight, about three-fourths of shell length, venter gently concave; posterior narrower than anterior and extended below, truncate above; LV overlaps RV, compressed; terminoventral areas compressed and bear narrow marginal rims; densely and coarsely pitted; L, holotype, 0.63; H, 0.33 mm (Swain, 1985).

Occurrence. Miocene, Esmeralda Formation, Black Springs road section, 2.85 miles (4.5 km) west of Black Spring, Black Spring 1:24000 Quadrangle, Nye County, Nevada, about 800 feet (244 m) above base of lake bed sequence (Swain, 1985); also in same section, 160 feet (49 m) above base (Swain, 1987c).

Genus Ilyocyprimorpha Mandelstam, 1956

(Ilyocyprididae-Ilyocypridinae)

Ilyocyprimorpha linonodosa (Swain)

Plate 23, figure 4

Ilyocypris arvadensis tuberculata Swain, 1949, p. 179, pl. 33, figs. 10-12 (non I. gibba tuberculata Brady, 1878).

Ilyocypris arvadensis linonodosa (Swain), Swain, 1951, p. 236.

Ilyocyprimorpha linonodosa (Swain), Swain, 1964, p. 275, pl. 43, fig. 11.

Subquadrate-subelliptical, highest one-fourth of length from anterior end; dorsum nearly straight, venter faintly concave; posterior narrower than anterior; LV larger than RV, moderately convex; bisulcate in anterodorsal third; median lobe rounded to bluntly pointed; large rounded nodes lie on either side of sulcate area; longitudinal row of irregular nodes in ventral third; densely and finely pitted general surface; L, 1.19; H, 0.69 mm (Swain, 1964).

Occurrence. Paleocene, roof of Roland Coal, SE$^{1}/_{4}$ sec. 13, T. 9 S., R. 45 E., Powder River County, Montana (Swain, 1949).

Ilyocyprimorpha? shafferi (Staplin)

Plate 23, figure 21

Ilyocypris shafferi Staplin, 1963, p. 1190, pl. 160, figs. 40, 41.

Large, subquadrate-subreniform, highest in anterior fourth; dorsum nearly straight to sinuous, venter concave; posterior narrower and more truncated than anterior; bisulcate in anterodorsal quadrant; median lobe poorly defined; several large nodes in dorsal third; numerous pustules in dorsal third, ventral fourth and in terminal areas; ventral pustules roughly aligned with venter; general surface densely and finely pitted; anterior and anteroventral margins spinose; LV slightly overlaps RV; radial canals numerous and short; inner lamellae narrow ("broader in RV"); LV hinge a groove with terminal slight sockets; adductor muscle scar a group of four spots; two "mandibular" frontal scars and an "antennular" scar near end of anterior sulcus; L, 1.40; H, 0.84 mm (Staplin, 1963).

Occurrence. Pleistocene, Tazewell Stage, Bloomington deposits, Illinois (Staplin, 1963) (see plate descriptions).

Ilyocyprimorpha variabilonodosa Swain, 1986a
(see Malheurocypris variabilonodosa)

Genus Ilyocypris Brady and Norman, 1889
(Ilyocyprididae-Ilyocypridinae)
Ilyocypris arvadensis tuberculata Swain, 1951
(see Ilyocyprimorpha linonodosa (Swain))

Ilyocypris biplicata (Koch)
Plate 23, figures 32, 44
Cypris biplicata Koch, 1838, species 16.
Ilyocypris biplicata (Koch), Scourfield, 1904, p. 34.
Jeicypris biplicata (Koch)(error), Dubowsky, 1927, p. 113.
Ilyocypris biplicata (Koch), Lister, 1975, p. 19, pl. 2, figs. 5, 6, TF 21.

Elongate, subquadrate-reniform, highest anteriorly; dorsum nearly straight, venter concave; anterior broader than posterior and more rounded, the latter truncate; terminal marginal small tubercles; may have two posterior hollow tubercles; bisulcate in anterodorsal quadrant; the median sulcus longer than anterior sulcus; surface densely and finely punctate; inner lamellae broadest anteriorly; hinge adont; L, 0.92 to 0.99; H, 0.52 mm (Lister, 1975).

Occurrence. Pleistocene, Lake Bonneville Group, Saltair core, SW1/4 sec. 25, T. 1 N., R. 3 W.; Section 28 core hole SE1/4 SW1/4 sec. 28, T. 1 N., R. 2 W., Salt Lake County, Utah (Lister, 1975). Holocene, Europe (Koch, 1838).

Ilyocypris bradyi Sars
Plate 23, figures 12-15
Ilyocypris bradyi Sars, 1891, p. 59. Swain, 1962,p. 807, pl. 95, fig. 6; pl. 96, figs. 13a, b, TF 4b (includes Recent synonymy). Gutentag and Benson, 1962, p. 42, pl. 1, figs. 8, 9, TF 12. Staplin, 1963, p. 1186, pl. 160, figs. 34, 35.
Ilyocypris bradii Sars, Furtos, 1933, p. 428, pl. 1, figs. 8-10.

Subquadrate, highest about one-fifth of length from anterior end; dorsum nearly straight, venter concave; anterior broader than posterior, extended below; LV slightly larger than RV, compressed; dorsomedian, deep pitlike sulcus, anterior to which lie a small node and a second short sulcus; centrally located rounded pit; densely and finely pitted general surface, scattered small pustules; free margin with fringe of spines; LV hinge with terminal elongated sockets and interterminal groove; adductor muscle scar of four spots; two frontal spots; inner lamellae broadest anteriorly, striated; vestibules present; radial canals numerous, terminating in spines on outer margin; normal canals numerous; L, 0.83; H, 0.40 mm (Swain, 1963).

Occurrence. Pleistocene, Gubik Formation, Arctic coastal plain, Alaska, Teshekpuk Lake shot holes, Party 43, line 1-48, shot point 47 at 60 feet (18.3 m); line 1A-48, shot point 27A at 60 feet (18.3 m); line 11-47, shot point 9 at 50-60 feet (15.2-18.3 m); Oumalik No. 2 well at 0-10 feet (3.05 m) (Swain, 1963). Illinoian and Wisconsin Stages, Odee Formation, Meade County, Kansas; Kansan or Yarmouthian Stages, Reno County, Kansas, below Pearlette Ash (Gutentag and Benson, 1962). Mankato Substage, Wisconsin Stage of Illinois; late Kansan Stage of Nebraska and Iowa (Staplin, 1963). Recent of Europe, North America, and Turkestan (authors).

Ilyocypris gibba (Ramdohr)
Plate 23, figures 16-20
Cypris gibba Ramdohr, 1808, p. 91,pl. 3, figs. 13, 14, 17.
Ilyocypris gibba (Ramdohr), Brady and Norman, 1889, p. 107, pl. 22, fig. 1-5. Gutentag and Benson, 1962, p. 44, pl. 1, figs. 6, 7, TF 13. Staplin, 1963, p. 1187, pl. 160, figs. 36, 37, 39. Swain, 1963,p. 806, pl. 96, fig. 14, TF 4a. Benson, 1967, p. 222, TF 6-5, p. 228, TF 10-3. Lister, 1975, p. 20, pl. 2, figs. 7, 8, TF 22.

Subreniform-subquadrate, highest in anterior fourth; dorsum nearly straight to slightly convex, venter concave; posterior narrower than anterior and more blunt; bisulcate in anterodorsal quadrant; pits lie ventral to sulci; small spines and tubercles in free marginal areas; inner lamellae broadest anteriorly; LV hinge a groove with posterior elongate socket; four adductor muscle scar spots; two "mandible adjustor" scars; L, female, 0.96 to 0.98; H, 0.51 to 0.54 mm (Staplin, 1963).

Occurrence. Pleistocene, late Kansan or Yarmouthian Stage, above Pearlette Ash, Reno County, Kansas, SE$^1/_4$ NE$^1/_4$ sec. 1, T. 25 S., R. 7 W., (Gutentag and Benson, 1962). Mankato Substage, Wisconsin Stage of Illinois; Sappa Silt, late Kansan of Kansas and Nebraska (Staplin, 1963). Gubik Formation, Arctic coastal plain, Alaska, Ikpikpuk Lake Shot Holes, line 6, shot point 8 at 30-40 feet (9.1 to 12.2 m); line 15-47, shot point 14 at 40-50 feet (12.2 to 15.2 m) (Swain, 1963). Yarmouthian Stage, Dickinson County, Kansas (Benson, 1967). Quaternary of Great Britain, Recent, Holarctic (authors).

Ilyocypris shafferi Staplin
(see Ilyocyprimorpha shafferi (Staplin))

Ilyocypris shawneetownensis Staplin
Plate 23, figure 22
Ilyocypris shawneetownensis Staplin, 1963, p. 1191, pl. 160, fig. 38.

Elongate, subquadrate, highest anteromedially; dorsum nearly straight, with a posterior projection, venter concave; posterior narrower and more truncate than anterior; LV slightly overlaps RV dorsally; bisulcate in anterodorsal quadrant; anterior sulcus (S1) the longer; median lobe narrow, with a dorsal node, another node posterior to median sulcus (S2) on bulging posterior lobe; ventromedian surface with longitudinal nodose ridge; surface densely and finely pitted; inner lamellae broadest anteriorly; vestibules present on free margins; LV hinge with terminal elongate weak sockets; L, 0.84 to 0.87; H, 0.48 to 0.50 mm (Staplin, 1963).

Occurrence. Pleistocene, Mankato Substage, Wisconsin Stage, Ohio Valley near Shawneetown, Illinois (Staplin, 1963).

Ilyocypris thousandspringsensis Swain
Plate 23, figures 23, 24
Ilyocypris thousandspringsensis Swain, 1986b, p. 213, pl. 4, figs. 15-18; pl. 5, fig. 33.

Elongate, subquadrate-subelliptical, highest anteromedially, dorsum nearly straight, five-eighths of length, venter slightly concave; posterior a little broader than anterior, the latter truncate above; LV overlaps RV dorsally and ventrally; compressed; two slightly anteromedian dorsal sulci reach venterad of midheight; anteriorad and posteriorad of sulci are two rounded nodes; coarsely pitted surface; L, holotype, 0.67; H, 0.38 mm (Swain, 1986b).

Occurrence. Miocene, Humboldt Formation, Gulf Refining Co., No. 1 Thousand Springs well, 350-2650 feet (107-808 m), Elko County, Nevada (Swain, 1986b).

Ilyocypris? sp., Sohn
Plate 23, figure 25
Ilyocypris? sp., Sohn, 1956, p. 119, pl. 1, fig. 1, not described.

Small, subovate, highest anteromedially, dorsum more convex than venter; bisulcate?, median sulcus longer than anteromedian sulcus; three to five small nodes in sulcate area; additional small nodes in posterior half; L, 0.50; H, 0.30 mm (from Sohn's illustrations, 1956).

Occurrence. Pliocene, Teewinot Formation, Jackson Hole, Wyoming (Sohn, 1956).

Genus Jonesina Ulrich and Bassler, 1908 (nomina dubia)

(Kloedenellidae? or Geisinidae?)

Jonesina? arcuata (Bean)

Plate 23, figures 26-29

Cypris arcuata Bean, 1836, p. 377, TF 55.

Beyrichia arcuata (Bean), Jones and Kirkby, 1886, p. 438, pl. 12, figs. 12-14.

Jonesina arcuata (Bean), Ulrich and Bassler, 1908, p. 324, pl. 44, figs. 17-19.

Jonesina? arcuata (Bean), Scott and Summerson, 1943, p. 672, pl. 1, figs. 12-15, 21; not described.

Subovate-quadrate to subpyriform, highest anteromedially; dorsum sinuous to nearly straight, venter convex; anterior broader than posterior and extended above in some specimens; bisulcate in anterodorsal quadrant; median sulcus long and comma-shaped; anterior sulcus short, oblique; marginal rim narrow; terminal margins compressed; smooth general surface; L, 0.92; H, 0.45 mm (data from Scott and Summerson, 1943). Species assigned to Jonesina have in recent years been placed in Family Geisinidae (Sohn in Moore, 1961), but present species lacks ventral concavity of Geisinidae. The species is closely similar to Hastifaba Cooper, 1945, and may belong to that genus.

Occurrence. Lower Pennsylvanian, Briceville Formation, Kanawha Series, along highway between Coalfield and Christmas, Morgan County, Tennessee; Hance Formation, on U.S. 19 just north of Molus, Harlan County, Kentucky (Scott and Summerson, 1943).

Jonesina? bispinosa Scott and Summerson

Plate 23, figures 37, 38

Jonesina? bispinosa Scott and Summerson, 1943, p. 673, pl. 2, figs. 4, 6, 9.

Subovate-subquadrate, highest anteromedially; dorsum nearly straight, venter convex; ends nearly equally rounded, but posterior is extended above; RV overlaps LV along free margins; bisulcate in anterodorsal quadrant; median sulcus extends to midheight; anterior sulcus short and oblique; lobe does not reach to dorsum; marginal rim; finely reticulate; a spine lies ventral to each cardinal angle; L, 1.10; H, 0.75 mm (Scott and Summerson, 1943). The species may belong in Hastifaba Cooper, 1946.

Occurrence. Lower Pennsylvanian, Kanawha Series, Mingo Formation, roof of coal mine about 250 feet (76 m) above road pass of U.S. 25 E, between Clear Lake and Little Cannon Creek, Bell County, Kentucky (Scott and Summerson, 1943).

Jonesina minnekahtensis Roth, 1933

(see Theriosynoecum minnekahtense (Roth))

Jonesina pahasapensis Roth, 1933

(see Theriosynoecum pahasapense (Roth))

Jonesina? sp., Scott and Summerson

Plate 23, figure 39

Jonesina? sp., Scott and Summerson, 1943, pl. 1, fig. 17 (not cited in text).

Subelliptical-subquadrate, highest anteromedially; height about half of length; dorsum and venter nearly straight; anterior truncated above, ends nearly equal in height; bisulcate in anterodorsal quadrant; median sulcus shorter than anterior sulcus; lobe not reaching to dorsum; smooth?; no dimensions given (from Scott and Summerson illustration).

Occurrence. Presumably Lower Pennsylvanian, but no locality given.

Genus Kassinina Mandelstam, 1960

(Cyprididae?-Cypridinae?)

Kassinina? gasspeakensis Swain

Plate 23, figures 11, 40-43

Kassinina? gasspeakensis Swain, 1987c, p. 426, pl. 2, figs. 15-21; pl. 3, figs. 1-3; pl. 5, fig. 2.

Elongate, subelliptical-subreniform, highest anteromedially; immature shells subpyriform; dorsum gently convex, straightened postmedially; venter nearly straight to slightly concave; posterior narrower than anterior, both ends extended below; LV overlaps RV most along venter and anterior; convexity moderate; midportion depressed as seen in edge view; finely and densely reticulate; tiny nodes superimposed on reticulate ridges; LV hinge with furrow for edge of RV; inner lamellae of moderate width, with vestibules; adductor muscle scar a median compact group of spont; L, holotype RV, 0.82; H, 0.43 mm (Swain, 1987c).

Occurrence. Miocene, Horse Spring Formation, Fossil Ridge, NW$^1/_4$ (unsurveyed) sec. 31, T. 17 S., R. 16 E., Clark County, Nevada, about 710 feet (216 m) above base of exposure (Swain, 1987c).

Kassinina mesoreticulata Swain and Wagoner

Plate 24, figures 1-4

Kassinina mesoreticulata Swain and Wagoner, 1985, p. 133, pl. 1, figs. 7, 8; pl. 4, figs. 1-5.

Subelliptical-subreniform, highest medially; dorsum gently convex, venter slightly concave; terminal margins equally rounded, slightly extended below; LV overlaps RV most anterodorsally and midventrally; moderately convex; medium-coarse bertillion reticulating ridges, concentric toward margins; surface slightly depressed anteromiddorsally; RV hinge beveled above and weakly rabbeted, overlapped by RV; adductor muscle scar a group of five spots; two frontal spots; inner lamellae moderate in width, broadest anteriorly; slight vestibule; L, holotype, 1.03; H, 0.58 mm (Swain and Wagoner, 1985).

Occurrence. Miocene, Esmeralda Formation, Blair Junction section, Esmeralda County, Nevada, sample C1N (Swain and Wagoner, 1985).

Kassinina microreticulata Swain and Wagoner

Plate 24, figures 5-7

Kassinina microreticulata Swain and Wagoner, 1985, p. 133, pl. 1, figures 4, 9-11, 13-16. Swain, 1985, pl. 428.

Resembles K. mesoreticulata Swain and Wagoner, but is more finely reticulate and greatest height is more anterior in position; male? shells less convex than females?; L, holotype, male, 1.07; H, 0.64 mm (Swain and Wagoner, 1985).

Occurrence. Miocene, Esmeralda Formation, Blair Junction section Esmeralda County, Nevada, samples C1A to C1D (Swain and Wagoner, 1985). Esmeralda Formation, Black Spring road section, 2.85 miles (4.5 km) west of Black Spring, Nye County, Nevada, about 800 feet (244 m) above base of lake bed sequence; 3.3 miles (5.3 km) east of Finger Rock Wash, Mineral County, Nevada, in Simon Mine area, in lower 100-150 feet (30-45 m) of lake beds sequence (Swain, 1985).

Kassinina paucireticulata Swain and Wagoner

Plate 24, figures 8-10

Kassinina paucireticulata Swain and Wagoner, 1985, p. 132, pl. 1, figs. 1-3, 6, 12.

Similar to K. microreticulata Swain and Wagoner but has a weaker pattern of surface ornamentation; L, 1.17; H, 0.69 mm (Swain and Wagoner, 1985).

Occurrence. Miocene, Esmeralda Formation, Blair Junction section, Esmeralda County, Nevada, sample 1A to 2A (Swain and Wagoner, 1985).

Genus Limnicythere Brady, 1868

(see Limnocythere Brady, 1868)

Limnicythere ? dibolba Swain 1947

(see Limnicytheropteron dibolba (Swain)

Limnicythere unicornis Swain, 1947
(see Limnicytheropteron unicornis (Swain))

Limnicythere williamsi Swain, 1947
(see Limnocythere williamsi (Swain))

Genus Limnicytheropteron Swain, 1986a
(Limnocytheridae-Limnocytherinae)
Limnicytheropteron cephalotoidea Swain
Plate 24, figures 11-13
Limnicytheropteron cephalotoidea Swain, 1986a, p. 101, pl. 4, figs. 10, 11, 15.

Males elongate-subquadrate to subtrapezoidal; females shorter and higher, subovate to subquadrate, highest postmedially; dorsum straight, venter gently convex except for projecting postmedian alae; subequal valves; LV slightly larger than RV; anterior margin broader than posterior; posteroventral valve surface expanded as spacious pointed alae in males, less spacious in females; coarsely but weakly reticulated general surface; LV with terminal elongate hinge sockets and interterminal rabbet furrow; adductor scar a submedian vertical row of four spots; at least one frontal spot; L, holotype, male 0.50; H, 0.30 mm (Swain, 1986a).

Occurrence. Pliocene, Glenns Ferry Formation, "Big Cut" locality, sec. 14, T. 4 S., R. 1 W., Owyhee County, Idaho (Swain, 1986a).

Limnicytheropteron dibolba (Swain)
Plate 24, figures 33-35
Limnicythere? dibolba Swain, 1947, p. 528, pl. 77, figs. 27-29.

Subpyriform, highest near anterior end; dorsum nearly straight, venter sinuous, converging with dorsum posteriorly; anterior broader than posterior; ventral half of valve surface strongly expanded, alate; expanded surface with two submedian nodes, the posterior node the larger; vertical median sulcus extends from dorsum to crest of alate swelling; anterior marginal rim; LV hinge with terminal small sockets; inner lamellae of LV moderate in width; L, holotype, 0.54; H, 0.30 mm (Swain, 1947).

Occurrence. Pliocene, Salt Lake Group, sec. 15, T. 13 N., R. 2 W., Cache County, Utah (Swain, 1947).

Limnocytheropteron typica Swain
Plate 24, figures 14-17
Limnicytheropteron typica Swain, 1986a, p. 101, pl. 5, figs. 4-16.

Elongate subquadrate-subelliptical, highest postmedially in mature shells, anteromedially in immature shells; dorsum nearly straight, venter concave except for projecting alae in some forms; posterior narrower than anterior; latter with compressed rim; LV a little larger than RV; LV hinge with terminal subtriangular notches and interterminal rabbeted edge; inner lamellae narrow, vestibules terminally; adductor muscle scar an anteromedian vertical row of four spots; prominent backward-projecting alate spine on each valve; reticulate surface; L, female? holotype, 1.00; H, 0.77 mm; L, male? 1.02, H, 0.49 mm (Swain, 1986a).

Occurrence. Pliocene, Glenns Ferry Formation, sec. 14, T. 4 S., R. 1 W., Owyhee County, Idaho (Swain, 1986a).

Limnicytheropteron unicornis (Swain)
Plate 24, figures 18-23
Limnicythere unicornis Swain, 1947, p. 526, pl. 77, figs. 22-26.
Limnocythere? unicornis (Swain), Swain, Becker and Dickinson, 1970, p. 465. Swain, 1986a, p. 102. Swain, 1987, p. 127, pl. 2, fig. 14.

Elongate, subquadrate-subelliptical, highest about one-fifth from anterior end; dorsum nearly straight, venter sinuous, concave medially; posterior narrower than anterior; ventral half alate, the latter have spine-like terminations posteriorly; compressed rim anteriorly and postventrally; smooth; shallow dorsomedian depression; LV hinge with terminal small sockets; inner lamellae of moderate width; L, holotype 0.54; H, 0.30 mm; dimorphic, some forms more elongate than others (Swain, 1947).

Occurrence. Pliocene, Salt Lake Group sec. 15, T. 13 N., R. 2 W., Cache County, Utah (Swain, 1947). Glenns Ferry Formation, sec. 14, T. 4 S., R. 1 W., Owyhee County, Idaho (Swain, 1986a). Salt Lake Group, Cache Valley Formation, T. 13 N., R. 3 W., southwest of Newton, Cache County, Utah (Swain, 1987).

Limnicytheropteron winecupensis Swain
Plate 24, figures 24-27
Limnocythere? sp. aff. L. unicornis (Swain), Swain, Becker and Dickinson, 1971, p. 486, pl. 5, figs. 6a-f.
Limnicytheropteron winecupensis Swain, 1986b, p. 214, pl. 4, figs. 20-25; pl. 5, figs. 13-15.

Elongate-subquadrate to wedge-shaped, highest near anterior end; dorsum nearly straight, about three-fourths of length, venter slightly concave, converging with dorsum posteriorly; anterior broader than posterior; valves subequal; long narrow elevated ridges along dorsum; alaform ridge in middle half of ventral part of each valve, terminating posteriorly in a knob or spine; median sulcus-like broad depression in dorsal half; anterior eighth compressed; coarsely reticulate surface; L, 0.47; H, 0.28 mm (Swain, 1986).

Occurrence. Miocene, lower part of Humboldt Formation, Gulf Oil Co, Thousand Springs well, center SE1/4 SW1/4 sec. 8, T. 40 n., R. 66 E., Elko County, Nevada, 1140-1150 feet (348-351 m) and 1890-1900 feet (548-579 m); also in conglomeratic limestone at base of fish-bearing beds, just above Triassic, Black Mountain section, Winecup Ranch, Elko County, Nevada (Swain, 1986b).

Limnicytheropteron? sp. aff. L. winecupensis Swain
not figured
Limnicytheropteron? sp. aff L. winecupensis Swain, Swain, 1990, p. 154, pl. 2, fig. 16.

LV internal mold subquadrate, highest one-fourth from anterior end; dorsum nearly straight, venter slightly convex, straightened medially, converging with dorsum posteriorly; posterior narrower than anterior; ventroposterior area of valve with low alaform expansion; L, 0.70; H, 0.42 mm (Swain, 1990).

Occurrence. Late Miocene?, Truckee Formation, two miles (3.2 km) WSW of Hazen, Churchill County, Nevada (Swain, 1990).

Genus Limnocypridea Lyubimova, 1956

(Trapezoidellidae)
Limnocypridea morrisonensis (Roth)
Plate 23, figures 30, 31
Plate 26, figures 16-20
Bairdiocypris morrisonensis Roth, 1933, p. 401, pl. 48, figs. 6a-d.
Limnocypridea morrisonensis (Roth), Sohn, 1979, p. 11.

Subtriangular-subreniform, dorsum strongly convex, umbonate; venter slightly concave; anterior margin rounded, extended below, posterior narrowly rounded, strongly extended below; valves compressed, minutely punctate; L, 2.9; H, 1.9 mm (Sohn, 1979).

Occurrence. Lower Cretaceous, Lakota Formation, sec. 28, T. 4 N., R. 6 E., Lawrence County, South Dakota (Sohn, 1979); referred to the Late Jurassic Morrison Formation by Roth (1933).

Genus Limnocythere Brady, 1868

(Limnocytheridae-Limnocytherinae)
Limnocythere ceriotuberosa Delorme

Plate 24, figures 28, 29

Limnocythere ceriotuberosa Delorme, 1967, p. 360, pl. 2, figs. 9-12. Lister, 1975, p. 24, pl. 4, figs. 13-16, TF 26. Cameron and Lundin, 1977, p. 349, pl. 1, figs. 15,1 6.

Subtrapezoidal-subrectangular, highest anteriorly (female), or more elongate, concave ventrally and with postventral downwarp, highest posteriorly (male); anterior broader than posterior (female), the opposite in male; strongly reticulate, busulcate in anterodorsal quadrant; midventral elongate alae; few sieve type normal pores; inner lamellae moderately broad anteriorly; radial canals few, simple; adductor muscle scar median to anteromedian row of four spots; two widely separated frontal spots; L, female, 0.65; H, 0.38 mm; L, male, 0.78; H, 0.43 mm (Lister, 1975).

Occurrence. Pleistocene, Lake Bonneville Group, Saltair core hole, SE1/4 sec 25, T. 1 N., R. 3 W.; Sec. 28 core hole, SE1/4 SW1/4 sec. 28, T. 1., R. 2 W., Salt Lake County, Utah (Lister, 1975). Late Pleistocene, Lake Cochise deposits, Willcox Playa, Cochise County, Arizona (Cameron and Lundin, 1977). Holocene of prairie lakes of Canada (Delorme, 1970); Minnesota (Smith, 1991).

Limnocythere chippewaensis Staplin

Plate 24, figures 30-32

Limnocythere chippewaensis Staplin, 1963, p. 1192, pl. 160, figs. 1-3.

Limnocythere friabilis Benson and McDonald, 1963, p. 24, pl. 3, figs. 1-4 (fide Lister, 1975).

Small, subreniform (female), subovate-subelliptical (male), highest anteromedially (female), to postmedially (male); dorsum slightly convex (female) to nearly straight (male); venter concave; posterior narrower than anterior in female, nearly equal in male but extended above; flattened border except midventrally; anteromiddorsal quadrant busulcate, with intervening node; surface finely reticulate; several small pits in anterior half; radial canals may be reflected as ridges across marginal zone; terminoventral margins may have spines; LV overlaps RV posteriorly; LV hinge with cardinal obscure sockets and interterminal edge that overlaps RV; adductor muscle scar with four spots in vertical row; two frontal scars; about eight dorsal scars; inner lamellae very narrow; L, 0.48 to 0.52; H, 0.27 to 0.30 mm (Staplin, 1963).

Occurrence. Pleistocene, Aftonian? Stage of Illinois; pre-Nipissing low water stage of Lake Michigan (Staplin, 1963). Holocene, Lake Erie (Benson and McDonald, 1963).

Limnocythere? dibolba Swain, 1947

(see Limnicytheropteron dibolba (Swain))

Limnocythere friabilis Benson and McDonald

Plate 24, figures 36-39

Limnocythere friabilis Benson and McDonald, 1963, p. 24, pl. 3, figs. 1-4.

Limnocythere chippewaensis Staplin, 1963, p. 1192, pl. 160, figs. 1-3 (fide Lister).

Female subpyriform, highest anteromedially, dorsum slightly convex, venter concave; anterior broader than posterior; reticulate surface; bisulcate in anterodorsal quadrant; median sulcus shallow, nearly reaching venter, anterior sulcus very weak; low wide node between sulci; postventrally are small marginal spines; scattered sieve type normal pores; adductor muscle scar an anteromedian row of four spots; two widely separated frontal spots; few dorsal spots; narrow inner lamellae, no vestibules; LV hinge with terminal tiny elongate sockets; L, 0.48 to 0.52; H, 0.27 to 0.28 mm (female); L, male, 0.53 to 0.58; H, 0.25 to 0.28 mm (Lister, 1975).

Occurrence. Pleistocene, Lake Bonneville Group, Saltair core hole, SE1/4 sec. 25, T. 1 N., R. 3 W.; Sec. 28 core hole, SE1/4 SW1/4 sec. 28, T. 1 N., R. W., Salt Lake County, Utah (Lister, 1975). Holocene, Lake Erie (Benson and McDonald, 1963). Because of uncertainty as to similarity of L. friabilis and L. chippewaensis, both species are here retained.

Limnocythere (Limnocytherina) herricki Staplin

Plate 24, figures 40, 41

Limnocythere herricki Staplin, 1963, p. 1193, pl. 160, figs. 9, 10.

Female elongate, subtriangular-reniform, highest near posterior end; dorsum nearly straight, venter concave; ends rounded, extended below, truncate above; reticulate surface; bisulcate in anterodorsal quadrant; small node separates sulci; ventromedian surface inflated; LV hinge with terminal obscure sockets and interterminal knife-edge projection underlapping RV; adductor muscle scar a tight vertical row of four spots; inner lamellae narrow; male unknown; L, 0.71 to 0.75; H, 0.36 to 0.39 mm (Staplin, 1963).

Occurrence. Pleistocene, Sankoty Sand (Nebraskan Stage) and late Mankato Substage, Wisconsin Stage of Illinois (Staplin, 1963).

Limnocythere huntingdonensis Dickinson and Swain

Plate 24, figures 42, 43; Plate 25, figure 1-4

Limnocythere huntingdonensis Dickinson and Swain, 1967, p. 344, pl. 37, figs. 16-24, TF 4. Swain, 1987a, p. 377, pl. 4, fig. 13.

Female shell subquadrate, highest anteriorly; dorsum nearly straight to slightly convex, venter weakly to moderately concave; anterior broader than posterior, both ends extended below; valves subequal, compressed terminally; a sulcus consisting of two elongate pits just anterad of middle; two pits anterior to sulcus; two large nodes anterad of sulcus; postventral alae on each valve; surface reticulate; LV hinge with terminal elongate depressions and intervening bar; inner lamellae broadest anteriorly, no vestibules; adductor muscle scar a vertical anteromedian row of four spots; one? frontal spot; male more elongate and equal-ended than female; L, 0.63; H, 0.33 mm (Dickinson and Swain, 1967).

Occurrence. Pliocene, Hay Ranch Formation (=Upper Humboldt), two miles north of Twin Bridges, Elko County, Nevada, sec. 24, T. 32 N., R. 55 E.; roadcut one mile south of Elko, Nevada on State Highway 46, sec. 19, T. 34 N., R. 55 E.; 20 miles north of Austin, Lander County, Nevada on State Highway 8A (Dickinson and Swain, 1967). Cache Valley Formation, Hoopes Creek loc., NW1/4 SE1/4 NE1/4 sec. 10, T. 12 S., R. 40 E., Franklin County, Idaho (Swain, 1987).

Limnocythere (Limnocytherina) illinoisensis (Sharpe)
Plate 25, figures 5-7
Limnicythere illinoisensis Sharpe, 1897, p. 425.
Limnocythere illinoisensis Sharpe, Müller, 1912, p. 231. Lister, 1975, p. 25, pl. 5, figs. 1-3, TF 28.

Elongate subquadrate-subpyriform-subreniform, highest anteriorly (female), subpyriform-elliptical, highest posteriorly (male); dorsum straight, venter concave; anterior broader than posterior in female; nearly equal to or narrower than posterior in male; anterior and postventral areas compressed; bisulcate in anterodorsal quadrant; median sulcus reaches to midheight; anteromedian sulcus short; nodes and pits lie anteriorad of median sulcus; surface coarsely reticulate; normal sieve-type pores scattered; midventral alate expansion; inner lamellae narrow, faintly striated, no vestibules; L, female, 0.71; H, 0.36; L, male, 0.78; H, 0.36 mm (Lister, 1975).

Occurrence. Pleistocene, Lake Bonneville Group, Saltair core hole, SE1/4 sec. 25, T. 1 N., R. 3 W.; Sec. 28 core hole, sec. 28, T. 1 N., R. 2 W., Salt Lake County, Utah (Lister, 1975). Holocene, Illinois (Sharpe, 1897).

Limnocythere (Limnocytherina) itasca Cole
Plate 25, figures 8-10
Limnocythere itasca Cole, 1949, p. 351. Lister, 1975, p. 26, pl. 4, figs. 5-8.
Limnocythere trapeziformis Staplin, 1963, p. 1199.
Limnocythere (Limnocytherina) itasca Cole, Delorme, 1971, p. 48, TF 68-80, 100, 101.

Female subtrapezoidal-subelliptical, highest anteromedially; dorsum straight, venter moderately convex; anterior broader than posterior, the latter bluntly truncate above; bisulcate, the median sulcus longer than the anterodorsal sulcus; rounded lobe separates the two sulci; ventral surface expanded, subalate; anterior area compressed; surface coarsely reticulate; a prominent ventrolateral spine points posteriorly; inner lamellae broadest anteriorly, striate, no vestibules; radial canals few, straight; L, 0.65 to 0.67; H, 0.32 to 0.33 mm; male more elongate, posterior broader than anterior, more concave ventrally; L, 0.66 to 0.67; H, 0.32 mm (Lister, 1975).

Occurrence. Pleistocene, Lake Bonneville Group, Saltair core hole, SE1/4 sec. 25, T. 1 N., R. 3 W.; Sec. 28 core hole, SE1/4 SW1/4 sec. 28, T. 1 N., R. 2 W., Salt Lake County, Utah (Lister, 1975). Late Kansan Stage, Sappa Silt, Lyons County, Iowa; Mankato Substage, Wisconsin Stage, Chicago area, Illinois; Tazewell Stage? of Ontario (Staplin, 1963, as L. trapeziformis). Holocene, north central United States and plains of Canada (Cole, 1949; Delorme, 1971; Swain, 1995).

Limnocythere ornata wabashensis Staplin

Plate 25, figures 11-13

Limnicythere ornata Furtos, 1933, (part?) p. 423, pl. 16, figs. 12-17.

Limnocythere ornata wabashensis Staplin, 1963, p. 1194, pl. 160, figs. 20-24.

Female elongate, subelliptical, highest anterodorsally; dorsum sinuous, slightly concave medially, venter concave; anterior end broader than posterior, both terminal areas compressed below; bisulcate, the median sulcus extends to midheight, the anteromedian sulcus shorter and oblique; a large node lies anterior and posterior to each sulcus; ventromedian surface with alate expansions, two nodes occur postmidventrally; general surface reticulate; LV overlaps RV posteriorly; LV hinge with terminal sockets; intervening edge underlaps RV; adductor muscle scar a vertical row of four spots; inner lamellae without vestibules; L, holotype, 0.69; H, 0.37 mm. Male larger and more elongate than female, nodes subdued; L, 0.80; H, 0.38 mm (Staplin, 1963).

Occurrence. Pleistocene, Tazewell to Mankato Substages, Wisconsin Stage of Illinois (Staplin, 1963). Holocene, Lake Erie and Chippewa Lake, Ohio (Furtos, 1933).

Limnocythere (Limnocytherina) paraornata Delorme

Plate 25, figures 14, 15

Limnocythere ornata (Furtos) Delorme, 1967, p. 1281 (non Furtos, 1933).

Limnocythere (Limnocytherina) paraornata Delorme, 1971, p. 50, TF 104-115, 165-168. Lister, 1975, p. 27, p. 4, figs. 1-4, TF 31.

Female subelliptical-subrectangular-reniform, highest postmedially; dorsum straight, venter concave; posterior a little broader than anterior; terminal marginal zones compressed;

bisulcate, the median sulcus longer than anteromedian sulcus; the two separated by a rounded node; bridge across median sulcus near midheight; surface densely reticulate; ventromedian alate swelling; posterior surface swollen; scattered sieve-type normal pores; inner lamellae broad anteriorly, striate; L, 0.85; H, 0.50 mm. Male more elongate than female; surface patterns weaker, larger than female; L, 0.95; H, 0.47 mm (Lister, 1975).

Occurrence. Pleistocene, Lake Bonneville Group, Saltair core hole, SE1/4 sec. 25, T. 1 N., R. 3 W.; Sec. 28 core hole, SE1/4 SW1/4 sec. 28, T. 1 N., R. 2 W., Salt Lake County, Utah (Lister, 1975). Holocene, in streams of SE corner of plains of Canada (Delorme, 1971).

Limnocythere parascutariense Delorme

Plate 25, figures 16, 17

Limnocythere parascutariense Delorme, 1971, p. 50, TF 116-128. Lister, 1975, p. 27, pl. 5, figs. 4-6, TF 30.

Female elongate, subrectangular-subreniform, highest postmedially; dorsum nearly straight to slightly convex, venter concave; posterior broader than anterior; postventral margin downwarped; compressed anterior marginal zone; bisulcate, median sulcus long, bridged near midheight; anteromedian sulcus oblique, short; reticulate surface; inner lamellae broad anteriorly; no vestibules; L, 0.65; H, 0.37 mm. Male more elongate, larger and with more enlarged posterior end than female; L, 0.84; H, 0.37 mm (Lister, 1975).

Occurrence. Pleistocene, Lake Bonneville Group, Saltair core hole, SE1/4 sec. 25, T. 1 N., R. 3 W.; Sec. 28 core hole, SE1/4 SW1/4 sec. 28, T. 1 N., R. 2 W,. Salt Lake County, Utah (Lister, 1975). Holocene, streams and lakes of SE interior plains of Canada (Delorme, 1971).

Limnocythere (Limnocytherina) posterolimba Delorme

Plate 25, figures 18, 19

Limnocythere posterolimba Delorme, 1967, p. 362, pl. 1, fig. 8.

Limnocythere (Limnocytherina) posterolimba Delorme, Delorme, 1971, p. 52, TF 131-144.

Limnocythere cf. posterolimbata (sic) Delorme, Swain, 1990, p. 153, pl. 1, figs. 18, 19. Swain, 1979, p. 313, pl. 1, figs. 14-16.

Subquadrate-subreniform, highest one-third of length from posterior end; dorsum nearly straight, about half of length, venter concave; posterior narrower than anterior, the latter extended below; LV larger than RV; compressed; bisulcate in anterodorsal quadrant, median sulcus reached midheight, oblique anterior sulcus reaches only to one-fourth of height; intervening rounded node has cross furrow ventrally; low submarginal ridge midventrally; free margins with compress rim; L, 0.85; H, 0.43 mm (Swain, 1990).

Occurrence. Pliocene, Idaho Group undifferentiated, just below lava flow, 15 miles (24 km) west of Burns, Oregon (Swain, 1990). Holocene, Pyramid Lake, Nevada (Swain, 1979); ponds on south-central plains of Canada (Delorme, 1971).

Limnocythere posterolimbata Delorme, Swain, 1979, 1995 (error)
(see Limnocythere posterolimba Delorme)

Limnocythere pseudoreticulata Staplin
Plate 25, figures 20-24
Limnocythere pseudoreticulata Staplin, 1963, p. 1195, pl. 160, figs. 7, 8. Lister, 1975, p. 28, pl. 5, figs. 11-14, TF 32.

Female small, subrectangular-subreniform, highest near anterior end; dorsum slightly convex, venter concave; anterior broader than posterior; flattened margin especially anteriorly; bisulcate in anterodorsal quadrant; the two sulci lie near midlength, have intervening node and subjacent transverse furrow; strongly reticulate surface; posterior area with scattered papillae; LV overlaps RV posteriorly; LV hinge with terminal obscure sockets and underlapping interterminal edge; adductor muscle scar a vertical row of four spots; two frontal spots; inner lamellae narrow; L, 0.51 to 0.60; H, 0.29 to 0.33 mm. Male a little larger and more elongate than female, subelliptical; posterior pointed; L, 0.58 to 0.61; H, 0.29 to 0.33 mm (Staplin, 1963).

Occurrence. Pleistocene, Tazewell Substage, Wisconsin Stage of Illinois (Staplin, 1963). Lake Bonneville Group, Saltair core hole, SE1/4 sec. 25, T. 1 N., R. 3 W.; Sec. 28 core hole, SE1/4 SW1/4 sec. 28, T. 1 N., R. 2 W., Salt Lake County, Utah (Lister, 1975).

Limnocythere pterygoventrata Dickinson and Swain
Plate 25, figures 25-27
Limnocythere pterygoventrata Dickinson and Swain, 1967, p. 345, pl. 38, figs. 1-4. Cameron and Lundin, 1977, p. 349, pl. 1, figs. 7, 8.

Female subpyriform, highest medially, dorsum arched, venter concave; anterior a little broader than posterior; valves subequal; anterior compressed; a slightly anteromedian sulcus with crossbar in dorsal half; postventral quadrant depressed, producing a rudder-like aspect; smooth. Male highest in posterior third; posterior broader than anterior; L, female, 0.78; H, 0.43 mm (Dickinson and Swain, 1967).

Occurrence. Pliocene, Hay Ranch Formation (=Upper Humboldt), two miles (3.2 km) north of Twin Bridges on Huntingdon Creek; one mile (1.6 km) south of Elko, Elko County, Nevada; 20 miles (32 km) north of Austin, Lander County, Nevada on Highway 8A; Gulf Oil Co., No. 1 Wilkins Ranch well, Elko County, Nevada, 957-3281 feet (292-1000 m) (Dickinson and Swain, 1967). Pleistocene, Lake Cochise beds, Willcox Playa, Cochise County, Arizona (Cameron and Lundin, 1977).

Limnocythere reticulata Sharpe

Plate 25, figures 28, 29

Limnicythere reticulata Sharpe, 1897, p. 427, pl. 29, figs. 1-7.

Limnocythere reticulata Sharpe, Hoff, 1942, p. 163, pl. 9, figs. 136-141. Staplin, 1963, p. 1196, pl. 160, figs. 25, 26.

Female subrectangular, highest anteromedially; dorsum nearly straight, venter concave; posterior narrower than anterior; anterior marginal area compressed, less compressed ventrally and posteriorly; anterodorsal quadrant bisulcate, the median sulcus the longer and both with crossbars medially; finely and evenly reticulate surface, except smooth anterodorsal area; rare papillae postmedially; LV slightly overlaps RV posteriorly; LV hinge with obscure cardinal sockets; medially, valve edge underlaps RV; inner lamellae very narrow; L, 0.60 to 0.69; H, 0.33 to 0.38 mm. Male more elongate, posterior pointed; L, 0.63 to 0.73; H, 0.31 to 0.37 mm (Staplin, 1963).

Occurrence. Pleistocene, Illinoian Stage and Tazewell Substage, Wisconsin Stage of Illinois (see plate description); late Cary or Mankato Substage, Wisconsin Stage of western Indiana; Mankato or late Cary of Story County, Iowa (Staplin, 1963).

Limnocythere robusta Delorme

Plate 26, figures 32, 33

Limnocythere robusta Delorme, 1967, p. 362, pl. 2, figs. 13-16; 1970, p. 54, pl. 10, figs. 158-161. Cameron and Lundin, 1977, p. 349, pl. 1, figs. 13, 14.

Elongate subelliptical, females shorter and narrower posteriorly than males; highest anteriorly; dorsum nearly straight, venter more or less concave; anterior broader than posterior; smooth to faintly reticulate; bisulcate, the median sulcus longer than anterior sulcus; non-alate; ventral longitudinal ridge; inner lamellae narrow, no vestibule; L, 0.65 to 0.75; H, 0.25 to 0.40 mm (After Delorme, 1970).

Occurrence. Pleistocene of Canada (Delorme, 1970); Late Pleistocene Lake Cochise deposits, Willcox Playa, Cochise County, Arizona (Cameron and Lundin, 1977).

Limnocythere cf. L. sanctipatricii Brady and Norman

Plate 25, figures 30, 31

Limnicythere sancti-patricii Brady and Robertson, 1869, p. 17, pl. 18, figs. 8-11; pl. 21, fig. 4. Brady, Crosskey and Robertson, 1874, p. 174, pl. 2, figs. 1-3.

Limnocythere sancti-patricii Brady and Robertson, Kaufmann, 1896, p. 342, pl. 13, figs. 20-41. Swain, 1963, p. 808, pl. 96, figs. 11a-d, TF 5a.

Female elongate subquadrate in adults, subpyriform in juveniles, highest medially to anteromedially; dorsum nearly straight, venter concave; anterior broader than posterior, anterior marginal zone compressed; anteromiddorsal sulcus represented by two deep pits separated by a narrow ridge; anterodorsal oblique sulcus shallower and defines a low anterodorsal node on its side; pitted; RV hinge with terminal small toothlike ridges; interterminal valve edge rabbeted below; L, 0.68; H, 0.35 mm. Male elongate-subquadrate with small nodes on either side of median sulcus (Swain, 1963).

Occurrence. Pleistocene, Gubik Formation, Arctic coastal plain, Alaska, Ikpikpuk Lake shot holes at depth of 40 feet (12 m) in 17 holes; Point Barrow core test No. 1 at 100-110 feet (30-33 m) and below (Swain, 1963). Holocene, northern Europe, United States, Canada; Pleistocene, Great Britain (authors).

Limnocythere sappaensis Staplin

Plate 25, figures 32, 33

Limnocythere sappaensis Staplin, 1963, p. 1197, pl. 160, figs. 4-6. Forester, 1996, p. 33.

Female subquadrate, subreniform, highest postmedially; dorsum slightly convex, venter concave; posterior broader than anterior, the latter extended below; anterior and posterior marginal zones compressed; anterodorsal quadrant bisulcate; median sulcus the longer, with cross-ridge, anterior sulcus oblique; rounded node between sulci; finely reticulate surface; scattered papillae; LV hinge with terminal obscure sockets, intervening valve edge underlaps RV; inner lamellae narrow; no vestibules; L, 0.53 to 0.60; H, 0.30 to 0.34 mm. Male larger and longer; posterior very broad; L, 0.60 to 0.70; H, 0.31 to 0.39 mm (Staplin, 1963).

Occurrence. Pleistocene, Sappa Silt (late Kansan Stage) of Iowa and Nebraska (see plate description) (Staplin, 1963); Pleistocene to Holocene, Owens Lake Basin, California (Forester, 1996).

Limnocythere scotti Staplin

Plate 25, figures 34, 35

Limnocythere scotti Staplin, 1963,p. 1197, pl. 160, figs. 17-19.

Female small, subrectangular-subreniform, highest anteromedially; dorsum nearly straight to gently convex, venter concave; ends about equally rounded, extended below; terminal marginal areas compressed; weakly bisulcate in anterodorsal quadrant, median sulcus the longer, anterior sulcus oblique; a small pit may lie between sulci; smooth and glistening surface;

postmidventral swelling; LV overlaps RV posteriorly; LV hinge with obscure cardinal sockets and intervening edge that underlaps RV; inner lamellae, slight vestibule; L, 0.60 to 0.70; H, 0.33 to 0.38 mm. Male more elongate than female, a rounded node posterior to median sulcus; prominent midventral swelling of surface and postjacent depression; L, 0.66 to 0.68; H, 0.32 to 0.33 mm (Staplin, 1963).

Occurrence. Pleistocene, Aftonian Stage of Illinois; Mankato Substage, Wisconsin Stage of Indiana and Illinois (Staplin, 1963) (see plate description).

Limnocythere sharpei Staplin
Plate 25, figures 36-38
Limnocythere sharpei Staplin, 1963, p. 118, pl. 160, figs. 27-29.

Female subquadrate-subreniform, highest postmedially; dorsum nearly straight to sinuous, venter concave; posterior more pointed than anterior, truncate above; terminal areas compressed, the anterior the broader; bisulcate in anterodorsal quadrant; the median sulcus reaches ventrad of midheight and has a crossbar medially; rounded lobes separates short anterodorsal sulcus from median sulcus; strongly reticulate; ventromedian ridgelike swelling parallel to margin; few papillae; LV hinge with cardinal small sockets, intervening valve edge underlaps RV; inner lamellae very narrow; L, 0.70 to 0.72; H, 0.38 to 0.40 mm. Male, posterior much higher than in females; ventroposterior broad marginal flange; L, 0.80 to 0.82; H, 0.40 to 0.42 mm (Staplin, 1963).

Occurrence. Pleistocene, Tazewell Substage, Wisconsin Stage of western Illinois; Mankato or late Cary Substage, Wisconsin Stage of western Indiana and southeastern Illinois (see plate description) (Staplin, 1963).

Limnocythere staplini Gutentag and Benson
Plate 25, figures 39-44
Limnocythere staplini Gutentag and Benson, 1962, p. 51, pl. 1, figs. 1-3, TF 15. Lister, 1975, p. 29, pl. 5, figs. 7-10, TF 33. Cameron and Lundin, 1977, p. 349, pl. 1, figs. 5, 6.

Female small, subquadrate-subreniform, highest anteromedially; dorsum arched, sinuate, venter concave; anterior broader than posterior; bisulcate in anterodorsal quadrant; median sulcus reaches ventrad of midheight; anterodorsal sulcus shorter; rounded node separates sulci; ventromedian longitudinal swelling; reticulate; LV hinge with cardinal elongate small sockets; inner lamellae broad, no vestibules; L, holotype, 0.52; H, 0.31 mm. Male, straighter dorsum, deeper ventral concavity, more pronounced anteromedian swelling than female; L, 0.63; H, 0.32 mm (Gutentag and Benson, 1962).

Occurrence. Pleistocene, Illinoian to Wisconsin Stages, Odee Formation, Meade County, Kansas; Jones Ranch beds, Meade County, Kansas; Nye Sink deposits, Beaver County, Oklahoma; Late Kansan to Yarmouthian Stage, Reno County, Kansas (Gutentag and Benson, 1962) (see plate description). Lake Bonneville Group, Saltair core hole SE$^{1}/_{4}$ sec. 25, T. 1 N., R. 3 W.; Sec. 28 core hole SE$^{1}/_{4}$ SW$^{1}/_{4}$ sec. 28, T. 1 N., R. 3 W., Salt Lake County, Utah (Lister, 1975). Late Pleistocene, Lake Cochise deposits Willcox Playa, Cochise County, Arizona (Cameron and Lundin, 1977).

Limnocythere trapeziformis Staplin, 1963
(see Limnocythere itasca Cole, 1949)

Limnocythere? sp. aff. L. unicornis (Swain), Swain, Becker and Dickinson, 1967
(see Limnicytheropteron winecupensis Swain)

Limnocythere varia Staplin
Plate 25, figures 47, 48; Plate 26, figure 1
Limnocythere varia Staplin, 1963, p. 1200, pl. 160, figs. 30-32.

Female elongate, subrectangular-subreniform, highest about one-fourth of length from anterior end; dorsum nearly straight, venter concave; posterior narrower than anterior, extended below; anterior marginal zone flattened, posterior less so; bisulcate in anterodorsal quadrant; median sulcus reaches to midheight, with crossbar; rounded node separates sulci; anterodorsal sulcus short with anterodorsal node; ventromedian longitudinal subalate swelling; reticulate; LV hinge with cardinal weak sockets; intervening valve edge underlaps RV; vestibules weak or absent; L, 0.90 to 0.98; H, 0.40 to 0.46 mm. Male larger and more elongate, posterior more rounded and with postventral flange; L, 0.97 to 1.03; H, 0.40 to 0.46 mm (Staplin, 1963).

Occurrence. Pleistocene, Tazewell to Mankato Substages, Wisconsin Stage of Illinois (see plate description) (Staplin, 1963); Bone Bed at Boney Spring, Benton County, Missouri, coll. by J. E. King unpublished.

Limnocythere verrucosa Hoff
Plate 25, figures 49, 50
Limnocythere verrucosa Hoff, 1942, p. 161, pl. 9, figs. 127-135. Staplin, 1963, p. 1201, pl. 160, figs. 11-14.

Female small, elongate, subrectangular, highest anteromedially; dorsum straight, venter concave; anterior broader than posterior; flattened marginal border broad anteriorly; bisulcate in anterior dorsal quadrant; median sulcus extends nearly to venter, with crossbar; rounded node separates that sulcus from short anterodorsal sulcus; swollen areas posterior to sulci; two ventromedian swollen areas; finely reticulate surface; radial canals numerous; LV hinge with cardinal small sockets; intervening valve edge underlaps RV; vestibules small or absent; L, 0.54 to 0.55; H, 0.30 to 0.31 mm. Male longer than female; posterior extended above; surface ornamentation subdued; L, 0.62 to 0.68; H, 0.38 to 0.41 mm (Staplin, 1963).

Occurrence. Pleistocene, Aftonian? Stage of Recent of Illinois; Mankato Substage, Wisconsin Stage of Indiana; Tazewell and Cary? Substages Wisconsin Stage of Ontario (Staplin, 1963). Recent of Illinois (Hoff, 1942).

Limnocythere? williamsi (Swain)
Plate 26, figures 2-4
Limnicythere? williamsi Swain, 1947, p. 527, pl. 77, figs. 30-33.
Limnocythere? williamsi (Swain), Swain, 1987a, p. 125.

Subquadrate-subpyriform, highest near anterior end; dorsum nearly straight, venter slightly concave, converging with dorsum posteriorly; anterior much broader than posterior, extended below; LV slightly larger than RV; free margins with narrow rim; ventroterminal areas compressed; bisulcate in anterodorsal quadrant; median sulcus reaches to midheight; rounded node separates median sulcus from short oblique anterodorsal sulcus; surface coarsely pitted; midventral longitudinal subalate swelling; LV hinge with small terminal sockets; interterminal edge beveled below to overlap RV; adductor muscle scar a vertical row of three or more elongate spots; inner lamellae fairly broad; narrow vestibules terminally; L, holotype RV 0.56; H, 0.38 mm (Swain, 1947).

Occurrence. Pliocene, Salt Lake Group, sec. 15, T. 13 N., R. 2 W., Cache County, Utah (Swain, 1947; Swain, 1987a).

Limnocythere? sp. 1, Sohn
Plate 26, figure 8
Limnocythere? sp. 1, Sohn, 1956, p. 120, pl. 1, fig. 2 (not described).

Subquadrate, highest submedially; dorsum slightly convex, venter nearly straight; posterior narrower than anterior, truncate above; median sulcus reaches venterad of midheight; anterodorsal sulcus shorter; ventral longitudinal swelling; postdorsal submarginal swelling; L, 0.48; H, 0.27 mm (from Sohn's photograph, 1956).

Occurrence. Pliocene, Teewinot Formation, Jackson Hole, Wyoming, USNM 562013, L-2-89, Unit 17 (Sohn, 1956).

Limnocythere? sp. 2, Sohn

Plate 26, figure 9

Limnocythere? sp. 2, Sohn, 1956, p. 121, pl. 1, figs. 3, 4 (not described).

Male? elongate, subquadrate, highest postmedially; posterior broader than anterior, extended above; postventral flangelike expansion; anterior border area flattened; bisulcate in anterodorsal quadrant; median sulcus reaches midheight; anterodorsal sulcus shorter; intervening node with furrow below; midventral longitudinal swelling; L, 0.60; H, 0.29 mm (from Sohn's photograph, 1956).

Occurrence. Pliocene, Teewinot Formation, Jackson Hole, Wyoming, USNM 562014, L-2-83 (Sohn, 1956).

Limnocythere? sp., Swain

Plate 25, figure 51

Limnocythere? sp., Swain, 1985, p. 435, pl. 5, fig. 23.

Broken valves subquadrate to subreniform, highest anteriorly; dorsum convex, venter concave, anterior broader than posterior; nodose and reticulate-pustulose surface; L, broken LV, 0.72; H, 0.45 mm (Swain, 1985).

Occurrence. Miocene, Esmeralda Formation, 3.3 miles (5.3 km) east of Finger Rock Wash, Simon Mine area, Mineral County, Nevada, in lower 100-150 feet (30-45 m) of lake beds sequence (Swain, 1985).

Limnocythere sp. 1, Swain

Plate 26, figures 11-13

Limnocythere sp. 1, Swain, 1986a, p. 102, pl. 4, figs. 13, 14, 17?, 18?

Male elongate, subquadrate-subreniform, highest about one-fourth of length from anterior end; dorsum straight, venter concave; anterior broader than posterior, the latter extended below, truncate above; LV larger than RV; terminal fourths compressed; oblique furrow anterodorsally; dorsomedian vertical sulcus nearly reaches midheight; lobe broken by two nodes between sulci; low alate expansions postventrally; smooth, L, 0.57; H, 0.32 mm, Female? shorter, nearly straight venter and lacks frill-like terminal compressions of male? dimorph (Swain, 1986a).

Occurrence. Pliocene, Glenns Ferry Formation, sec. 14, T. 4 S., R. 1 W., Owyhee County, Idaho ("Big Cut") (Swain, 1986a).

Limnocythere? sp. 3, Sohn

Plate 26, figure 10

Limnocythere? sp. 3, Sohn, 1956, p. 121, pl. 1, figs. 5, 6 (not described).

Elongate, subquadrate-subreniform, highest anteromedially; dorsum slightly convex, venter concave; posterior narrower than anterior, truncate above and below; bisulcate, median sulcus reaches venterad of midheight, anterior sulcus shorter; intervening node has defining furrow below; anterior marginal area compressed; midventral longitudinal swelling; L, 0.68; H, 0.35 mm (from Sohn's photograph, 1956).

Occurrence. Pliocene, Teewinot Formation, Jackson Hole, Wyoming, USNM 562015 (Sohn, 1956).

Limnocythere sp. indet. A, Staplin

Plate 26, figure 15

Limnocythere sp. indet A, Staplin, 1963, p. 1202, pl. 160, fig. 33.

Subquadrate, highest posteriorly, dorsum nearly straight to slightly concave, venter concave; posterior much broader than anterior; obliquely bisulcate in anterodorsal quadrant; median sulcus the longer, reaching nearly to midheight; two nodes separate the two sulci; two other nodes lie posterior to median sulcus; midventral longitudinal subalate swelling; surface reticulate; L, RV, 0.77; H, 0.43 mm (Staplin, 1963).

Occurrence. Pleistocene, Tazewell Substage, Wisconsin Stage, northwestern Illinois (see plate description) (Staplin, 1963).

Limnocythere sp., Swain

Plate 26, figure 5

Limnocythere sp., Swain, 1956, p. 137, pl. 3, fig. 17. Swain, 1964a, p. 278, pl. 44, fig. 3.

Elongate-subovate to subquadrate, highest near anterior end; dorsum convex, venter concave; anterior broader than posterior; ventro-terminal borders compressed; subcentral vertical sulcus extends nearly the height of valve; L, 0.61; H, 0.33 mm (Swain, 1964).

Occurrence. Eocene, lower approximately 965 feet (294 m) of Green River Formation, Anvil Point section, southwest of Rifle, Moffat County, Colorado along road from U. S. Bureau of Mines Demonstration Plant to Anvil Point (Swain, 1964a).

Limnocythere sp., Benson

Plate 26, figures 6, 7

Limnocythere sp., Benson, 1969, p. 108, pl. 21, figs. 1, 2, 6, 7, 8 (not Pl. 22, figures 1-3, =Cyprideis sp.).

Elongate-subpyriform, male more elongate than female, highest anteromedially; dorsum gently convex to nearly straight, venter concave; anterior broader than posterior; terminoventral marginal areas compressed; bisulcate in anterodorsal quadrant; median sulcus extends venterad of midheight; coarsely reticulate; L, female, 0.61; H, 0.31 mm (Benson, 1969).

Occurrence. Early Pleistocene, Rita Blanca Lake Beds near Channing, Texas, (Benson, 1969).

Limnocythere sp., Swain and Wagoner
not figured
Limnocythere sp., Swain and Wagoner, 1985, p. 143 (not described or figured).

Occurrence. Miocene, Esmeralda Formation, Blair Junction section, Esmeralda County, Nevada, samples 1D, 1Q, 1R?, and 2C (Swain and Wagoner, 1985).

Limnocythere sp. 2, Swain
Plate 26, figure 14
Limnocythere sp. 2, Swain, 1986a, p. 103, pl. 5, fig. 17.

Elongate, sublanceolate, highest about one-fourth of length from anterior end; dorsum sinuous but nearly straight, venter slightly concave; anterior broader than posterior; valves strongly convex; terminal area compressed; shallow oblique sulcus anterodorsally; anteromedian sulcus does not reach midheight; large node posterior to median sulcus; ventromedian surface subalaform; weak coarse reticulations postdorsally; L, 0.77; H, 0.40 mm (Swain, 1986a).

Occurrence. Pliocene, Glenns Ferry Formation, sec. 14, T. 4 S., R. 1 W., "Big Cut", Owyhee County, Idaho (Swain, 1986a).

Limnocythere? sp. 1, Swain
Plate 26, figure 34
Limnocythere? sp. 1, Swain, 1982, p. 123, pl. 1, figs. 11-13.

Subquadrate (female?) to elongate-subelliptical (male?), highest anteriorly; dorsum convex to straight, venter sinuous to concave; anterior broader than posterior; anteromedian and posteromedian surface of each valve swollen, more so posteriorly; median sulcus nearly reaches venter, broadens near dorsum; terminal areas compressed; smooth; L, 0.67; H, 0.40 mm (after Swain, 1982).

Occurrence. Lower Cretaceous, Washitan Stage, in deep wells in Hendry and Pinellas Counties, Florida (Swain, 1982).

Limnocythere? sp. 2, Swain
Plate 26, figure 35

Limnocythere? sp. 2, Swain, 1982, p. 126, pl. 2, fig. 14.

Subquadrate, highest one-fourth from posterior end; dorsum straight; venter sinuate; posterior broader than anterior and extended above, truncate below; LV larger than RV; anteromedian sulcus narrow; two low nodes anterior to sulcus, one node posterior to sulcus; ventral ridge forms low ala; terminal fifths of shell compressed; L, 0.70; H, 0.40 mm (after Swain, 1982).

Occurrence. Lower Cretaceous, Washitan Stage, in well in Palm Beach County, Florida (Swain, 1982).

Genus Lineocypris Zalanyi, 1929

(Candonidae)

Lineocypris? sp., Sohn

Plate 26, figures 22, 23

Lineocypris? sp., Sohn, 1956, p. 120, pl. 1, figs. 7-10 (not described).

Elongate, subreniform-subelliptical, highest medially to postmedially; dorsum gently convex, venter concave; anterior a little narrower than posterior, both ends extended below; smooth; inner lamellae narrow, broadest anteriorly; L, 0.65; H, 0.34 mm (from Sohn's photograph).

Occurrence. Pliocene, Teewinot Formation, Jackson Hole, Wyoming NW1/4 NW1/4 sec. 18, T. 42, N., R. 115 W., Teton County, Wyoming; NW1/4 sec. 36, T. 42 N., R. 116 W.; secs. 1, 2, T. 43 N., R. 115 W.; sec. 21, 22, 28, 29, T. 42 N., R. 115 W. (Sohn, 1956).

Genus Longispinella Sohn, 1979

(Cyprideidae)

Longispinella asymmetrica Sohn

Plate 13, figures 6, 7; Plate 26, figures 24-26

Longispinella asymmetrica Sohn, 1979, p. 18, pl. 4, figs. 7-20; pl. 5, figs. 1-7, 13-16.

Subtriangular, highest medially; dorsum strongly convex, angulated at position of greatest height; venter slightly convex; anteroventral rostrum (beak), alveolus (notch and furrow) well developed; cyathus (angular post-ventral projection in LV) triangular; LV overlaps RV; robust lateral spine in center of posterior third of each valve; densely punctate except in alveoli; LV with anterodorsal sulcus; smooth ridge separates sulcus from alveolus; hinge margin incised; L, 0.56; H, 0.40 mm (Sohn, 1979).

Occurrence. Lower Cretaceous, Lakota Formation, Chilson Member, Fall River County, South Dakota, USGS 26098, 31130 (Sohn, 1979).

Longispinella longispina (Peck)

Plate 26, figures 27-29; Plate 27, figures 44, 45

Cypridea longispina Peck, 1941, p. 300, pl. 43, figs. 6-9; 1951, p. 312, pl. 48, figs. 12-15; Sohn, 1958, p. 123, pl. 1, figs. 1-4.

Longispinella longispina (Peck), Sohn, 1979, p. 19, pl. 4, figs. 1-5, pl. 5, figs. 5-12, 17-23.

Subtriangular, highest just anterior to middle; dorsum strongly arched, subumbonate; venter nearly straight; anteroventral beak, hood and notch well developed; posterior narrower than anterior, both ends extended below; LV overlaps RV; strong posteromedian spine on both valves; smooth surface; hinge line deeply indented; L, 0.9; H, 0.55 mm (Peck, 1941).

Occurrence. Lower Cretaceous, Kootenai Formation, near Great Falls, Montana (Peck, 1941); Cloverly Formation near Lander, Wyoming; Lakota Formation, 3.2 miles (5.1 km) SE of Hot Springs, Wyoming (Peck, 1951); Lakota Formation, NW1/4 sec. 29, T. 51 N., R. 65 W., Crook County, Wyoming (Sohn, 1958). Lower Kootenai Formation, T. 18 N., R. 4 E., Cascade County, Montana; Lakota Formation, Fall River County, South Dakota, U.S.G.S. coll. 31154, 31153 (Sohn, 1979).

Genus Looneyella Peck, 1951

(Hutsoniidae)

Looneyella monticula (Jones), Peck

Plate 26, figures 30, 31; Plate 27, figures 1-4, 42, 43

Cythere monticula Jones, 1893, p. 389, pl. 15, figs. 13a, b. White, 1895 (reproduction of Jones' plate).

Looneyella monticula (Jones), Peck, 1951, p. 576, pl. 80, figs. 1-6.

Subtriangular, highest near anterior end; dorsum nearly straight, concave medially, venter convex; anterior much broader than posterior, extended below; posterior caudate, pointed, extended above; LV overlaps RV; alate ventrally, with strong median longitudinal ridge; bisulcate in anterodorsal quadrant; rounded node or lobe separates sulci; anteromarginal ridge; prominent rounded elevation dorsomedially; reticulate surface; flattened venter with narrow parallel ridges; LV hinge with terminal elongate taxodont sockets; L, 0.95; H, 0.55 mm (Peck, 1951).

Occurrence. Upper Cretaceous, Bear River Formation on Sulphur Creek and in Shell Hollow, near Evanston, Uinta County, Wyoming (Peck, 1951).

Looneyella quadrispina Peck

Plate 27, figures 5-7

Looneyella quadrispina Peck, 1951, p. 576, pl. 80, figs. 7-9.

Trigonal, highest medially, dorsum slightly convex, venter sinuous owing to projection of midventral ala, converging with dorsum toward posterior; anterior broader than caudate posterior and extended below; posterior extended above; middorsal large spine does not project beyond margin, whereas ventral ala does so; roughly reticulate surface; L, 1.01; H, except for ala, 0.75 mm (Peck, 1951).

Occurrence. Upper Cretaceous, Bear River Formation, Shell Hollow, 7 miles (11.26 km) north of Evanston, Uinta County, Wyoming (Peck, 1951).

Genus Lutkevichinella Schneider, 1956

(Cytherissinellidae)

Lutkevichinella sp., Kietzke

Plate 26, figure 21

Limnocytheriinae, gen. et sp., indet, Kietzke, 1987, p, 124, TF 2H, I

Lutkevichinella sp., Kietzke, 1989a, p. 223.

Subelliptical, highest medially, dorsum slightly convex, venter nearly straight, anterior a little narrower than posterior; LV overlaps RV; a single vertical submedian sulcus; finely reticulated, becoming coarser terminally; L, 0.60 to 0.64; H, 0.28 to 0.32 mm (Kietzke, 1987).

Occurrence. Upper Triassic, Chinle Formation, upper shale member, Bull Canyon, Santa Rosa County, New Mexico; Redonda Member, Shark Tooth Hill, Tucumcari County, New Mexico (Kietzke, 1987).

Genus Malheurocypris Swain, 1990

(Ilyocyprididae)

Malheurocypris variabilonodosa (Swain)

Plate 27, figures 8-12

Ilyocyprimorpha variabilonodosa Swain, 1986a, p. 99, pl. 3, figs. 19-21; pl. 4, figs. 1, 2.

Malheurocypris variabilonodosa (Swain), Swain, 1990, p. 152.

Subquadrate-subtrapezoidal, highest one-fifth of length from anterior end; dorsum nearly straight except for nodose protuberances, venter gently concave; posterior narrower than anterior, both ends extended below, subtruncate above; LV overlaps RV; strongly convex; prominent pustulose ornamentation; anteriorly and anteroventrally is a submarginal curved row of pustules together with subsidiary flanking pustules; a row of smaller pustules lies close to anterior margin; submedially is a set of several pustules forms an elevated area; postmedially is an elongated elevated area surmounted by several pustules; anterodorsally is a set of five or six pustules of which two or three project beyond dorsal margin; general surface finely and densely papillose; LV hinge margin rabbeted expanded terminally; inner lamellae narrow but variable; vestibules present; L, holotype, 0.67; H, 0.46 mm (Swain, 1986a).

Occurrence. Pliocene, Glenns Ferry Formation, sec. 14, T. 4 S., R. 1 W., "Big Cut", Owyhee County, Idaho (Swain, 1986a).

Malheurocypris sp., Swain
Plate 27, figures 13-15
Malheurocypris sp., Swain, 1990, p. 152, pl. 2, figs. 1-3.

Elongate subquadrate-subreniform, highest one-fifth of length from anterior end; dorsum nearly straight, venter strongly concave; anterior broader than posterior, both ends extended below; LV larger than RV, moderately convex; surface features include anteromedian strongly inflated node with two smaller nodes on its crest, anteroventral short anteroventrally directed high ridge, rounded high median node; posteromedian ventrally directed ridge, posterodorsal high node, 15-20 other smaller nodes; general surface finely and densely pustulose; LV hinge a furrow supported by ridgelike thickening on shell interior; inner lamellae of moderate width terminally, narrow vestibules: L, 0.93; H, 0.52 mm (Swain, 1990).

Occurrence. Pliocene, Idaho Group, NE1/4 SW1/4 sec. 9, T. 22 S., R. 46 E., Malheur County, Oregon (Swain, 1990).

Genus Metacypris Brady, 1870
(Limnocytheridae-Timiriasevinae)
"Metacypris" angularis (Peck, 1941)
(see Dryelba angularis (Peck))
"Metacypris" bradyi Jones
Plate 27, figures 30, 31
Metacypris bradyi Jones, 1986, p. 146, pl. 4, figs. 2a-c
?Theriosynoecum bradyi (Jones), Sohn, 1982, p. 311 [Not Theriocynoecum by Sohn.]

Female? subelliptical, highest postmedially; dorsum and venter subparallel, each concave medially; posterior broader than anterior; valves strongly inflated posteriorly; anteromedian sulcus, postmedian tubercle and other scattered tubercles; L, 1.13; H, 0.6 mm (after Jones, 1886). May be Cytheridella by comparison with C. todiltensis (Swain).

Occurrence. Upper Jurassic, Morrison Formation, "Atlantosaurus beds", near Canyon City, Colorado (Jones, 1886).

Metacypris? clarkstonensis Swain

Plate 27, figures 22, 23

Metacypris? clarkstonensis Swain, 1987a, p. 125, pl. 2, figs. 6-9.

Subovate-subquadrate, highest one-fourth of length from anterior end; dorsum nearly straight, venter slightly sinuous, slightly convex, converging posteriorly with dorsum; anterior broader than posterior; dimorphic; males less convex and narrower in posterior third than females; LV overlaps RV; weak irregular nodelike elevations and broad pits on surface; slight anterodorsal depression; narrow terminal compressed rims; L, holotype, female, 0.73; H, 0.43 mm (Swain, 1987).

Occurrence. Pliocene, Salt Lake Group, Cache Valley Formation, South Clarkston section, T. 13 N., R. 12 W., Cache County, Utah; T. 13 N., R. 3 W., southwest of Newton, Cache County, Utah (Swain, 1987).

"Metacypris" consobrina Jones

Plate 27, figures 34, 35

Metacypris consorbina Jones, 1893, p. 388, pl. 15, figs. 1a-c

Subelliptical, highest in posterior third; dorsum and venter both concave medially, subparallel; ends nearly equally rounded; valves strongly inflated posteriorly; slightly anteromedian transverse sulcus; irregularly pitted; L, 0.45; H, 0.25 mm (after Jones, 1893). May be a Cytheridella (see C. todiltensis (Swain))

Occurrence. Lower Cretaceous, Lower Bear River Formation or Gannett Group, Cokeville, Wyoming (Jones, 1893).

"Metacypris" cunieformis Jones

Plate 27, figures 40, 41

Metacypris cunieformis Jones, 1893, p. 389, pl. 15, figs. 3a-c

Oblong, highest anteriorly, dorsum and venter subparallel, nearly straight to sinuous; ends broadly rounded, the anterior the broader, the posterior truncate below; anterior fourth compressed; smooth; L, 0.65; H, 0.35 mm (after Jones, 1893). Generic status uncertain.

Occurrence. Lower Cretaceous, Lower Bear River Formation, Cokeville, Wyoming (Jones, 1893).

"Metacypris" forbesi Jones

Plate 27, figures 28, 29

Metacypris forbesi Jones, 1885, p. 345, pl. 8, figs. 11, 16; 1886, p. 146, pl. 4, figs. 1a-c
?Theriosynoecum forbesi (Jones), Sohn, 1982, p. 311 [Not Theriosynoecum by Sohn.]

Female? elongate, subpyriform, highest postmedially; dorsum and venter both slightly concave medially, converging toward anterior; valves strongly inflated posteriorly; anterodorsal transverse sulcus; scattered pustules; L, 1.0; H, 0.53 mm (after Jones, 1886). May be Cytheridella, see C. todiltensis (Swain).

Occurrence. Upper Jurassic, Morrison Formation, "Atlantosaurus Beds", near Canyon City, Colorado (Jones, 1886); Purbeck Beds of England (Jones, 1885).

Metacypris paracordata Swain
Plate 27, figures 24-27
Metacypris? sp., Swain, 1956, p. 137, pl. 3, figs. 15a-e.
Metacypris paracordata Swain, 1964a, p. 278, pl. 24, figs. 5a-f.

Subelliptical, highest medially; dorsum moderately convex, venter nearly straight to slightly concave; anterior narrower than posterior, both ends extended below; valves subequal, strongly convex posteriorly; faint oblique anterodorsal sulcus, finely reticulate; L, holotype, 0.56; H, 0.30 mm (Swain, 1964).

Occurrence. Eocene, Colton-Green River transition beds, 8 miles (12.8 km) west of Rangely, Colorado on north side of White River, sec. 3, T. 1 N., R. 103 W., Rio Blanco County, Colorado)Swain, 1964).

Metacypris persulcata Peck, 1941
(see Dryelba? persulcata (Peck))

"Metacypris" simplex Jones
Plate 27, figures 38, 39
Metacypris simplex Jones, 1893, p. 389, pl. 15, figs. 9a-c

Suboblong, highest anteriorly; dorsum slightly convex, venter straight, converging posteriorly with dorsum; venter overhangs margin; ends truncate; finely reticulate surface; L, 0.4; H, 0.2 mm (after Jones, 1893). May be Trapezoidella.

Occurrence. Lower Cretaceous, Lower Bear River Formation or Gannett Group, Cokeville, Wyoming (Jones, 1893).

"Metacypris" subcordata Jones

Plates 27, figures 36, 37

Metacypris subcordata Jones, 1893, p. 388, pl. 15, figs. 2a-c

?Theriosynoecum subcordata (Jones), van Morkhoven, 1963; Sohn, 1982

[Not Theriosynoecum by Sohn.]

Suboblong, highest medially; dorsum and venter slightly convex, the former sinuous anteriorly due to oblique anterodorsal sulcus; valves strongly inflated posteriorly; reticulate pattern of surface pitting; L, 0.6; H, 0.35 mm (after Jones, 1893).

Occurrence. Lower Cretaceous, Lower Bear River Formation or Gannett Group, Cokeville, Wyoming (Jones, 1893).

Metacypris todiltensis Swain, 1946
(see Cytheridella todiltensis (Swain))

"Metacypris" whitei Jones
Plate 27, figures 32, 33

Metacypris whitei Jone, 1886, p. 146, pl. 4, figs. 3a-c

Male? oblong, highest postmedially, dorsum slightly concave medially, venter nearly straight; posterior narrow than anterior, the latter truncate above; valves moderately convex posteriorly; bisulcate in anterior half; long median vertical sulcus with a crossbar; smooth?; L, 1.13; H, 0.6 mm; slight anteroventral notch? (Jones, 1896). May be a Cypridea if notch is present or a Cytheridella if no notch is present.

Occurance. Upper Jurassic, Morrison Formation, "Atlantosaurus beds", near Canyon City, Colorado (Jones, 1886).

Metacypris sp., Swain, 1964a
(see Timiriasevia heylmuni Swain)

"Metacypris"?, sp.
Plate 28, figures 1, 2
Betty Kellett Nadeau, 1945, unpublished photographs

Elongate-ovate, highest medially; dorsum convex, venter slightly concave; anterior slightly broader than posterior, both ends extended below; valves subequal, strongly convex posteriorly; flattened venter, weak anterodorsal sulcus; finely pitted, venter striated longitudinally; dimensions unknown (from Nadeau photographs).

Occurrence. Upper Jurassic or Lower Cretaceous, Morrison Formation?, Canyon City?, Colorado.

Genus Monoculus Linne, 1758, nomina dubia

Monoculus candidus (Müller), Jurine, 1870

(see Candona candida (O. F. Müller))

Genus Morrisonia Branson, 1935, invalid

Morrisonia wyomingensis Branson, 1935

(see Theriosynoecum wyomingense (Branson))

Genus Neolimnocythere Delachaux, 1928

(Limnocytheridae-Limnocytherinae)

Neolimnocythere? prestonensis Swain

Plate 28, figures 3-6, 7-11

Neolimnocythere? prestonensis Swain, 1986c, p. 326, pl. 4, figs. 8-11. Swain, 1990, p. 154, pl. 2, figs. 4-9.

Elongate, subquadrate-sublanceolate, highest one-fourth of length from anterior end; dorsum nearly straight to slightly sinuous, venter concave; anterior broader than posterior and extended below; valves subequal; shallow anteromedian sulcus in dorsal third; anterodorsal surface is ridgelike and bears two nodes; middorsal surface nodelike, extending more or less beyond dorsal margin; ventromedian surface has three longitudinally elongate nodes or short ridges; anterior fifth of shell compressed; coarsely reticulate surface; RV hinge with anterior elongate tooth; postjacent socket, interterminal bar and posterior elongate tooth; inner lamellae narrow, small vestibules; adductor muscle scar an anteromedian row of four elongate spots; one or more frontal spots; L, holotype, 0.59; H, 0.30 mm (Swain, 1086).

Occurrence. Pliocene, Cache Valley Formation, NW$^{1}/_{4}$ SE$^{1}/_{4}$ NE$^{1}/_{4}$ sec. 10, T. 12 S., R. 40 E., Franklin County, Idaho (Swain, 1986). Idaho Group, sec. 9, T. 22 S., R. 46 E., Malheur County, Idaho (Swain, 1990).

Genus Oreanocypris Swain, 1986

(Candonidae-Cyclocypridinae)

Oreanocypris acutispinosus Swain

Plate 28, figures 12, 13

Oreanocypris acutispinosus Swain, 1986a, p. 93, pl. 3, figs. 11, 14.

Subquadrate-subovate, highest one-third of length from anterior end; dorsum gently convex, truncate postmedially, venter gently concave; anterior broader than posterior; LV overlaps RV especially along venter; convexity moderate; postmedian long hollow spine on each valve, curves backward; smooth to finely papillose; LV hinge a weakly rabbeted furrow; inner lamellae narrow, small vestibules terminally; radial canals widely spaced?; adductor muscle scar a median group of four spots; one or two anteroventral spots; L, holotype RV 0.72; h, 0.43 mm (Swain, 1986a).

Occurrence. Pliocene, Glenns Ferry Formation, sec. 14, T. 4 S., R. 1 W., Owyhee County, Idaho, "Big Cut" (Swain, 1986a).

Oreanocypris ascioformis Swain

Plate 28, figures 14, 15

Oreanocypris ascioformis Swain, 1986a, p. 94, pl. 3, figs. 15, 16.

Subquadrate-subtrapezoidal, highest postmedially; dorsum gently convex, venter nearly straight to slightly concave; posterior a little broader than anterior, both ends extended below; LV overlaps RV, most so along venter; convexity of valves low; large blunt to pointed-ended, wedgelike elevation postmedially; smooth surface; LV hinge with rabbet-groove; inner lamellae narrow, vestibules weakly developed; adductor muscle scar a submedian group of four small spots; one or two frontal spots; L, holotype RV 0.60; H, 0.42 mm (Swain, 1986a).

Occurrence. Pliocene, Glenns Ferry Formation, sec. 14, T. 4 S., R. 1 W., Owyhee County, Idaho (Swain, 1986a).

Oreanocypris subtriangularis Swain

Plate 28, figures 16-18

Oreanocypris subtriangularis Swain, 1986a, p. 93, pl. 3, figs. 13, 17, 18.

Subtrapezoidal (mature) to subtriangular (immature), highest one-third to one-fourth of length from anterior end; dorsum moderately convex, straightened postmedially; venter weakly concave (mature) to convex (immature); anterior broader than posterior, both ends extended below; LV larger than RV; convexity moderate; rounded postmedian elevation on each valve; smooth; LV hinge rabbeted; inner lamellae of moderate width, broadest anteriorly, terminal vestibules; radial canals widely spaced; adductor muscle scar a submedian group of four spots; one or two frontal spots; L, holotype, 0.56; H, 0.36 mm (Swain, 1986a).

Occurrence. Pliocene, Glenns Ferry Formation, sec. 14, T. 4 S., R. 1 W., Owyhee County, Idaho (Swain, 1986a).

Genus Pactolocypris Swain, 1985

(Cyprididae-Herpetocyprellinae)

Pactolocypris biprojectus Swain

Plate 28, figures 19-23

Tuberocypris sp. 1, Swain, Becker and Dickinson, 1971, p. 478, pl. 1, figs. 3a-e.

Pactolocypris biprojectus Swain, 1985, p. 427, pl. 3, figs. 4, 7, 8a, 9, 9a; pl. 4, figs. 12-16. Swain, 1986b, p. 208, pl. 3, fig. 18; pl. 5, fig. 18.

Elongate, subquadrate-subovate, highest medially to postmedially; dorsum nearly straight to gently convex, venter slightly convex to sinuous; anterior nearly equal to or slightly narrower than posterior, truncate above; LV larger than RV; two large, hollow nodes on ventromedian surface; the more anterior node is the smaller of the two (see Swain, 1986b, p. 3, figs. 18); surface weakly and densely pitted; LV hinge rabbeted; adductor muscle scar a group of four or five spots; two anteroventral frontal spots, inner lamellae very narrow; L, holotype, 0.81; H, 0.49 mm (Swain, 1986b).

Occurrence. Miocene, Esmeralda Formation, 2.85 miles (4.5 km) by road west of Black Spring, Black Spring 1:24000 Quadrangle, Nye County, Nevada; about 800 feet (244 m) above base of lake bed sequence; 3.3 miles (5.3 km) east of Finger Rock Wash on road to Simon Mine, Stewart Spring 1:24000 Quadrangle, Mineral County, Nevada in lower 100-150 feet (30-45 m) of lake bed sequence (Swain, 1985); Humboldt Formation, section at Cobre, Elko County, Nevada, 80-84 feet (24-26 m) above base of exposure (Swain, 1986b). Humboldt Formation, Medicine Creek section, Winecup Ranch, T. 39 N., R. 65 E., Elko County, Nevada (Swain et al., 1971).

Pactolocypris cancellatus (Swain and Wagoner)

Plate 28, figures 26-30

Tuberocypris cancellatus Swain and Wagoner, 1985, p. 137, pl. 3, figs. 1-8.

Pactolocypris cancellatus (Swain and Wagoner), Swain, 1990, p. 149, pl. 2, figs. 10, 12.

Not Pactolocypris cancellatus (Swain and Wagoner), Swain, 1986b, p. 219, pl. 2, fig. 22 (=Pactolocypris centronotus (Swain, 1964a)).

Elongate-subreniform to subellitpical, highest medially to anteromedially; dorsum convex, venter nearly straight to slightly concave; ends nearly equally rounded, extended below; LV overlaps RV; tubercle present postmidventrally to postmedially; surface with fine reticulating pattern of Bertillion-like ridges arranged concentrically terminally and ventrally; adductor muscle scar an anteroventral group of five spots; two frontal spots; normal pores on low tubercles; L, holotype, 0.98; H, 0.53 mm (Swain and Wagoner, 1985).

Occurrence. Miocene, Esmeralda Formation, Blair Junction section, Esmeralda County, Nevada, samples 1A, 2A, 2C?, 4-7? (Swain and Wagoner, 1985). Truckee Formation, late Miocene?, 2 miles (3.2 km) WSW of Hazen, Churchill County, Nevada (Swain, 1990).

Pactolocypris cancellatus pronodosus Swain

Plate 28, figures 24, 25

Pactolocypris cancellatus pronodosus Swain, 1986b, p. 209, pl. 3, figs. 22, 23.

Resembles P. cancellatus (Swain and Wagoner, 1985) but the distinctive submedian node is anteromedian rather than postmedian in position and the surface cancellation is more subdued than in P. cancellatus; L, holotype, 0.82; h, 0.50 mm (Swain, 1986b).

Occurrence. Miocene, lower part of Humboldt Formation, Gulf Oil Co., No. 1 Thousand Springs well, center SE1/4 SW1/4 sec. 8, T. 40 N., R. 66 E., Elko County, Nevada, 1610-2250 feet (491-686 m) (Swain, 1986b).

Pactolocypris centronotus (Swain)

Plate 28, figure 31; Plate 31, figures 15-17

Candona? n., sp., Swain, 1956, p. 136, pl. 3, figs. 12a, b.

Tuberocypris centronotus Swain, 1964a, p 274, pl. 44, figs. 2a-e.

Pactolocypris cf. cancellatus (Swain and Wagoner), Swain, 1986b, pl. 2, fig. 22.

Pactolocypris cf. centronotus (Swain), Swain, 1987, p. 223, pl. 2, fig .24.

Subelliptical-subreniform, highest anteromedially; dorsum convex, venter nearly straight to slightly concave; posterior broader than anterior, the latter extended below, truncate above; LV overlaps RV; relatively compressed valves; large, high, pointed node on each valve about two-fifths of length from posterior end; finely pitted; L, holotype 0.90; H, 0.51 mm (Swain, 1964).

Occurrence. Eocene, Colton-Green River transition beds, northeastern part of Uinta Basin, near west end of Raven Ridge, Rio Blanco County, Colorado (Swain, 1964). Elko Oil Shale, Catlin Works (abandoned), Elko, Nevada (Swain, 1986b, 1987).

Pactolocypris laevus Swain

Plate 28, figures 32-35

Pactolocypris laevus Swain, 1986b, p. 208, pl. 2, figs. 20, 21, 23; pl. 3, figs. 16-17.

Subelliptical-subreniform, highest medially; dorsum moderately convex, venter slightly concave; anterior narrower than posterior, both ends extended below; LV overlaps RV along dorsal slopes and venter, moderately convex; each valve bears a low ridgelike elevation slightly posterior and ventral to middle; smooth or very weakly punctate; L, holotype, 0.88; H, 0.52 mm (Swain, 1986b).

Occurrence. Miocene, Humboldt Formation, Medicine Creek section T. 39 N., R. 65 E., Winecup Ranch, Elko County, Nevada, 1883-1921 feet (574-586 m) below top of section (Swain, 1986b).

Pactolocypris pactolensis Swain

Plate 29, figures 1-4

Pactolocypris pactolensis Swain, 1985, p. 426, pl. 3, figs. 1-3, 5, 6, 10-12, 15; pl. 4, fig. 7. Swain, 1990, p. 149, pl. 2, figs. 11, 13.

Elongate, subovate-subquadrate, highest medially to anteromedially; dorsum gently convex, venter nearly straight to slightly concave; anterior a little narrower than posterior, both ends extended below; LV slightly overlaps RV; moderately convex; low rounded nodelike elevation posteroventrad of middle; weakly and densely pitted medially, nearly smooth terminally; adductor muscle scar a compact group of four spots; two frontal spots; LV hinge with rabbet groove; inner lamellae narrow, very small terminal vestibules; few radial canals; L, holotype, 0.94; H, 0.60 mm (Swain, 1985).

Occurrence. Miocene, Esmeralda Formation, 2.85 miles (4.5 km) west of Black Spring, Black Spring 1:24000 Quadrangle, Nye County, Nevada, about 800 feet (244 m) above base of lake beds sequence; 7.6 miles (12.1 km) east of Finger Rock Wash, Mineral County, Nevada on road to Pactolus Mine, Goldyke 1:24000 Quadrangle, about 1000 feet (305 m) above base of lake beds sequence; Finger Rock Wash, 2 miles (3.2 km) west of Stewart Spring, Mineral County, Nevada, Stewart Spring 1:24000 Quadrangle, about 900 feet (275 m) above base of lake beds sequence (Swain, 1985). Truckee Formation, late? Miocene, 2 miles (3.2 km) WSW of Hazen, Churchill County, Nevada (Swain, 1990).

Pactolocypris suborbicularis Swain

Plate 29, figures 5-9, 12 (aff.)

Tuberocypris sp. 2, Swain, Becker and Dickinson, 1971, pl. 1, figs. 15a-c.

Pactolocypris suborbicularis Swain, 1985, p. 428, pl. 3, figs. 13; pl. 4, figs. 10, 11. Swain 1986b, p. 208, pl. 2, figs. 24, 25; pl. 3, figs. 19, 20; pl. 5, figs. 6, 7, 28.

Subovate, highest slightly anterad of midlength, venter nearly straight, posterior slightly broader on average than anterior; the latter extended below; LV overlaps RV; moderately convex; large low rounded node lies ventral and posterior to middle; smooth or with fine Bertillion markings medially and dense pustules marginally; L, holotype 1.07; H, 0.78 mm (Swain, 1985).

Occurrence. Miocene, Esmeralda Formation, 2.85 miles (4.5 km) west of Black Spring, Nye County, Nevada, Black Spring 1:24000 Quadrangle, about 800 feet (244 m) above base of lake beds sequence (Swain, 1985). Humboldt Formation, Medicine Creek section, T. 39 N., R. 65 E., Winecup Ranch, Elko County, Nevada, conglomeratic limestone at base of exposure; Gulf Oil Co., No. 1 Thousand Springs well, center SE1/4 SW1/4 sec. 8, T. 40 N., R. 66 E., Elko County, Nevada, 800-840 feet (244-256 m) (Swain, 1986b). ?Oligocene, Chadron Formation, Helmville, Nebraska (R. Konizeski, unpublished).

Pactolocypris suborbicularis attenuatus Swain

Plate 29, figure 10, 11

Pactolocypris suborbicularis attenuatus Swain, 1986b, p. 207, pl. 3, figs. 14, 15; pl. 5, fig. 28.

Relatively more elongate than suborbicularis; LV hinge rabbeted; inner lamellae narrow; adductor muscle scar an anteromedian group of four or five spots; two frontal spots; L, holotype, 0.82; H, 0.50 mm (Swain, 1986b).

Occurrence. Miocene, Humboldt Formation, Gulf Oil Co., No. 1 Thousand Springs well, center SE1/4 SW1/4 sec. 8, T. 40 N., R. 66 E., Elko County, Nevada, 1680-1700 feet (515-518 m); Cobre surface section, Elko County, Nevada, 186-196 feet (57-60 m) and 975-987 feet (297-301 m) below top of exposure (Swain, 1986b).

Genus Paracandona Hartwig, 1899

(Candonidae-Candoninae)

Paracandona euplectella (Brady and Norman)

Plate 29, figure 13

Candona euplectella Brady and Norman, 1889, p. 105, pl. 9, figs. 7, 8.

Paracandona euplectella (Brady and Norman), Hartwig, 1899, p. 309. Staplin, 1963, p. 791, pl. 94, figs. 13, 14.

Elongate, subelliptical, highest postmedially; dorsum nearly straight, venter slightly concave; anterior slightly broader than posterior; LV barely overlaps RV; finely and densely reticulate and with scattered papillae, especially in terminal and ventral areas; inner lamellae narrow, broadest anteriorly; LV hinge with groove to receive RV edge; male slightly larger than female; valves strongly convex; L, female, 0.61; H, 0.32 mm; L, male, 0.77 mm (Staplin, 1963).

Occurrence. Pleistocene, Mankato or late Cary Substage, Wisconsin Stage, western Indiana (Staplin, 1963). Recent, Holarctic (authors).

Genus Paracypria Sars, 1910

(Candonidae-Thalassocypridinae)

Paracypria? acuminata (Roth)

Plate 29, figures 18, 19

Paracypris acuminata Roth, 1933, p. 404, pl. 48, figs. 5a-c; Sohn, 1957, 1958.

Sublanceolate, highest slightly anteromedially, dorsum convex, venter slightly concave; posterior narrower than anterior, acuminate, both ends extended below; LV overlaps RV; smooth; hingement consists of simple valve contact; L, 1.08; H, 0.48 mm (Roth, 1933).

Occurrence. Lower Cretaceous, Lakota Formation, sec. 28, T. 4 N., R. 6 E., Lawrence County, South Dakota (Roth, 1933; Sohn, 1957, 1958).

Paracypria declivoidea Swain

Plate 29, figures 14,1 5

Paracypria declivoidea Swain, 1986a, p. 91, pl. 1, figs. 6-10.

Sublanceolate-subreniform, highest medially to anteromedially; dorsum moderately convex, venter concave; anterior broader than posterior, both extended below; LV overlaps RV especially midventrally, anteriorly and on dorsal slopes; smooth; LV hinge grooved to receive RV edge; inner lamellae broad anteriorly, vestibules present terminally; adductor muscle scar a median group of four spots; two anteroventral frontal spots; L, holotype, 0.60; H, 0.31 mm (Swain, 1986a).

Occurrence. Pliocene, Glenns Ferry Formation, sec. 14, T. 4 S., R. 1 W., Owyhee County, Idaho at "Big Cut" (Swain, 1986a).

Paracypria? sp., Swain and Wagoner

Plate 29, figure 16

Paracypria? sp., Swain and Wagoner, 1985, p. 139, pl. 4, fig. 9.

Elongate-sublanceolate, highest medially, dorsum convex, venter slightly concave; anterior broader than posterior, both ends extended below; LV overlaps RV except middorsally; smooth; L, 0.62; H, 0.33 mm (Swain and Wagoner, 1985).

Occurrence. Miocene, Esmeralda Formation, Blair Junction section, Esmeralda County, Nevada, sample 1C2 (Swain and Wagoner, 1985).

Genus Paracypridea Swain, 1946

(Cyprideidae)

"Paracypridea" sp., Sohn

Plate 29, figure 17

"Paracypridea" sp., Sohn, 1969, p. 137, pl. 2, figs. 17-20.

Elongate, subelliptical-subreniform, highest postmedially; dorsum nearly straight to slightly convex, venter slightly concave with weak anteroventral beak (rostrum) and notch (alveolus); anterior equal to or narrower than posterior; smooth?; broad anterior inner lamellae; adductor muscle scar a submedian group of spots (after Sohn's illustration, 1979); L, 2.12; H, 1.21 mm.

Occurrence. Lower Cretaceous?, Aptian?, Newark Canyon Formation?, W1/2 NW1/4 sec. 21, T. 29 N., R. 52 E., Eureka County, Nevada (Sohn, 1979).

Genus Paracypridopsis Kaufmann, 1900

(Cypridopsidae-Cypridopsinae)

Paracypridopsis? curta Swain

Plate 29, figures 20-22

Paracypridopsis? curta Swain, 1986b, p. 209, pl. 4, figs. 2, 3; pl. 5, fig. 29.

Elongate-trapezoidal to subtriangular, highest about two-fifths of length from anterior end; dorsum moderately convex, truncate posteriorly, venter nearly straight; posterior narrower than anterior, strongly extended below; LV overlaps RV, but very little terminally; relatively compressed; smooth; L, 0.52; H, 0.26 mm (Swain, 1986b).

Occurrence. Miocene, Humboldt Formation, Gulf Oil Co., No. 1 Thousand Springs well, center SE1/4 SW1/4 sec. 8, T. 40 N., R. 66 E., Elko County, Nevada, 15-2475 feet (4.6-755 m); Gulf Oil Co., No. 2 Wilkins Ranch well, sec. 9, T. 39 N., R. 69 E., Montello area, Elko County, Nevada, 3130-3160 feet (954-963 m) and below; Cobre surface section, Elko County, Nevada 921-1011 feet (281-308 m) above base of exposure; in conglomerate limestone at base of Medicine Creek surface section, Winecup Ranch, Elko County, Nevada, just above Triassic. Oligocene or Miocene, Indian Well or Humboldt Formation, Palisade section, north end of Pine Valley, two miles (3.2 km) south of Palisade, Eureka County, Nevada (Swain, 1986b).

Paracypridopsis? fouchi Swain

Plate 29, figures 23-36

Paracypridopsis? fouchi Swain, 1987b, p. 224, pl. 1, fig. 8; pl. 3, figs. 3-5, 15-17, 19.

Elongate subpyriform, highest two-fifths of length from anterior end; dorsum convex, truncated fore and aft of position of greatest height; posterior much narrower than anterior, subacuminate, truncate above, strongly extended below; LV overlaps RV; compressed; smooth except for five small nodes on anteroventral and postventral surfaces, each of which has a normal pore in its crest; LV hinge with furrow to receive edge of RV; adductor muscle scar a submedian group of four or five spots, obliquely oriented; two anteroventral frontal spots; L, holotype, 1.15; H, 0.60 mm (Swain, 1987b).

Occurrence. Eocene, Twin Bridges Limestone, center sec. 36, T. 32 N., R. 55 E., Elko County, Nevada; Sheep Pass Formation, Johnson Canyon section, Grant Range, Nye County, Nevada. Paleocene or Eocene, Sheep Pass Formation, Sheep Pass Canyon section, Egan Range, Eureka County, Nevada, lower carbonate unit (Swain, 1987b).

Paracypridopsis sp. (Lister)

Plate 8, figures 48, 49

Cypridopsis sp., Lister, 1975, p. 16, pl. 3, figs. 9, 10, TF 15.

Elongate, subtrapezoidal-subreniform, highest anteromedially; dorsum straightened postmedially; both dorsal slopes truncated; posterior acuminate, anterior rounded; both ends extended below; LV larger than RV; smooth; inner lamellae very broad terminally; radial canals numerous; adductor scar of three large and three small spots; two frontal spots; L, 0.66; H, 0.32 mm (Lister, 1975). Elongate subtrapezoidal outline fits Paracypridopsis better than Cypridopsis.

Occurrence. Pleistocene, Lake Bonneville Group, Saltair core hole, SE1/4 sec. 25, T. 1 N., R. 3 W., and Sec. 28 core hole, SE1/4 SW1/4 sec. 28, T. 1 N., R. 2 W., Salt Lake County, Utah (Lister, 1975).

Paracypridopsis sp., Swain

Plate 29, figure 29

Paracypridopsis sp., Swain, 1987c, p. 430, pl. 2, fig. 22; pl. 4, figs. 1-4.

Elongate, subelliptical-subreniform, highest slightly anteromedially; dorsum convex, venter concave; posterior narrower than anterior, both ends extended below; LV overlaps RV most along anterior and venter; compressed; smooth; LV hinge furrowed to receive RV edge; inncr lamellae broadest anteriorly; L, 0.88; H, 0.43 mm (Swain, 1987c).

Occurrence. Miocene, Horse Camp Formation, sec. 12, T. 10 N., R. 58 E., Nye County, Nevada, about 3.8 miles (6.24 km) east of Currant (Swain, 1987c).

Genus Paracypris Sars, 1866

(Candonidae-Paracypridinae)

Paracypris acuminatus Roth, 1933

(see Paracypria acuminata (Roth))

Genus Pelocypris Klie, 1939

Pseudoilyocypris Ferguson, 1967

(Cyprididae-Pelocypridinae)

Pelocypris tuberculatus (Ferguson)

Plate 29, figures 30, 31

Pseudoilyocypris tuberculatum Ferguson, 1967, p. 244.

Pelocypris tuberculatus (Ferguson), Lister, 1975, p. 21, pl. 2, figs. 9, 10, TF 23.

Subquadrate, highest anteromedially; dorsum nearly straight to slightly sinuous, venter slightly concave; anterior broader than posterior; median sulcus shallow, reaching to midheight; densely punctate; tubercles on each side of sulcus; midventral tubercle; free margins denticulate; valves subequal?; inner lamellae broadest anteriorly; vestibules terminally; numerous radial canals; adductor muscle scar a median group of five spots; two frontal spots; L, 1.93; H, 1.12 mm (Lister, 1975).

Occurrence. Pleistocene, Lake Bonneville Group, Saltair core hole, SE1/4 sec. 25, T. 1 N., R. 3 W.; Sec. 28 core hole, SE1/4 SW1/4 sec. 28, T. 1 N., R. 2 W., Salt Lake County, Utah (Lister, 1975). Recent, Texas and New Mexico lakes (Ferguson, 1967).

Genus Petrobrasia Krommelbein, 1965

(Cyprididae?)

Petrobrasia? sp., Sohn

Plate 29, figures 32, 33

Petrobrasia? sp., Sohn, 1969, p. B7, pl. 2, figs. 11, 12, 16.

Subtriangular, highest medially or anteromedially; dorsum convex, venter nearly straight; anterior a little narrower than posterior; adductor muscle scar a median group of four spots; two frontal spots; two antennal? spots dorsal to main group; L, 1.31+; H, 0.87+ mm (Sohn, 1979).

Occurrence. Lower Cretaceous?, late Aptian?, Newark Canyon Formation, W1/2 NW1/4 sec. 21, T. 20 N., R. 53 E., Elko County, Nevada (Sohn, 1969).

Genus Phacocypris Cai and Huang, 1978

(Cyprididae-Cypridinae)

Phacocypris? cf. pisiformis Hou and Shan

Plate 31, figures 18, 19

Phacocypris? cf. pisiformis Hou and Shan, 1978 in Hou et al., 1988, Swain, 1985, p. 422, pl. 5, figs 19, 10.

Elongate-ovate, highest medially; dorsum convex, venter nearly straight to slightly convex; ends nearly equally rounded, extended below; LV overlaps RV; moderately convex; smooth; L, 0.95; H, 0.58 mm (Swain, 1985).

Occurrence. Miocene, Esmeralda Formation, lower shale sample, Finger Rock Wash, 0.5 mile (0.8 km) south of Rawhide Ranch, Granny Goose 1:24000 Quadrangle, 1400-1500 feet (425-455 m) above base of lake bed sequence, Mineral County, Nevada; Black Spring Road section, 2.85 miles (4.5 km) west of Black Spring, Nye County, Nevada, Black Spring 1:24000 Quadrangle (Swain, 1985).

Phacocypris? sp. aff. P. tongbinensis Hou and Shan

Plate 31, figures 20, 21

Phacocypris? sp. aff. P. tongbinensis Hou and Shan, 1978 in Hou et al., 1988, Swain, 1985, p. 422, pl. 5, figs. 1, 2, 21, 22.

Elongate, subovate-subreniform, highest medially to postmedially; dorsum convex, venter slightly concave; posterior slightly broader than anterior, both ends extended below; LV overlaps RV most along anterior and ventral margins; moderately convex; smooth; L, 0.87; H, 0.50 mm (Swain, 1985).

Occurrence. Miocene, Esmeralda Formation, Finger Rock Wash, 0.5 mile (0.8 km) south of Rawhide Ranch, Granny Goose 1:24000 Quadrangle, Mineral County, Nevada, 1400-1500 feet (425-455 m) above base of lake bed sequence; 2 miles (3.2 km) west of Stewart Spring, Stewart Spring 1:24000 Quadrangle, Mineral County, Nevada, 1400-1500 feet (425-455 m) above base of lake bed sequence; 2 miles (3.2 km) west of Stewart Spring, Stewart Spring 1:24000 Quadrangle, Mineral County, Nevada, about 900 feet (275 m) above base of lake bed sequence; 7.6 miles (12.1 km) east of Finger Rock Wash on road to Pactolus Mine, Goldyke 1:24000 Quadrangle, Mineral County, Nevada, about 1000 feet (305 m) above base of lake bed sequence (Swain, 1985).

Genus Physocypria Vavra, 1897

(Cypridopsidae-Cypridopsinae)

?Physocypria fadeewi Dubowsky

Plate 29, figures 34, 35

Cypria (Physocypria) fadeewi Dubowsky, 1927, p. 84, TF 10-18.

Physocypria fadeewi Dubowsky, Staplin, 1963, p. 1178, pl. 159, figs. 6, 7.

Suboval, highest medially; dorsum strongly convex, venter gently convex; anterior narrower than posterior; LV overlaps RV except along dorsum; RV with a row of tubercles ventroterminally and terminally; LV hinge edge fits into RV flange groove terminally; medially hinge edge recessed; inner lamellae broadest anteriorly; smooth or with scattered pits; L, 0.55 to 0.65; H, 0.40 to 0.45 mm (Staplin, 1963).

Occurrence. Pleistocene, latest Mankato Substage, Wisconsin Stage, Chicago, Illinois region (Staplin, 1963). Recent, Russia, Florida (Dubowsky, 1927; Furtos, 1933).

Physocypria pustulosa (Sharpe)

Plate 30, figures 1-4

Cypria pustulosa Sharpe, 1897, p. 461, pl. 48, figs. 6-10.

Cypria (Physocypria) pustulosa Sharpe, Sharpe, 1918, p. 821, TF 1284a-c.

Physocypria pustulosa (Sharpe), G. W. Müller, 1912, p. 134; Staplin, 1963, p. 1178, pl. 159, figs. 1-5; Lister, 1975, p. 12, pl. 3, figs. 5, 6, TF 11. Cameron and Lundin, 1977, p. 349, pl. 2, figs. 17, 18.

Physocypria globula Furtos, 1933,p. 468, pl. 16, figs. 1-9 (fide Staplin, 1963).

Small, subovate to subtriangular, highest postmedially; dorsum moderately to strongly convex, venter straight to slightly convex; anterior narrower than posterior, extended below; LV overlaps RV strongly along dorsum and venter; RV with row of marginal tubercles at terminal and rarely, ventral margins; LV hinge edge fits under RV terminally and is depressed medially; smooth or with scattered pits; six adductor muscle scar spots; two frontal spots; inner lamellae and vestibules broadest anteriorly; L, 0.50 to 0.65; H, 0.33 to 0.50 mm (Staplin, 1963).

Occurrence. Pleistocene, Yarmouth Stage to Recent of Illinois and Kansas; late Kansan Sappa Silt of Kansas and Nebraska (Staplin, 1963). Lake Bonneville Group, Saltair core hole SE1/4 sec. 25, T. 1 N., R. 3 W., Sec. 28 core hole, SE1/4 SW1/4 sec. 28, T. 1 N., R. 2 W., Salt Lake County, Utah (Lister, 1975). Late Pleistocene, Lake Cochise deposits, Willcox Playa, Cochise County, Arizona (Cameron and Lundin, 1977). Recent, widespread in the USA (authors).

Genus Pontoniella Mandelstam, 1960

(Cyprididae or Candonidae-Disopontocypridinae)

Pontonella Mandelstam, 1956

Pontoniella catlinensis Swain

Plate 29, figures 27, 28; Plate 30, figures 5-7

Pontoniella sp., Swain, Becker and Dickinson, 1969, p. 481, pl. 4, figs. 12 a, b.

Pontoniella? catlinensis Swain, 1987b, p. 227, pl. 2, figs. 14-16, 20-23.

Subtriangular-sublanceolate, highest anteromedially; dorsum strongly convex, straightened medially, venter nearly straight to slightly concave; posterior margin acuminate, strongly produced below; RV slightly larger than LV; compressed; smooth; L, holotype 0.90; H, 0.50 mm (Swain, 1987).

Occurrence. Eocene and/or Oligocene, Elko Formation, oil shale just above old Catlin Works, southeast edge of Elko, Nevada; Eocene, Twin Bridges Limestone, near center sec. 36, T. 32 N., R. 55 E., Elko County, Nevada (Swain, 1987).

Pontoniella? sp., Swain and Wagoner
Plate 30, figure 8
Pontoniella? sp., Swain and Wagoner, 1985, p. 139, pl. 5, fig. 9.

Very small, sublanceolate, highest about one-third of length from anterior end; dorsum moderately convex, venter gently convex; posterior much narrower than anterior; LV slightly larger than RV; weakly ornamented by coarse reticulating ridges; L, 0.40; H, 0.23 mm (Swain and Wagoner, 1985).

Occurrence. Miocene, Esmeralda Formation, Blair Junction section, Esmeralda County, Nevada, sample 1o (Swain and Wagoner, 1985).

Pontoniella sp., Swain
Plate 30, figure 9
Pontoniella sp., Swain, 1985, pl. 4, fig. 21.

Sublanceolate, highest three-eighths of length from anterior end; dorsum strongly convex, venter concave; posterior narrower than anterior; both ends extended below; LV overlaps RV; moderately convex; original specimens furrowed by filamentous algae on surface; L, 0.75; H, 0.40 mm (Swain, 1985).

Occurrence. Miocene, Esmeralda Formation, Black Spring road section, 2.85 miles (4.5 km) west of Black Spring, Nye County, Nevada, Black Spring 1:24000 Quadrangle, about 800 feet (244 m) above base of lake bed sequence; 3.3 miles (5.3 km) east of Finger Rock Wash, Simon Mine area, Mineral County, Nevada, Stewart Spring 1:24000 Quadrangle, in lower 100-150 feet (30-45 m) of lake bed sequence (Swain, 1985).

Pontoniella sp., Swain
Plate 30, figure 10
Pontoniella sp., Swain, 1986a, p. 92, pl. 1, fig. 11.

Elongate subtriangular, highest about two-fifths of length from anterior end; dorsum moderately convex, truncate on either side of point of greatest height; venter concave, posterior narrower than anterior, both ends extended below; LV slightly larger than RV; compressed; smooth; RV hinge edge fits into furrow in LV; inner lamellae of moderate width anteriorly; vestibules terminally; radial canals few and short; L, RV, 0.58; H, 0.32 m (Swain, 1986a).

Occurrence. Pliocene, Glenns Ferry Formation, sec. 14, T. 4 S., R. 1 W., Owyhee County, Idaho (Swain, 1986a).

Pontoniella sp., Swain
Plate 30, figure 11
Pontoniella sp., Swain, 1986b, p. 210, pl. 3, fig. 21.

Subpyriform-acuminate, highest anteromedially; dorsum moderately convex, venter slightly concave; anterior much broader than posterior, the latter strongly extended below; LV slightly larger than RV; compressed; smooth; L, 0.75; H, 0.43 mm (Swain, 1986b).

Occurrence. Miocene, Humboldt Formation, Gulf Oil Co., No. 2 Wilkins Ranch well, sec. 19, T. 39 N., R. 69 E., Elko County, Nevada in Montello area, 3130-3160 feet (954-963 m) and below (Swain, 1986b).

Genus Potamocypris Brady, 1870
(Cypridopsidae-Cypridopsinae)

"Potamocypris" affinis Jones
Plate 31, figures 22, 23
Potamocypris affinis Jones, 1893, p. 387, pl. 15, figs. 8a, b

Suboblong, highest anteriorly; dorsum nearly straight to slightly convex; venter slightly concave; anterior broader than posterior; both ends subtruncate; valves compressed; smooth; L, 0.7; H, 0.4 mm (after Jones, 1893). Generic status uncertain.

Occurrence. Lower Cretaceous, Lower Bear River Formation or Gannett Group, Cokeville, Wyoming (Jone, 1893).

Potamocypris cf. comosa Furtos
Plate 30, figure 16
Potamocypris comosa Furtos, 1933, p. 435, pl. 7, figs. 11-20.
Potamocypris cf. comosa Furtos, Swain, 1987a, p. 368, pl. 1, fig. 1.

Subtriangular, highest medially; dorsum strongly convex, venter gently concave; anterior narrower than posterior; compressed marginal flange anteroventrally, both ends extended below; LV overlaps RV; compressed; smooth; L, 0.61; H, 0.36 mm (LV) (Swain, 1987).

Occurrence. Pliocene, Cache Valley Formation, Hoopes Creek locality, NE$^1/_4$ SE$^1/_4$ sec. 10, R. 12 S., R. 40 E., Franklin County, Idaho (Swain, 1987a).

Potamocypris sp. aff. P. fulva Brady, Swain, 1964b
(see Advenocypris? concinna Swain)

Potamocypris granulosa Daday
Plate 30, figure 17
Potamocypris granulosa Daday, 1902,p. 293. Lister, 1975, p. 16, pl. 3, figs. 11, 12, TF 16. Cameron and Lundin, 1977, p. 338, pl. 1, figs. 3, 4.

Elongate-reniform, highest anteromedially; dorsum convex, venter concave; anterior a little broader than posterior; LV longer than RV, RV higher than LV; granular-pitted surface; inner lamellae narrow, broadest anteriorly; L, 0.53 to 0.58; H, 0.33 to 0.34 mm (Lister, 1975).

Occurrence. Pleistocene, Lake Bonneville Group, Saltair core hole, SE1/4 sec. 25, T. 1 N., T. 3 W.; Sec. 28 core hole, SE1/4 SW1/4 sec. 28, T. 1 N., R. 2 W., Salt Lake County, Utah (Lister, 1975). Late Pleistocene, Lake Cochise deposits, Willcox Playa, Cochise County, Arizona (Cameron and Lundin, 1977). Holocene, Europe, Canada (authors).

Potamocypris minuta Staplin
Plate 30, figures 18, 19
Potamocypris minuta Staplin, 1963, p. 1184, pl. 159, figs. 47, 48.

Small, subtriangular, highest two-fifths of length from anterior end; dorsum convex, venter concave, anterior broader than posterior; LV longer than RV but RV overlaps LV dorsally and ventrally; vestibules present; radial canals short and numerous; LV edge fits into groove on RV; adductor muscle scar a group of five or six spots; two frontal scars; L, LV, 0.44 to 0.59; H, 0.26 to 0.37 m; L, RV 0.43 to 0.52; H, 0.28 to 0.30 mm (Staplin, 1963).

Occurrence. Pleistocene, Mankato or late Cary Substage, Wisconsin Stage of southeastern Illinois; Mankato Substage of northwestern Illinois (Staplin, 1963).

Potamocypris? moorei Swain
Plate 30, figures 20-22
Potamocypris? moorei Swain, 1987c, p. 430, pl. 2, figs. 6, 23, 24; pl. 4, figs. 5, 6.

Subtriangular, highest submedially; dorsum strongly convex, venter nearly straight to slightly convex; ends narrowly rounded and extended below, the posterior slightly the narrower; LV overlaps RV most strongly along dorsum and venter; dorsal border of LV forms a broad rimlike expansion; smooth, L, holotype, 0.88; H, 0.58 mm (Swain, 1987c). Dorsal LV overlap unlike typical Potamocypris.

Occurrence. Miocene, Horse Camp Formation, NW1/4 sec. 2, T. 10 N., R. 58 E., Nye County, Nevada, near Currant (Swain, 1987c).

Potamocypris pallida Alm

Plate 30, figure 23

Potamocypris pallida Alm, 1914, p. 470. Lister, 1975, p. 17, pl. 2, figs. 17, 18, TF 17.

Elongate-reniform, highest anteromedially; dorsum convex, venter concave; anterior a little broader than posterior, both ends extended below; compressed; smooth; inner lamellae broadest posteriorly (atypical); vestibules occur terminally; L, LV 0.63 to 0.64; H, 0.32 to 0.33 mm (Lister, 1975).

Occurrence. Pleistocene, Lake Bonneville Group, Saltair core hole, SE1/4 sec. 25, T. 1 N., R. 3 W.; Sec. 28 core hole, SE1/4 SW1/4 sec. 28, T. 1 N., R. 2 W., Salt Lake County Utah, (Lister, 1975).

Potamocypris? pellucidus Swain

Plate 30, figures 24, 25

Potamocypris? pellucidus Swain, 1963, p. 800, pl. 95, figs. 1a, b, TF 2A.

Subtriangular, highest medially; dorsum strongly convex, venter nearly straight; anterior slightly broader than posterior, both ends extended below; RV overlaps LV dorsally; valves compressed; smooth; RV hinge a rabbet groove to receive edge of LV; inner lamellae narrow; adductor muscle scar a median group of four spots; two frontal spots; L, holotype RV 0.65; H, 0.46 mm; L, LV, 0.67; H, 0.43 mm (Swain, 1963).

Occurrence. Pleistocene, Gubik Formation, freshwater facies, Alaska, Ikpikpuk Lake Shot Holes, line 12, shot points 6 and 15, depth 30-40 feet (9-12 m); line 10, shot point 20, depth 0-10 feet (0.3 m) (Swain, 1963).

Potamocypris smaragdina (Vavra)

Plate 30, figures 26-29

Cypridopsis smaragdina Vavra, 1891, p. 80, TF 26.

Potamocypris smaragdina (Vavra), Daday, 1900. Gutentag and Benson, 1962, p. 29, pl. 1, figs. 4, 5, TF 7. Staplin, 1963, p. 1185, pl. 159, figs. 40-43, 46, 49. Lister, 1975, p. 17, TF 18.

Potamocypris smaragdina compressa Furtos, 1933, p. 435, pl. 6, figs. 10-14.

Females elongate subtriangular, highest anteromedially; dorsum arched, venter concave; posterior narrower and more pointed than anterior; LV flanged and extends beyond RV terminally; RV overlaps LV dorsally and ventrally; punctate, normal pore canals numerous; RV hinge grooved to receive LV edge; inner lamellae broad anteriorly, directed outward postventrally in LV; L, LV, 0.55 to 0.79; H, 0.30 to 0.37 m; L, RV, 0.50 to 0.56; H, 0.30 to 0.40 mm (Staplin, 1963).

Occurrence. Pleistocene, late Kansan Stage to Wisconsin Stage, Mead and Reno Counties, Kansas, and Beaver County, Oklahoma (Gutentag and Benson, 1962). Illinoian Stage and late Mankato Substage, Wisconsin Stage of Illinois; Sappa Silt (late Kansan Stage of Kansas (Staplin, 1963). Recent of Europe, North America from Mexico to Canada (authors).

Potamocypris unicaudata Schafer
Plate 30, figure 30
Potamocypris unicaudata Schafer, 1943, p. 212. Lister, 1975,p. 18, pl. 2, fig. 16, TF 19.

Subtriangular-subreniform, highest anteromedially; dorsum broadly arched, venter concave; posterior narrower and more pointed than anterior; RV overlaps LV dorsally and ventroterminally; smooth; inner lamellae of moderate width, vestibules terminally; L, 0.76; H, 0.42 mm (Lister, 1975).

Occurrence. Pleistocene, Lake Bonneville Group, Saltair core hole, SE$^{1}/_{4}$ sec. 25, T. 1 N., R. 2 W., Salt Lake County, Utah (Lister, 1975).

"Potamocypris" unisulcatus Jones
Plate 31, figures 24, 25
Potamocypris unisulcatus Jones, 1893, p. 387, pl. 15, figs. 10a, b

Elongate subpyriform, highest anteriorly; dorsum and venter nearly straight, converging posteriorly; transverse sulcus medially; valves compressed; smooth; L, 1.05; 0.65 mm (after Jones, 1893).

Occurrance. Upper Cretaceous, Montana Formation, Coalville, Utah (Jones, 1893); Osborn Series, Isle of Wight, England.

Potamocypris variegata (Brady and Norman)
Plate 30, figures 31, 32
Cypridopsis variegata Brady and Norman, 1889, p. 91, pl. 8, figs. 20, 21.
Candonella variegata (Brady and Norman), Vavra, 1898, p. 12.
Potamocypris variegata (Brady and Norman), Daday, 1900, p. 193. Staplin, 1963, p. 1185, pl. 59, figs. 44, 45.

Elongate-subreniform, thick shelled, highest anteromedially; dorsum convex, angulated in LV, venter concave; anterior broader and less pointed than posterior; LV overlaps RV and is flanged terminally; RV overlaps LV dorsally and ventrally; inner lamellae narrower in RV than in LV; vestibules present; pitted; RV hinge with groove for reception of LV edge; adductor muscle scar of five or six spots; two anteroventral mandible adjustor spots and two anterodorsal antennal scars; L, RV, 0.49 to 0.58; H, 0.30 to 0.36 mm (Staplin, 1963).

Occurrence. Pleistocene, Mankato Substage, Wisconsin Stage of Illinois; Tazewell Substage?, Wisconsin Stage of Ontario. Recent of Europe, Iran, Ohio (Staplin, 1963).

Potamocypris? williamsi Swain
Plate 30, figure 12-15
Potamocypris sp., Swain, 1956, p. 136, pl. 3, figs. 9, 10.
Potamocypris williamsi Swain, 1964a, p. 273, pl. 43, figs. 1a-h.

Subquadrate-subreniform, highest postmedially; dorsum straightened medially, venter nearly straight to concave; anterior slightly narrower than posterior; LV larger than RV and overlaps it strongly dorsally and ventrally (atypical of Potamocypris); not strongly convex; anterior border compressed; smooth; LV hinge furrowed to receive edge of RV; inner lamellae narrow; L, holotype, 1.06; H, 0.66 mm (Swain, 1964).

Occurrence. Eocene, Wasatch-Green River transition beds along northern side of Uinta Basin, Utah, and central Piceance Basin, Colorado; lower Green River Formation, Fossil, Lincoln County, Wyoming (elongate variety) (see plate description) (Swain, 1964a).

Potamocypris? sp., Swain and Brown
Plate 30, figure 33
Potamocypris? sp., Swain and Brown, 1972, p. 15, pl. 1, fig. 15.

Subtriangular, highest submedially; dorsum convex, venter nearly straight to slightly concave; posterior slightly narrower than anterior, subacuminate; both ends strongly extended below; LV overlaps RV dorsally (atypical of Potamocypris) and ventrally; smooth?; L, 0.65; H, 0.34 mm (Swain and Brown, 1972).

Occurrence. Upper Jurassic or (probably) Lower Cretaceous, Unit H, NC-DA-OT-12 well, Dare County, North Carolina, 7836-7846 feet (2388-2391 m) (Swain and Brown, 1972).

Potamocypris sp., Lister
Plate 30, figure 34
Potamocypris sp., Lister, 1975, p. 18, pl. 3, figs. 13, 14, TF 20.

Subtriangular-subreniform, highest anteromedially; dorsum convex, venter concave; posterior narrower than anterior; minutely and irregularly pitted; inner lamellae narrow, may be immature forms; L, 0.58 to 0.60; H, 0.33 mm (Lister, 1975).

Occurrence. Pleistocene, Lake Bonneville Group, Saltair core hole, SE1/4 sec. 25, T. 1 N., R. 3 W.; Sec. 28 core hole, SE1/4 SW1/4 sec. 28, T. 1 N. R. 2 W., Salt Lake County, Utah (Lister, 1975).

Potamocypris n. sp. and var. Swain, 1956
(see Advenocypris umbonata Swain)

Genus Prionocypris Brady and Norman, 1896
(Cyprididae-Eucypridinae)
Prionocypris sp., aff. P. glacialis (Sars)
Plate 30, figure 35
Prionocypris sp. aff. P. glacialis (Sars), Dickinson and Swain, 1967, p. 342, pl. 36, fig. 18.

Large, elongate subelliptical, highest one third of length from anterior end; dorsum convex, venter nearly straight to slightly sinuous; posterior narrower than anterior; valves subequal?, moderately convex; smooth; LV hinge with slight groove; inner lamellae narrow posteriorly; L, 1.73; H, 1.0 mm (Dickinson and Swain, 1967).

Occurrence. Pliocene, Hay Ranch Formation (=Upper Humboldt), roadcut one mile (1.6 km) south of Elko, Nevada on State Highway 46, sec. 19, T. 34 N., R. 55 E., Elko County, Nevada (Dickinson and Swain, 1967).

Prionocypris pigra (Fischer)
Plate 30, figure 36
Cypris pigra Fischer, 1851, p. 158, pl. 9, figs. 11-16.
Cypris tumefacta Brady and Robertson, 1870, p. 13, pl. 14, figs. 4-6.
Erpetocypris tumefacta (Brady and Robertson), Brady and Norman, 1889, p. 87, pl. 8, figs. 5-7; pl. 13, fig. 18.
Herpetocypris tumefacta (Brady and Robertson), Sars, 1891, p. 62.
Prionocypris tumefacta (Brady and Robertson), Kaufmann, 1900, p. 284, pl. 16, figs. 13-15; pl. 21, figs. 9-13.
Eucypris pigra (Fischer), Müller, 1912, p. 173.
Prionocypris pigra (Fischer), Sars, 1928, p. 132, pl. 51, fig. 2. Swain, 1963, p. 801, pl. 96, figs. 3, 7, 8a, b, TF 2b.

Elongate, subreniform, highest medially; dorsum convex, venter slightly concave; posterior narrower than anterior, both ends extended below; LV overlaps RV on free margins but middorsally RV overlaps LV; moderately convex; smooth; LV hinge? a groove to receive edge of RV?; inner lamellae of moderate width, vestibules terminally; about 25 radial canals anteriorly; adductor muscle scar a median group of five spots; L, 0.76; H, 0.43 mm (Swain, 1963).

Occurrence. Pleistocene, Gubik Formation, non-marine facies, Arctic coastal plain, Alaska, Ikpikpuk Lake shot holes, line 15-47, shot point 14 at 40-50 feet (12-15 m); Party 47 shot holes, line 8, shot point 10 at 40-50 feet (12-15 m) line 10, shot point 31 at 40-50 feet (12-15 m) (Swain, 1963). Recent, Europe (authors).

Prionocypris tumefacta (Brady and Robertson), Kaufmann, 1900
(see Prionocypris pigra (Fischer))

Genus Procyprois Swain, 1964
(Cyprididae-Notodromadinae?)
Procyprois? gracilis Swain
Plate 30, figure 37
Procyprois? gracilis Swain, 1985, p. 431, pl. 1, fig. 4.

Subtriangular-subovate, highest slightly postmedially; dorsum moderately convex, venter nearly straight to slightly concave; anterior narrower than posterior, both ends extended below; LV overlaps RV except anteriorly; compressed; smooth; L, holotype, 0.92; H, 0.61 mm (Swain, 1985).

Occurrence. Miocene, Esmeralda Formation, Black Spring road, 2.85 miles (4.5 km) west of Black Spring, Nye County, Nevada, about 800 feet (244 m) above base of lake beds sequence; 3.3 miles (5.3 km) east of Finger Rock Wash on road to Pactolus Mine, Mineral County, Nevada, Goldyke 1:24000 Quadrangle, estimated 100-150 feet (30-45 m) above base of lake bed sequence (Swain, 1985).

Procyprois ravenridgensis Swain
Plate 30, figure 38
Cyprois n., sp., Swain, 1956, p. 135, figs. 5a-j.
Procyprois ravenridgensis Swain, 1964a, p. 271, pl. 42, figs. 3a-j.

Subtriangular-subreniform, highest postmedially; dorsum strongly arched, venter slightly concave; anterior narrower than posterior, both ends extended below; LV overlaps RV except middorsally; convexity low to moderate; smooth to very finely pitted; females more convex posteriorly than males; LV hinge grooved to receive RV edge; inner lamellae broadest anteriorly but generally narrow; vestibules lie anteriorly; adductor muscle scar a median group of several spots; L, holotype, 0.99; H, 0.63 mm (Swain, 1964).

Occurrence. Eocene, upper Wasatch/Colton Formations and Colton-Green River transition beds, Uinta and Piceance Creek Basins, Utah and Colorado (Swain, 1964).

Procyprois twinbridgensis Swain

Plate 31, figures 1-4

Procyprois twinbridgensis Swain, 1987b, p. 226, pl. 1, figs. 1-3, 5-7; pl. 2, figs. 11a, b.

Thin and fragile, subovate, highest medially to postmedially; dorsum moderately convex, venter nearly straight to convex; posterior broader than anterior, both ends extended below; LV overlaps RV most along dorsum; compressed; fine polygonal surface ornamentation, coarser medially; scattered nodes terminally; adductor muscle scar a median group of four spots; two frontal spots; L, holotype, 1.03; H, 0.70 mm (Swain, 1987b).

Occurrence. Eocene, Twin Bridges Limestone, west side Huntingdon Creek, center sec. 36, T. 32 N., R. 55 E., Twin Bridges, Elko County, Nevada (Swain, 1987b).

Procyprois? sp., Swain

Plate 31, figures 5, 6

Procyprois? sp., Swain, 1985, p. 431, pl. 4, figs. 8, 9.

Elongate, subovate, highest about one-third of length from posterior end; dorsum gently convex, venter nearly straight; anterior broader than posterior, both ends extended below; LV overlaps RV most along venter; compressed; smooth; L, 0.80; H, 0.53 (Swain, 1985).

Occurrence. Miocene, Esmeralda Formation, Black Spring road, 2.85 miles (4.5 km) west of Black Spring, Black Spring 1:24000 Quadrangle, Nye County, Nevada, about 800 feet (244 m) above base of lake bed sequence (Swain, 1985).

Procyprois? sp., Swain

Plate 31, figure 7

Procyprois? sp., Swain, 1986b, p. 210, pl. 2, fig. 26.

Subtriangular-subovate, highest submedially; dorsum strongly convex, venter nearly straight; anterior narrower than posterior, both ends extended below; LV overlaps RV most along venter and anterodorsal slopes; compressed; smooth; L, 0.96; H, 0.63 mm (Swain, 1986b).

Occurrence. Miocene, Humboldt Formation, lower part, Gulf Oil Co., No. 1 Thousand Springs well, center SE$^{1}/_{4}$ SW$^{1}/_{4}$ sec. 8, T. 40 N., R. 66 E., Elko County, Nevada, 1850-1860 feet (564-567 m) and 2400-2450 feet (732-737 m) (Swain, 1986b).

Genus Pruvostina Scott and Summerson, 1943

(Carbonitidae by Sohn, 1985)

(Darwinuloidacea-Darwinuloididae by Molostovskaja, 1990)

Pruvostina wanlessi Scott and Summerson

Plate 31, figures 8-12

Pruvostina wanlessi Scott and Summerson, 1943, p. 671, pl. 1, figs. 1-5.

Subovate-subquadrate, highest postmedially; dorsum convex, may be straightened medially, venter nearly straight; anterior narrower than posterior; RV typically overlaps LV, most so along venter, but reversal of overlap has been noted; finely reticulate surface; hinge one-third of shell length, deeply impressed; L, 1.20; H, 0.75 mm (Scott and Summerson, 1943).

Occurrence. Lower Pennsylvanian, Scott Formation, Kanawha Series, old WPA (Work Projects Administration) road northwest of Newcomb, $^{1}/_{4}$ mile SE of road pass between Burnt Cove and Hatfield Branches, Williamsburg Quadrangle, Campbell County, Tennessee (Scott and Summerson, 1943).

Pruvostina? sp. 1, Sohn

Plate 31, figures 13, 14

Pruvostina? sp. 1, Sohn, 1985, p. 455, figs. 4.15-4.19 (not described).

Subquadrate-subovate, highest postmedially; dorsum nearly straight, venter concave; anterior narrower than posterior, extended below; RV overlaps LV along free margins; hinge impressed; smooth; L, 0.85; H, 0.60 mm (from Sohn's photographs, 1985).

Occurrence. Mississippian, Bluestone Formation, Bramwell Member, USGS coll. 1291-PC, Crumpler Quadrangle, Wyoming County, West Virginia (Sohn, 1985).

Genus Pseudoeucypris Schneider, 1957

(Cyprididae-Eucypridinae)

Pseudoeucypris pagei (Swain)

Plate 32, figures 8, 9, 11-13

Candona pagei Swain, 1949, p. 174, pl. 32, figs. 1-4, 6, 7.

Cypris pagei (Swain), Swain, 1956, p. 135, pl. 3, figs. 7a-e.
Pseudoeucypris pagei (Swain), Swain, 1964a, p. 269, pl. 42, figs. 4a-e; pl. 44, fig. 7.

Subovate-subtriangular, highest near midlength; dorsum strongly arched, venter straight to slightly convex; posterior narrower than anterior, both ends extended below; LV slightly larger than RV, overlapping along free margins; valves very strongly convex; very finely pitted in reticulate pattern; LV hinge margin with groove to receive edge of RV; inner lamellae broadest anteriorly; adductor muscle scar a median group of four or five spots; two frontal spots; L, 1.36; H, 0.83 mm (Swain, 1964).

Occurrence. Eocene, upper Wasatch or Colton Formation, Wasatch-Green River transition beds and lower Green River Formation, Uinta, Piceance Creek and southern Green River Basins, Utah, Colorado and Wyoming; also in western Green River and Fossil Basins, Wyoming; and in Wasatch-Green River interbeds, 17 miles (27 km) north of Kemmerer, Wyoming on U.S. Highway 189 (Swain, 1964).

Pseudoeucypris pagei (Swain) subsp.
Plate 32, figures 14, 15
Pseudoeucypris pagei var. Swain, 1964a, p. 269, pl. 42, figs. 7a-c; pl. 43, fig. 2.

Similar to P. pagei in outline and overlap relationships but is less convex; there is a nearly complete gradation with P. pagei; L, 1.18; H, 0.78 mm (Swain, 1964a).

Occurrence. Eocene, middle part of Wasatch-Green River transition beds, Dripping Rock Gulch section, Raven Ridge, Rio Blanco County, Colorado (Swain, 1964a).

Pseudoeucypris sp.
Plate 32, figure 16
Pseudocucypris sp., Swain, 1986b, p. 200, pl. 4, fig. 7.

Subovate, highest about two-fifths of length from anterior end; dorsum convex, subtruncate anterior to point of greatest height, venter concave; posterior narrower than anterior, both ends extended below; LV larger than RV; strongly convex; finely and densely reticulate, except in area of adductor muscle scar; L, 0.83; H, 0.49 mm (Swain, 1964).

Occurrence. Miocene, Humboldt Formation, Medicine Creek section, T. 39 N., R. 65 E., Winecup Ranch 1:24000 Quadrangle, Elko County, Nevada, 161-199 feet (49-61 m) above base of exposure (Swain, 1986a).

Genus Pseudoilyocypris Ferguson, 1967
(Ilyocyprididae)

Pseudoilyocypris tuberculatum Ferguson, 1967
(see Pelocypris tuberculatus (Ferguson)

Genus Reticulocypris Staplin, 1963
(Candonidae-Cyclocypridinae)
Reticulocypris tuberculata Staplin
Plate 32, figures 17, 18
Reticulocypris tuberculata Staplin, 1963, p. 1182, pl. 150, figs. 55, 56, 59.

Elongate, subreniform, highest anteromedially; dorsum convex, venter slightly concave; anterior broader than posterior, each end spinose, fewer at posterior end; densely reticulate; LV barely overlaps RV; RV hinge with groove to receive flange edge of LV; flange overlaps terminally not middorsally; inner lamellae broadest antero- and postventrally; adductor muscle scar not known; L, holotype, 0.48; H, 0.27 mm (Staplin, 1963).

Occurrence. Pleistocene, Mankato or late Cary Substage, Wisconsin Stage, near confluence of Wabash and Ohio Rivers (Staplin, 1963).

Genus Scabriculocypris Anderson, 1940
(Cypridacea?)
Scabriculocypris n. sp. Kietzke and Lucas
Plate 32, figures 19-21
Scabriculocypris n. sp. Kietzke and Lucas, 1995, p. 23, TF 2a-d

Elongate subrectangular, highest anteromedially; dorsum straight, venter slightly concave, converging with dorsum posteriorly; anterior broader than posterior; LV overlaps RV; weakly bisulcate?; densely reticulate, spinose in rows along dorsum and scattered over surface; L, 0.73; H, 0.35 mm (Kietzke and Lucas, 1995).

Occurrence. Lower Jurassic, Kayenta Formation, SW$^{1}/_{4}$ sec. 35, T. 27 N., R. 12 E., Coconino County, Arizona (Kietzke and Lucas, 1995).

Genus Scottia Brady and Norman, 1889
(Cyprididae-Scottinae)
Scottia sp. aff. S. browniana (Jones), Swain, 1949
(see Scottia subquadrata Swain)

Scottia? eaglespringsensis Swain
Plate 32, figure 24
Scottia? eaglespringsensis Swain, 1964a, p. 272, pl. 41, fig. 11.

Very small, elongate subquadrate-subreniform, highest medially to postmedially; dorsum slightly convex, venter concave; anterior broader than posterior, truncate below; valves subequal; smooth; L, 0.43; H, 0.25 mm (Swain, 1964).

Occurrence. Late Cretaceous or Paleocene, Sheep Pass Formation, Shell Oil Co., No. 1 Eagle Springs well, Nye County, Nevada, 7450 feet (2271 m) (Swain, 1964a).

Scottia? sp. aff. S.? eaglespringsensis Swain

Plate 32, figure 22

Scottia? sp. aff. S.? eaglespringsensis Swain, Swain, 1987b, p. 223, pl. 1, fig. 14.

Subquadrate, subtrapezoidal, highest submedially; dorsum slightly convex, straightened medially, venter gently concave; posterior narrower than anterior, both ends extended below; LV overlaps RV most along venter; compressed; finely and densely pitted; L, 0.70; H, 0.35 mm (Swain, 1987b).

Occurrence. Eocene, Twin Bridges Limestone, near center sec. 36, T. 32 N., R. 55 E., Twin Bridges, Elko County, Nevada (Swain, 1987b).

Scottia? sublanceolata Swain

Plate 32, figures 25, 26

Scottia? sublanceolata Swain, 1964a, p. 272, pl. 44, figs. 8a-b.

Very small, elongate sublanceolate-subreniform, highest medially; dorsum convex, venter concave; ends narrow, the anterior slightly the broader; valves subequal, smooth; L, holotype 0.35; H, 0.19 mm (Swain, 1964).

Occurrence. Late Cretaceous or early Tertiary, Sheep Pass Formation, Shell Oil Co., No. 1 Eagle Springs well, Nye County, Nevada, 7534-7535 feet (2296-2297 m) (Swain, 1964a).

Scottia subquadrata Swain

Plate 32, figure 23, 27

Scottia sp. aff. S. browniana (Jones), Swain, 1949, p. 176, pl. 32, fig. 19.

Scottia subquadrata Swain, 1964a, p. 272, pl. 42, fig. 5.

Subquadrate, highest medially; dorsum arched, straightened medially for about half of length, venter nearly straight; anterior slightly broader than posterior, both ends extended below; LV overlaps RV; smooth; L, holotype, 0.83; H, 0.52 mm (Swain, 1964).

Occurrence. Eocene, Green River Formation, 11 miles (18 km) north of Castlegate, Utah, lower part of formation above lowest red sandstone of the exposure (Swain, 1949).

Colton-Green River transition beds, 155 feet (47 m) above base, Dripping Rock Gulch section, Rio Blanco County, Colorado, sec. 12, T. 2 N., R. 104 W., (Swain, 1964).

Scottia sp. aff. S. subquadrata Swain

Plate 32, figure 28

Scottia sp. aff. S. subquadrata Swain, Swain, 1985, p. 422, pl. 1, fig. 6.

Subelliptical-subtrapezoidal, highest submedially; dorsum convex, venter nearly straight; posterior slightly narrower than anterior; LV overlaps RV, least anteriorly; compressed; smooth; L, 0.81; H, 0.49 mm (Swain, 1985).

Occurrence. Miocene, Esmeralda Formation, Black Spring road section, 2.85 miles (4.5 km) west of Black Spring, Black Spring 1:24000 Quadrangle, Nye County, Nevada, about 800 feet (244 m) above base of lake bed sequence (Swain, 1985).

Scottia? trapezoidalis Swain

Plate 32, figure 29

Scottia? trapezoidalis Swain, 1964a, p. 272, pl. 41, fig. 12.

Very small, subtrapezoidal, highest postmedially; dorsum slightly convex, venter concave; posterior narrower than anterior, both ends extended below, truncate above; valves subequal; L, 0.43; H, 0.25 mm (Swain, 1964).

Occurrence. Late Cretaceous or early Tertiary, Sheep Pass Formation, Shell Oil Co., No. 1 Eagle Springs well, Nye County, Nevada, 7450 feet (2271 m) (Swain, 1964a).

Genus Sollenella Abusik, 1960

(Lepertidiidae)

Sollenella? sp.

Plate 32, figure 10

Sollenella ? sp., Friedman and Lundin, 1998, p. 485, TF 5

Subovate, highest anteromedially, dorsum nearly straight, slightly sinuous, cardinal angles nearly equally obtuse; venter convex; anterior margin broader than posterior margin; anterodorsal small eye tubercle; free marginal narrow rim; median adductor muscle scar are with dorsal venous lines; weakly granulate surface; L, 4.3; H., 2.75 mm (after Friedman and Lundin, 1998).

Occurence. Middle Devonian, Gilboa Formation, East Windham, Greene County, New York (Friedman and Lundin, 1998). (1)

Genus Spirocypris Sharpe, 1903, invalid

Spirocypris tuberculatus Sharpe, 1908

(see Cypricercus tuberculatus (Sharpe))

Genus Stenocypris Sars, 1889

(Cyprididae-Herpetocypridinae)

Stenocypris? horsecampensis Swain

Plate 32, figures 30-34

Stenocypris? horsecampensis Swain, 1987c, p. 428, pl. 2, figs. 5, 7, 11, 12; pl. 3, figs. 10, 11, 15-18.

Elongate-sublanceolate to subtriangular, highest submedially; dorsum strongly convex, venter concave; posterior a little narrower than anterior, both ends extended below; LV overlaps RV most along dorsum and venter; valve flanks flattened medially in edge view; smooth; LV hinge a furrow to receive RV edge; inner lamellae broad terminally, concentrically striated anteriorly; L, holotype 0.99; H, 0.54 mm (Swain, 1987c).

Occurrence. Miocene, Horse Camp Formation, sec. 12, T. 10 N., R. 58 E., 3.8 miles (6.24 km) east of Currant, Nye County, Nevada (Swain, 1987c).

Stenocypris cf. sinuata Müller

Plate 32, figure 35

Stenocypris sinuata Müller, 1898, p. 273, pl. 17, figs. 9-13, 15-20, 23.

Stenocypris cf. sinuata Muller, Swain, 1964a

Elongate-sublanceolate, highest slightly postmedially; dorsum convex, venter slightly concave; posterior narrower and more acuminate than anterior; LV overlaps RV along dorsal slopes and venter; compressed; smooth; L, 0.90; H, 0.41 mm (Swain, 1964).

Occurrence. Eocene, Colton-Green River transition beds, Nine Mile or Minnie Maud Canyon, center east line sec. 15, T. 12 S., R. 13 E., Carbon County, Utah and eastward down Minnie Maud Creek (Swain, 1964a).

Genus Theriosynoecum Branson, 1936

(Dryelbidae)

Theriosynoecum? minnekatense (Roth)

Plate 31, figures 28-30

Jonesina minnekahtensis Roth, 1933, p. 399, pl. 48, figs. 1a-g

?Jonesina pahasapensis Roth, 1933, p. 401, pl. 48, figs. 2a-q

Theriosynoecum minnekahtensis (Roth), Van Morkhoven, 1963.

Elongate subovate, highest postmedially; dorsum slightly convex, sinuous, venter weakly convex; anterior narrower than posterior; LV slightly larger than RV; median sulcus extends venterad of midheight; short anterodorsal sulcus is oblique toward anterior; venter with weak narrow longitudinal ridges; surface with meshwork of polygonal pits; posteriorly are several small tubercles; hinge of LV a flange fitting into furrow on RV and with posterior socket that receives tooth on RV; L, 1.00; H, 0.56 mm; valves tumid, greatest width postmedian (this form may be female of T. pahasapensis (Roth, 1933).

Occurrence. Lower Cretaceous, Lakota Formation by (Sohn, 1958); Upper Jurassic, Morrison Formation by Roth (1933), sec. 28, T. 4 N., R. 6 E., Lawrence County, South Dakota (Roth, 1933).

Theriosynoecum? pahasapense (Roth)

Jonesina pahasapensis Roth, 1933, Plate 23, figures 33-36, p. 401, pl. 48, figs. 2a-g

Theriosynoecum pahasapense (Roth), Van Morkhoven, 1963.

This form seems identical to T.? minnekahtense (Roth) in all respects except the greater width of the latter. The two forms are probably male and female dimorphs as was suggested by Roth (1933) and T.? minnekahtense has page priority.

Occurrence. The form has the same occurrence as T.? minnekahtense.

Theriosynoecum wyomingense (Branson)

Plate 33, figures 1-5.

Morrisonia wyomingensis Branson, 1935, p. 521, pl. 51, figs. 17-21. (non Morrisonia Grote, 1874).

Theriosynoecum wyomingense (Branson), Branson, 1936, p. 323; Sohn, 1982.

Theriosynoecum sp., Sohn, 1958, pl. 1, figs. 19, 20.

Subquadrangular, highest submedially; dorsum and venter nearly straight and subparallel; ends of nearly equal curvature, posterior slightly broader than anterior, both ends extended below; anterior fifth compressed, bisulcate; median sulcus (S_2) extends to ventrad of midheight; anterodorsal sulcus (S_1) short and oblique; two dorsal nodes posterior to S_2, the more median node extends beyond hingeline; midventrally a short longitudinal ridge the anterior end of which turns dorsad and ends in a node; marginal ridge or ridges occur along free margins, except postdorsally; surface except in sulci, pitted; LV slightly overlaps RV; LV hinge with ridge, RV with hinge groove; L, 1.3; H, 0.9 mm (modified from Branson, 1936, who reversed valve orientation).

Occurrence. Upper Jurassic, Morrison Formation, Johnson County, Wyoming, as well as at other localities in Wyoming, Colorado, Montana and South Dakota, in Salt Wash Member or its equivalent (Sohn, 1958, 1982).

Theriosynoecum sp., Sohn, 1958
(see Theriosynoecum wyomingense (Branson))

Theriosynoecum? sp., Swain
Plate 33, figures 6, 7
Theriosynoecum? sp., Swain, 1986a, p. 103, pl. 5, figs. 2, 3.

Subovate-subquadrate, highest postmedially; dorsum sinuous, venter convex; anterior broader than posterior, extended below, whereas posterior is extended above; LV slightly larger than RV; strongly convex; narrow anterior marginal compressed zone; anteromedially in dorsal half is a shallow sulcus (S_2); anterodorsal sulcus (S_1) scarcely developed; anterior to S_2 is a nodelike elevation that projects beyond dorsal margin; midventral surface expanded as an ala in each valve; these end abruptly about one-fourth from posterior end; general surface smooth; L, 0.80; H, 0.49 mm (Swain, 1986a).

Occurrence. Pliocene, Glenns Ferry Formation, sec. 14, T. 4 S., R. 1 W., Owyhee County, Idaho, at "Big Cut" (Swain, 1986a).

Theriosynoecum? sp.
Plate 33, figure 8
Betty Kellett Nadeau, 1945, unpublished photograph.

Subquadrate, highest postmedially; dorsum and venter nearly straight, the latter slightly sinuous, converging toward anterior; posterior broader than anterior, the latter truncate above; LV slightly larger than RV; bisulcate in anterodorsal quadrant, median sulcus (S_2) reaches to midheight; anterodorsal sulcus (S_1) short and oblique; swellings of shell on either side of S_2; prominent postmedian rounded node; surface densely reticulate; dimensions unknown (after Nadeau photograph).

Occurrence. Upper Jurassic or Lower Cretaceous, Morrison Formation?, near Canyon City, Colorado? (Nadeau, unpublished).

Genus Timiriasevia Mandelstam, 1947
(Limnocytheridae-Timiraseviinae)
Timiriasevia? heylmuni Swain
Plate 33, figures 9-12, 13-16

Metacypris sp., Swain, 1964a, p. 279, pl. 44, figs. 11a-c.

Timiriasevia? heylmuni Swain, 1987b, p. 230, pl. 1, fig. 17; pl. 2, fig. 10. Swain, 1987c, p. 438, pl. 4, figs. 8, 14-17.

Subovate to subpyriform, highest anteromedially; dorsum gently convex, venter gently concave to nearly straight; anterior broader than posterior in males, narrower in females; valves subequal, inflated, females more so than males in posterior half; ornamented by 25-27 narrow, closely spaced, concentrically arranged ridges, medially sinuous, branching and anastomosing, more or less longitudinally arranged; adductor muscle scar an anteromedian? vertical row of four elongate spots, two frontal spots; hinge of left valve with terminal elongate furrows, rabbeted above and an interterminal narrow furrow, rabbeted below; L, holotype, 0.55; H, 0.27 mm (Swain, 1987b, c).

Occurrence. Eocene, Twin Bridges Limestone, near center sec. 36, T. 32 N., R. 55 E., Elko County, Nevada (Swain, 1964, 1987b, c).

Genus Trapezoidella Sohn, 1979

(Trapezoidellidae)

Trapezoidella rothi Sohn

Plate 33, figures 17-19

Trapezoidella rothi Sohn, 1979, p. 9, pl. 1, figs. 5-10; pl. 2, figs. 7-29, 33, 34-36.

Subhemicircular, highest slightly postmedially; dorsum strongly convex, venter straight to slightly concave; posterior broader than anterior; LV overlaps RV, umbonate along dorsum; LV with curved longitudinal ridge near ventral margin, and with narrow groove middorsally; RV trapezoidal in outline, with longitudinal swelling ventrad of hinge margin; smooth; L, holotype, 1.37; H, 0.97 mm (Sohn, 1979).

Occurrence. Lower Cretaceous, Lakota Formation, Chilson Member, SE1/4 sec. 11, T. 8 S., R. 3 E., Flint Hill Quadrangle, Fall River County, South Dakota; also at other localities in the county (Sohn, 1979).

Trapezoidella trapezoidalis (Roth)

Plate 31, figures 26, 27; Plate 33, figures 20-22

Bythocypris (Bairdiocypris) trapezoidalis Roth, 1933, p. 402, pl. 48, figs. 6a-d. Harper and Sutton, 1935, p. 628, pl. 76, figs. 16, 17.

Trapezoidella trapezoidalis (Roth), Sohn, 1979, p. 9, pl. 1, figs. 6-9, 12-17; pl. 2, figs. 30-32.

Subtrapezoidal, highest medially; dorsum nearly straight, venter concave; anterior broader than posterior, the latter truncate above; both ends extended below; LV overlaps RV; LV with midventral longitudinal ridge and middorsal submarginal furrow; smooth to minutely pitted; L, 1.03; H, 0.73 mm (Sohn, 1979).

Occurrence. Lower Cretaceous, Lakota Formation sec. 28, T. 4 N., R. 6 E., Lawrence County, South Dakota; also 3 miles (4.8 km) south of Piedmont, Meade County, South Dakota; U.S.N.M. 242839, Crook County, Wyoming; and other localities in Fall River, Lawrence and Butte Counties, South Dakota and Crook County, Wyoming (Sohn, 1979).

Trapezoidella sp. aff. T. trapezoidalis (Roth)
Plate 33, figures 23, 24
Betty Kellett Nadeau, 1945, unpublished photographs.

A form resembling T. trapezoidalis but with greatest height postmedian rather than anteromedian; dimensions unknown.

Occurrence. Upper Jurassic or Lower Cretaceous, Morrison? Formation, near Canyon City?, Colorado (from Nadeau photographs).

Genus Tuberocypris Swain, 1947
(Candonidae-Candoninae)
Tuberocypris acuminatus Swain
Plate 33, figures 25-28
Tuberocypris acuminatus Swain, 1947, p. 523, pl. 77, figs. 1-8. Swain, 1986a, p. 95. Swain, 1987a.

Subreniform-subquadrate, highest medially; dorsum convex, straightened medially, venter concave; anterior rounded, extended below, posterior acuminate, strongly extended below; LV overlaps RV; rounded median swelling, highest at its posterior edge; a depression containing adductor muscle scar lies within swelling; some specimens more convex posteriorly than others may be females; smooth general surface; LV hinge with groove defined ventrally by a weak ridge process arising from interior valve surface; inner lamellae of moderate width; vestibules terminally; radial canals widely spaced; adductor muscle scar a curved hooklike row of five or more spots, open postdorsally; L, holotype LV, 1.08; H, 0.57 mm (Swain, 1947).

Occurrence. Pliocene, Salt Lake Formation, sec. 15, T. 13 N., R. 2 W., on boundary between Cache and Box Elder Counties, Utah (Swain, 1947). Glenns Ferry Formation, sec. 14, T .4 S., R. 1 W., Owyhee County, Idaho (Swain, 1986a); Hay Ranch Formation, Tanner Reservoir overflow canal, sec. 9, T. 11 N., R. 19 W., Box Elder County, Utah (Swain, 1987a).

Tuberocypris brighti Swain

Plate 33, figures 29-33

Tuberocypris brighti Swain, 1986a, p. 95, pl. 1, figs. 12-16; pl. 2, figs. 5, 18; pl. 3, figs. 10a, b.

Subquadrate (adult) to subpryriform (juvenile), highest about one-fourth of length from posterior end in mature forms; dorsum nearly straight, venter concave, that of RV bears a small anteromedian spine; anterior broadly rounded; posterior higher than anterior but truncated above and below, and angulated ventrally; LV overlaps RV except possibly dorsally; a large slightly anteromedian hollow node (adults) varies from rounded to pointed and posteriorly directed; general surface smooth; LV hinge with groove to receive RV edge; inner lamellae broadest anteriorly, radial canals numerous anteriorly; adductor muscle scar lies in pit that represents external node; it consists of four spots in compact group; at least one frontal spot; L, holotype, 0.70; H, 0.37 mm (Swain, 1986a).

Occurrence. Pliocene, Glenns Ferry Formation, sec. 14, T. 4 S., R. 1 W., Owyhee County, Idaho (Swain, 1986a).

Tuberocypris cancellatus Swain and Wagoner, 1985
(see Pactolocypris cancellatus (Swain and Wagoner))

Tuberocypris centronotus Swain, 1964
(see Pactolocypris centronotus (Swain))

Tuberocypris ovalis Swain
Plate 34, figures 4, 5
Tuberocypris ovalis Swain, 1947, p. 524, pl. 77, figs. 9-11.

Subovate-reniform, highest just postmedially; dorsum convex, venter concave; anterior narrower than posterior, both ends extended below; LV slightly overlaps RV; dorsoventrally, elongated, slightly postmedian swelling; general surface smooth; LV hinge a groove with ventrally defining ridge; inner lamellae narrow; vestibules terminally; L, holotype RV, 1.19; H, 0.65 mm (Swain, 1947).

Occurrence. Pliocene, Salt Lake Formation, sec. 15, T. 13 N., R. 2 W., along boundary between Cache and Box Elder Counties, Utah (Swain, 1947).

Tuberocypris quadralis Swain
Plate 34, figures 6-13
Tuberocypris quadralis Swain, 1947, p. 523, pl. 77, figs. 12-17; Swain, 1987, p. 124, pl. 1, figs. 12-15.

Subquadrate, highest anteromedially; dorsum nearly straight to slightly convex, venter slightly concave; anterior broader than posterior, both ends extended below; LV overlaps RV; prominent median swelling with depressed pellucid central portion, where adductor muscle scar lies, swelling highest posteriorly; minutely punctate surface; LV hinge a furrow that is defined ventrally by a ridge process arising from shell interior; RV hinge edge compressed to fit into LV groove; inner lamellae broadest anteriorly; vestibules terminally; adductor muscle scar a hook-shaped row open dorsally, of about five spots; L, holotype, 0.90; H, 0.51 mm (Swain, 1947).

Occurrence. Pliocene, Salt Lake Formation, sec. 15, T. 13 N., R. 2 W., on boundary between Cache and Box Elder Counties, Utah (Swain, 1947). Cache Valley Formation, Salt Lake Group, T. 13 N., R. 3 W., southwest of Newton, Cache County, Utah (Swain, 1987).

Genus Tuberocyproides Swain, 1947

(Candonidae-Candoninae)

Tuberocyproides dipleura Swain

Plate 34, figures 14-16

Tuberocyproides dipleura Swain, 1947, p. 525, pl. 77, figs. 18-21.

Subreniform-acuminate, highest anteromedially; dorsum nearly straight, venter sinuate, concave medially, anterior broadly rounded, posterior acuminate, both ends extended below, posterior slightly concave above; anterior marginal low ridge; surface with two oblique strong ridges near valve middle that diverge ventrally toward termino-ventral extremities, each with abrupt bulbous termination; elevated knob lies between ridges on ventral slope; small depression medially; surface studded with coarse scattered pustules; LV hinge a narrow groove defined ventrally by a ridge process arising from interior valve surface; anterodorsal and postdorsal slopes flanged to overlap RV; inner lamellae broad anteriorly; L, holotype LV, 1.32; H, 0.73 mm (Swain, 1947).

Occurrence. Pliocene, Salt Lake Formation, sec. 15, T. 13 N., R. 2 W., along boundary between Cache and Box Elder Counties, Utah (Swain, 1947).

Tuberocyproides oreanensis Swain

Plate 34, figures 17-24

Tuberocyproides oreanensis Swain, 1986a, p. 95, pl. 1, figs. 17-20; pl. 2, figs. 1-3.

Subtrapezoidal (adults) to subpyriform (juveniles), highest anteromedially; dorsum nearly straight (adults) to convex (juveniles), venter concave; anterior broadly rounded, posterior acutely pointed, strongly extended below; LV overlaps RV; strongly convex; each valve with strongly elevated short, oblique cleatlike ridges anteroventrally and a longer one postventrally, trending toward those respective margins; ridge surfaces with or without nodes;

general surface with clusters or single nodes; LV hinge with terminal grooves and an interterminal rabbet groove to receive RV edge; inner lamellae broad anteriorly; vestibule anteriorly; radial canals widely spaced; adductor muscle scar a submedian group of five spots; two frontal spots; L, holotype 1.26; H, 0.78 mm (Swain, 1986a).

Occurrence. Pliocene, Glenns Ferry Formation, sec. 14, T. 4 S., R. 1 W., at "Big Cut", Owyhee County, Idaho (Swain, 1986a).

Genus Turkmenella Schneider, 1963

(Disopontocypridinae?)

Turkmenella laeva Swain

Plate 34, figures 25, 26

Turkmenella laeva Swain, 1987c, p. 431, pl. 1, figs. 7, 8.

Subtrapezoidal, highest one-fourth of length from anterior end; dorsum nearly straight, about half of length, venter slightly concave; posterior broader than anterior, truncate above; both ends extended below; LV overlaps RV; compressed; smooth; L, holotype, 0.77; H, 0.43 mm (Swain, 1987c).

Occurrence. Miocene, Esmeralda Formation, Warrior Mine measured section, 7650 feet (2332 m) S. 26° E. of NW corner Simon 1:24000 Quadrangle in unsurveyed area, Nye County, Nevada, 400 feet (122 m) above base of section (Swain, 1987c).

Turkmenella warriorensis Swain

Plate 34, figures 27-29

Turkmenella warriorensis Swain, 1987c, p. 432, pl. 1, figs. 9-12.

Subquadrate to subtrapezoidal, highest about one-fourth of length from anterior end; dorsum and venter both concave, the former owing to elevated cardinal angles; posterior narrower than anterior, both ends extended below; LV overlaps RV; relatively compressed; smooth; L, holotype, 0.80; H, 0.49 mm (Swain, 1987c).

Occurrence. Miocene, Esmeralda Formation, Warrior Mine measured section, 7650 feet (2332 m) S. 26° E. of NW corner Simon 1:24000 Quadrangle in unsurveyed area, Nye County, Nevada, 110 feet (34 m) and 400 feet (122 m) above base of section (Swain, 1987c).

Turkmenella? sp. Swain

Plate 34, figure 30

Turkmenella? sp., Swain, 1987c, p. 423, pl. 3, fig. 12.

Elongate, subtrapezoidal, highest medially; dorsum straightened medially for about half of shell length, venter concave; ends about equally rounded; LV overlaps RV most along middorsum; compressed; smooth; L, 0.88; H, 0.51 mm (Swain, 1987c).

Occurrence. Miocene, Horse Camp Formation, sec. 12, T. 10 N., R. 58 E., about 3.8 miles (6.24 km) east of Currant, Nye County, Nevada (Swain, 1987c).

Genus Ussuriocypris Mandelstam, 1956

(Cyprididae?-Cypridinae?)

Ussuriocypris? subreniformis Swain and Wagoner

Plate 21, figures 21-24

Ussuriocypris? subreniformis Swain and Wagoner, 1985, p. 135, pl. 5, figs. 6, 14; Swain, 1985, p. 421, pl. 3, figs. 14, 15

Subquadrate subovate reniform, highest one third of length from anterior end; dorsum convex, venter slightly concave to straight; ends nearly equal, the posterior slightly broader, both are extended below; LV larger than RV, relatively compressed; smooth; adductor muscle scar an anteromedian group of several spots; one or two frontal spots; L, 1.09; H, 0.62 mm (after Swain and Wagoner, 1985)

Occurence. Miocene, Esmeralda Formation, Black Spring road section, Black Spring Quadrangle, Nye County, Nevada; Blair Junction section, Esmeralda County, Nevada (Swain and Wagoner, 1985; Swain, 1985).

Genus Whipplella Holland, 1934)

(Suchonellidae by Sohn, 1985) (Darwinuloidacea-Darwinuloididae by Molostovskaja, 1990)

Whipplella carbonaria Scott

Plate 34, figures 34, 35

Whipplella carbonaria Scott, 1944, p. 143, pl. 24, figs. 3-5.

Carbonita humilis (Jones and Kirkby), Bless and Pollard, 1975, p. 111)

Elongate-ovate, highest medially; dorsum convex, venter slightly concave; anterior narrower than posterior; RV overlaps LV; hinge area about one-third of shell length, impressed into narrow slitlike depression; coarsely pitted; some pits are elongate, concentric, and in rows; L, 0.82; H, 0.45 mm (Scott, 1944).

Occurrence. Pennsylvanian, McLeansboro Group, Cohn Cyclothem, Webster Ferry, White County, Illinois (Scott, 1944).

Whipplella cunieformis Holland

Plate 34, figures 36-39

Whipplella cunieformis Holland, 1934, p. 344, pl. 25, figs. 5a-c. Scott, 1944, p. 143, pl. 24, figs. 18-20.

Carbonita inflata (Jones and Kirkby), Cooper, 1946, p.66, pl. 8, figs. 40 - 42.

Ovate, highest postmedially; dorsum convex, venter straight to gently concave; anterior narrower than posterior; RV overlaps LV; inflated valves; hinge line in middle third, depressed in a groove as seen in dorsal view; densely pitted in posterior half, less so in anterior half; L, 0.84; H, 0.59 mm (Holland, 1934; Scott, 1944).

Occurrence. Permian, Dunkard Series, Greene Group, upper part, Nineveh Formation; Limestone Hill, SW part of Wood County, NW West Virginia (Holland, 1934).

Whipplella deltoidea Holland, 1934
(see Gutschickia deltoidea (Holland))

Whipplella depressa Holland

Plate 32, figures 1-3

Whipplella depressa Holland, 1934, p. 345, pl. 23, figs. 7a-c. Scott, 1944, p. 143, pl. 24, figs. 21-23.

Carbonita inflata (Jones and Kirkby), Cooper, 1946, p. 66, pl. 8, figs. 40-42.

Ovate, highest postmedially; anterior narrower than posterior; dorsum convex, venter straight to gently concave; RV overlaps LV; inflated posteriorly; hinge area in middle third of shell lies in a depressed groove seen dorsally; coarsely pitted in concentric fashion; L, 0.76; H, 0.50 mm (Holland, 1934; Scott, 1944).

Occurrence. Permian, Dunkard Series, Greene Group, upper part, Nineveh Formation, Limestone Hill, southernmost Wood County, NW West Virginia (Holland, 1934).

Whipplella ninevehensis Holland, 1934
(see Gutschickia ninevehensis (Holland))

Whipplella ovata Holland, 1934
(see Cypridopsis ovata (Holland))

Whipplella parvula Holland, 1934
(see Cypridopsis parvula (Holland))

Whipplella sp., Sohn

Plate 31, figures 31, 32

Whipplella sp., Sohn, 1977, p. 136, TF 1g, 2a, b, h-k (illustrated but not described).

Elongate subreniform, subelliptical, highest medially; dorsum convex, venter concave to sinuous; anterior narrower than posterior; RV overlaps LV; smooth? or weakly ornamented; adductor muscle scar a double darwinulid-like row of 11 spots and a more anteroventral ridgelike spot; L, 1.25; H, 0.35 mm (after Sohn's photographs, 1977).

Occurrence. Permian, Greene Formation, upper Rockport Limestone Member, Wetzel County, West Virginia, USGS coll. 12901-PC; Windy Gap Limestone Member, Wetzel County, West Virginia, U.S.G.S. coll. 12905-PC (Sohn, 1977).

Whipplella? sp., Sohn

Plate 32, figures 4-6

Whipplella? sp. 1, Sohn, 1985, p. 455, TF 4.10-4.14, 4.21, not described.

Specimens preserved as steinkerns (internal molds); ovate-subreniform, highest medially; dorsum convex, venter concave; anterior narrower than posterior; RV overlaps LV; adductor muscle scar an anteromedian double row of 10-12 spots, an elongate ridgelike more anteroventral spot, and a more postdorsal spot; L, 0.71; H, 0.50 mm (Sohn illustration, 1985).

Occurrence. Upper Mississippian, Bluestone Formation, Bramwell Member, Crumpler Quadrangle, U.S.G.S. coll. 12981-PC, Wyoming County, West Virginia (Sohn, 1985).

Whipplella? sp. 2, Sohn

Plate 31, figure 33

Whipplella? sp. 2, Sohn, 1985, p. 435, TF 5.5-5.7 (not described in detail).

Cited as resembling Whipplella? sp. 1, Sohn (1985) but with larger and fewer adductor muscle scar spots; L, 0.45; H, 0.30 mm (Sohn, 1985). Specimens preserved as internal molds (steinkerns).

Occurrence. Upper Mississippian, Bluestone Formation, Bramwell Member, Crumpler Quadrangle, Wyoming County, West Virginia, U.S.G.S. coll. 12981-PC (Sohn, 1985).

Genus Xiyingia Li and Shi, 1978

(Cyprididae-Huabeinidinae)

Xiyingia? sp. aff. X. elliptica Hou and Shi

Plate 32, figure 7

Xiyingia? sp. aff. X. elliptica Hou and Shi, 1978 in Hou et al., 1988; Swain, 1985, p. 430, pl. 2, fig. 12.

Elongate-subquadrate, highest about one-fourth of length from posterior end; dorsum nearly straight, venter slightly concave; anterior a little narrower than posterior, the latter extended below; LV overlaps RV; compressed; anterodorsally a weak sulcus-like depression in RV; smooth; L, 1.07; H, 0.50 mm (Swain, 1985).

Occurrence. Miocene, Esmeralda Formation, Finger Rock Wash, Granny Goose 1:24000 Quadrangle, 0.5 mile (0.8 km) south of Rawhide Ranch, Mineral County, Nevada, 1400-1500 feet (425-455 m) above base of lake bed sequence (Swain, 1985).

REFERENCES

Adamson, R. D., Hardy, C. T. and Williams, J. S., 1955. Tertiary rocks of Cache Valley, Utah and Idaho. Utah Geological Society Guidebook, 10, 1-22

Albers, J. P. and Steward, J. H., 1972. Geology and mineral deposits of Esmeralda County, Nevada. Nevada Bureau of Geology and Mines, 78, 80 p.

Alm, G., 1914. Beiträge zur Kenntnis der nördlichen und arktischen Ostracodenfauna. Ark. Zool. 9, 1-20.

__________ 1915. Monographie der schweidischen Süsswasser-Ostracoden nebst systematischen Besprechungen der Tribus Podocopa. Zool. Bidr., Uppsala, 4, 1-248.

American Association of Petroleum Geologists, COSUNA Chart, Great Basin, Tulsa.

Anderson, N. R., 1965. Upper Cenozoic stratigraphy of the Oreana, Idaho 15 minute Quadrangle. University of Michigan Doctoral Dissertation, 212 p. University Microfilm 65-10, 183.

Anderson, F. W., 1970. Carboniferous Ostracoda - the Genus Carbonita Strand. Geological Survey of Great Britain, Bulletin 32, 69-121.

Axelrod, D. J., 1956. Pre-Miocene floras from west-central Nevada. California University Publications, Geol. Sci. 33, 99-102, 141-142.

Baird, W., 1835. List of Entromostraca found in Berwickshire. Berwickshire Nat. Club, 1, 95-100.

__________, 1837-38. The natural history of the British Entomostraca. Mag. Zool. Bot., 1, 35-41, 514-526, 132-144.

__________, 1845. Arrangement of the British Entomostraca. Hist. Berwickshire Nat. Club, 2, 145-158.

__________, 1850. Natural history of the British Entomostraca. London, Ray Soc., 364 p.

Baker, A. A., Dane, C. H. and Reeside, J. B., Jr. 1936. Correlation of the Jurassic formations of part of Utah, Arizona, New Mexico and Colorado. U. S. Geological Survey, Prof. Paper 183, 66 p.

Becker, J., 1969. The freshwater Ostracoda of the Lower Humboldt Formation (Miocene) in northern Nevada, northern Utah and southeastern Idaho. M. S. Thesis, University of Minnesota, 55 p.

Benson, R. H., 1967. Muscle scar patterns of the Pleistocene (Kansan) ostracodes. (In) *Essays in Paleontology and Stratigraphy*, C. Teichert and E. L. Yochelson eds. Department of Geology, University of Kansas, Spec. Publ. 2, 211-241.

__________, 1969. Ostracodes of the Rita Blanca Lake deposits. (In) *Paleoecology of an early Pleistocene lake on the High Plains of Texas*, R. Y. Anderson and D. N. Kirkland, eds. Geological Society of America, Memoir 113, 107-113.

__________, and McDonald, H. D., 1963. Postglacial (Holocene) ostracodes from Lake Erie. University of Kansas, Paleontological Contr., Arthropoda, art. 4, 1-26.

Bless, M. J. M. and Pollard, J. E., 1975. Quantitative analysis of dimorphism in *Carbonita humilis* (Jones and Kirkby). (In) *Biology and Paleobiology of Ostracoda*, F. M. Swain, L. S. Kornicker and R. F. Lundin, eds. Bulletin of American Paleontology, 65, 109-122.

Bradley, W. H., 1931. Origin and microfossils of the oil shale of the Green River Formation of Colorado and Utah. U. S. Geological Survey, Prof. Paper 168, 58 p.

Brady, G. S., 1854. On species of Ostracoda new to Britain. Ann. Mag. Nat. Hist., ser. 13, 59-64.

__________, 1868. A monograph of the Recent British Ostracoda. Trans. Linn. Soc., London, 26, 353-496.

__________, 1870. Notes on Entomostraca taken chiefly in the Northumberland and Durham district (1869). Nat. Hist. Soc. Northumberland and Durham, Trans. 3, 361-373.

__________, and Norman, A. M., 1889. Monograph of the marine and freshwater Ostracoda of the North Atlantic and northwestern Europe. i. Podocopa. Trans. Roy. Dublin Soc., ser 2, 4, 61-270.

__________, Crosskey, W. H., and Robertson, D., 1874. A monograph of the post-Tertiary Entomostraca of Scotland. Palaeontogr. Soc., London, 28, 1-232.

__________, and Robertson, D., 1872. On the distribution of British Ostracoda. Ann. Mag. Nat. Hist., ser. 4, 9, 48-70.

__________, and Robertson, D., 1869. Notes on a week's dredging in the west of Ireland. Ann. Mag. Nat. Hist., ser 4, 3, 353-374.

__________, and Robertson, D., 1870. The Ostracoda and Foraminifera of tidal rivers. Ann. Mag. nat. Hist., ser. 4, 6, 1-33.

Branson, C. C., 1945. Freshwater invertebrates from the Morrison (Jurassic) of Wyoming. Jour. Paleontology, 9, 514-522.

Branson, C. C., 1936. New name for Morrison ostracode genera. Jour. Paleontology, 10, 323.

Brontstein, Z. C., 1930. (Description of the Ostracoda fauna of Lake Baikal). Trudy Kom. po izuch. oz Baikala, 3, 117-157.

__________, 1947. (Ostracoda of fresh-water) Fauna U.S.S.R., Crustacea 2, Zool. Inst. Akad. Nauk S.S.S.R., n. s., no. 31, 339 p.

Brouwers, E., 1992. Cypris, newsletter for Ostracodologists, Denver, p. 30.

Brown, P. M., Miller, J. A. and Swain, F. M., 1972. Structural and stratigraphic framework and spatial distribution of permeability of the Atlantic coastal plain, North Carolina to New York. U. S. Geological Survey, Prof. Paper 796, 79 p.

Calahan, L.. W., 1939. Diagnostic fossils of the Ark-La-Tex area. Shreveport Geological Society Guidebook, 14th Ann. Field Trip, p. 36-56.

Cameron, S. P. and Lundin, R. F., 1977. Environmental interpretation of the ostracode succession in late Quaternary sediments of pluvial Lake Chochise, southeastern Arizona. (In) Aspects of the Ecology and Zoogeography of Recent and Fossil Ostracodes, H. Loeffler and D. Danielopol eds. Dr. W. Junk D.V., the Hauge, p. 325-352.

Carter, C., 1997. Ostracoda in Owens Lake core OL-92; alternation of saline and freshwater forms through time. Geological Society of America Spec. Paper 317, p. 113-119.

Chapman, F., 1896. On some Pliocene Ostracoda from near Berkeley, California. University of California, Bull. of the Department of Geology, 2, 93-100.

Cole, G. A., 1949. A new cytherid ostracod from Minnesota. Amer. Microscop. Soc. Trans., 68, 350-354.

Cooper, C. L., 1946. Pennsylvanian ostracodes of Illinois. Ill. State Geol. Survey, Bull. 70-177. p.

Croneberg, A., 1895. Beiträge zur Ostracodenfauna der Umgegend von Moscau, Bull. Soc. Nat. Moscau, 5, 284-306.

Daday, E., 1900. Ostracoda Hungariae, Budapest, 320 p.

__________, 1882. Koeslémenyek a Kkolozvâri m. Kir. Tudomány-egyetem allat-és oeszszshasonlet-oboncztani intezeleboel, part 2 Adatok Kolozsvar es koernyeke Crustacae-Faunajanak ismeretehez, Orvos Termeszettadomanyi Ertesitoe, 7, 211-258.

__________, 1902. Mikroscopische Süsswasser-Tiere aus Patagonien, Termeszettadomanyi Füzetek, 25, 201-310.

__________, 1985. Untersuchungen uber die Süsswassern-Mikrofauna Paraguays. Zoologica, 18.

Dane, C. H., 1954. Strategraphic and facies relationships of upper part of Green River Formation and lower part of Uinta Formation in Duchesne, Uintah, and Wasatch counties, Utah. American Association of Petroleum Geologists, Bull., 38, 404-425.

Delorme, L. D., 1967. New freshwater Ostracoda from Saskatchewan, Canada. Can. Jour. Zool. 45, 357-363.

__________, 1968. Pleistocene freshwater Ostracoda from Yukon, Canada. Ibid., 46, 859-876.

__________, 1970a. Freshwater ostracodes of Canada. Ibid., 48, 153-168.

__________, 1970b. Freshwater ostracodes of Canada, part II: Sub-family Cypridopsinae and Herpetocypridinae and Family Cyclocyprididae. Ibid, 48, 253-266.

__________, 1970c. Freshwater ostracodes of Canada, part III: Family Candonidae, Ibid, 48, 1099-1127.

__________, 1970d. Freshwater ostracodes of Canada, part IV: Families Ilyocyprididae, Notodromadidae, Darwinulidae, Cytherideidae and Entocytheridae. Ibid., 48, 1251-1279.

__________, 1971. Freshwater ostracodes of Canada, part V: Families Limnocytheridae, Loxoconchidae, Ibid., 49, 43-64.

Dickinson, K. A., 1959. The Ostracoda and Cladocera of the Upper Humboldt Formation, northeastern Nevada. M. S. Thesis, University of Minnesota, Minneapolis.

__________, and Swain, F. M., 1967. Late Cenozoic freshwater Ostracoda from northeastern Nevada. Jour. Paleontology, 41, 325-350.

Dolgoff, A., 1963. Volcanic stratigraphy of the Pahrangat area, Las Vegas Range, southeastern Nevada. Geololgical Society of America Bull. 74, 875-900.

Dubowsky, N. W., 1927 (Contribution toward the recognition of the freshwater Ostracoda fauna of the Ukraine). Trudy Khark. Tov., Dosl. prep., 50.

Eardley, A. J., 1944. Tertiary history of north-central Utah. Utah Geological Society Guidebook, Geology of Utah, 10, 37-44

Eaton, J. E., 1939. Ridge Basin, California. American Association of Petroleum Geologists, Bull., 33, 517-558.

Ebanks, W. J., 1965. Structural geology of the Gass Peak area, Las Vegas Range, Nevada (abst.). Houston Geological Society, Bull., 8, 22.

__________, 1986. Written Communication.

Ekman, S., 1908. Ostracoden aus den nördschwedischen Hochgebirges. Naturw. Unters. Sarekgeb. Sch. Lappl., IV, Zoologi.

Ekren, E. B., Byers, F. M., Jr. and Hardyman, R. E., 1980. Stratigraphy, preliminary petrology and some structural features of Tertiary volcanic rocks in Gabbs Valley and Gillis Ranges, Mineral County, Nevada. U. S. Geological Survey, Bull., 1464, 54 p.

Englund, K. J., Henry, W. T. and Cecil, C. B., 1981. Upper Mississippian and Lower Pennsylvanian depositional environments, southwestern Virginia and southeastern West Virginia. (In) T. G. Roberts, ed., Geological Society of America Cincinnati 1981 Field Trip Guidebooks, v. 1, Amer. Geol. Inst., Falls Church, VA.

Ferguson, E., Jr., 1967. New ostracodes from the playa lakes of eastern New Mexico and western Texas. Amer. Micros. Soc.,Trans. 86, 244-250.

Ferguson, H. G., Muller, S. W. and Cathcart, S. W. 1953. Geology of the Coaldale Quadrangle, Nevada. U. S. Geological Survey Quadrangle Map GO-23.

Fisher, C. A., 1909. Geology of the Great Falls coal field, Montana. U. S. Geological Survey, Bull. 56.

Fischer, S., 1851. Über das Genus Cypris und desen in der Umbegung von St. Petersburg vorkommenden Arten. Mem. sav.etr. St. Petersburg, 7, 127-167.

Forester, R., 1983. Relationship of two lacustrine ostracode species to solute composition and salinity; implications for paleohydrochemstry. Geology, 11, 435-438.

__________, 1985. Limnocythere bradburyi, n. sp.: a modern ostracode from central Mexico and a possible Quaternary paleoclimatic indicator. Jour. Paleontology, 59, 8-20.

__________, 1987. Late Quaternary paleoclimatic records from lacustrine ostracodes. (In) W. Ruddiman and H. E. Wright, eds., North America and adjacent oceans during the last deglaciation, The Geology of North America K-3, Geol. Soc. Amer. Boulder, CO, 261-276

__________, 1991. Pliocene climate history of the western United States derived from lacustrine ostracodes. Quat. Sci. Rev., 10, 136-146.

__________, 1996. Comments in Cypris, newsletter for ostracodologists, Denver, CO, p. 33.

__________, and Smith, A. J., 1994. Levels and style of precipitation during the late glaciation reconstructed from the ostracode fossil record. Technology Integration: above and below. Am. Nuclear Soc. Int. High Level Waste Management Conf., Las Vegas, Nevada, p. 2552.

Fouch, T. D., 1979. Character and distribution of Upper Cretaceous (?) and Paleocene nonmarine sedimentary rocks in east central Nevada. (in) Armentrout, J. M., et al., eds. Cenozoic Paleogeography of the Western United States. Soc. Econ. Paleont. and Mineral., Pacific Section, p. 97-111.

Friedman, G. M. and Lundin, R. F., 1988. Freshwater ostracodes from upper Middle Devonian fluvial facies, Catskill Mountains, New York. Jour. Paleontology, 72, 485-490.

Frolich, A. J. and Olsen, R. E., 1984. Newark Supergroup, a revision of the Newark Group of eastern North America. U. S. Geological Survey, Bull, 1537-A, p. 55-58.

Furtos, N., 1933. The Ostracoda of Ohio. Ohio Biol. Survey, 5, 411-524.

Galeeva, L. I., 1955. Ostracody melovykh otlozhenii Mongol'skoi Narodnoi Republikii, Moscow, Vses. Nauchno-Issled. Geol Razved. Neft. Inst. (VNIGNI) 94 p. (Gostoptechizat).

Goldich, S. S. and Elms, M. A., 1949. Stratigraphy and petrology of the Buck Hill Quadrangle, Texas. Geol. Soc. Amer. Bull., 60, 1133-1182.

Gore, P. J. W., 1994. Triassic-Jurassic lacustrine deposits in the Culpeper Basin, Virginia. (In) Global Geological Record of Lake Basins, v. 1, E. Gierlowski-Kordesch and K. Kelts, eds. Cambridge University Press, 155-163.

Gutentag, E. D., 1972. Analyses of ostracodes from the Sand Draw faunal localities. Am. Mus. Hat. Hist., Bull. 148 (art. 1), 37-39.

__________, and Benson, R. H., 1962. Neogene (Plio-Pleistocene) freshwater ostracodes from the central High Plains. Geological Survey of Kansas, Bull. 157 (4), 60 p.

__________, and Galli-Olivier, C., 1969. New Locations of Pleistocene (Kansan) molluscan and ostracode faunas, Dickinson County, Kansas. U. S. Geological Survey, Prof. Paper 650-D, D148-D154.

Harper, F. and Sutton, A. H., 1935. Ostracoda of the Morrison Formation from the Black Hills. Jour. Paleontology, 9, 623-628.

Hartmann, G., 1971. Discussion, in Paleoecologie des Ostracodes, H. J. Oertli, ed., Pau, France, Bull., Centre de Rech., SNPA, 5, suppl. p. 338.

__________, and Puri, H. S., 1974. Summary of Neontological and Paleontological Classification of Ostracoda. Mitt. Hamburg Zool. Mus. Inst., 70, 7-73.

Hayden, F. V., 1869. U. S. Geological Survey of Territories, 1st, 2nd and 3rd Ann. Rept., 192. p.

Holland, W. C., 1934. The ostracods of the Nineveh limestones of Pennsylvania and West Virginia. Carnegie Museum Ann., 22, (2-4), 343-350.

Holmes, J. A., 1996. Trace element and stable isotope geochemistry of non-marine ostracod shells in Quaternary paleoenvironmental reconstruction. Jour. Paleolimnology, 15, 223-235.

Hoskins, D. M., Stratigraphy and paleontology of the Bloomsburg Formation of Pennsylvania and adjacent states. Penn. Geol. Survey Bull.., ser. 4, G 36.

Hou You-tang and others, 1978. Early Tertiary ostracode fauna from the coastal region of Bohai. Palaeont. and Geol. Inst. Nanjing Acad. Sci., Science Press, Beijing, China, 205 p.

__________, 1988. On the change of the authors' names of new genera and species in the monograph "Early Tertiary Ostracode Fauna from the Coastal Region of Bohai" (1978). Acta Micropaleontologica Sinica, 5, 209-218.

Howe, H. V. and Laurencich, L., 1958. Introduction of the study of Cretaceous Ostracoda. Baton Rouge, Louisiana State University Press, 366 p.

Humphrey, F. L., 1960. Nevada Bur. Mines, Bull. 57, p. 41-42.

Jones, T. R., 1859. Description of the Entomostraca of the Pleistocene beds of Newburg, Copford, Clacton and Grays. Ann. Mag. Nat. Hist., ser. 2 (31), 25-28.

__________, 1956. A monograph of the Tertiary Entomostraca. Palaeontol. Soc. London, 9, 1-70.

__________, 1962. Monograph of the Fossil Estheridae. Palaeontogr. Soc. London, 1-134.

__________, 1870. On the bivalved Entomostraca from the Coal Measures of South Wales. Geol. Mag., 7, 218, 219.

__________, 1878. Ostracoda from the Subwealden of Sussex. Geol. Mag. n. ser., 5, 100-110.

__________. 1885. On the Ostracoda of the Purbeck Formation with notes on the Wealden species. Geol. Soc. London, Quart. Jour., 41, 311-353.

__________, 1886. Notes on the distribution of the Ostracoda of the Carboniferous formations of the British Isles. Ibid. 42, 496.

__________, 1886. On some fossil Ostracoda from Colorado. Geol. Mag. n. ser., 3, (4), 145-148.

__________, 1893. On some fossil Ostracoda from S. W. Wyoming and from Utah. Geol. Mag., decade 3, 10, 385-391.

__________, and Kirkby, J. W., 1867. On the Entomostraca of the Carboniferous rocks of Scotland. Geol. Soc., Glasgow, Trans. 2, 213-228.

__________, 1879. Notes of the Paleozoic bivalved Entromostraca, No. 12. Some Carboniferous species belonging to the Genus Carbonia Jones. Ann. Mag. Nat. Hist., ser. 5, 4, 28-39.

__________, 1886. Notes on the distribution of the Ostracoda of the Carboniferous of the British Isles. Geol. Soc. London, Quart. Jour., 42, 498-514.

__________, and Brady, G. S., 1874-1884. A monograph of the British fossil bivalved Entomostraca from the Carboniferous formations. Part 1 The Cypridinidae and their allies. Mon. Palaeontogr. Soc., 1874-1884, p. 1-56, 1874; p. 57-82, 1884.

Jurine, L. 1820. Histoire des monocles, qui se trouvent aus environs de Geneve, Geneve.

Kaufman, A., 1896. Die schweizerischen Cytheriden und ihre nachsten Verwandten. Rev. Suisse Zool., IV.

__________, 1900. Cypriden und Darwinuliden der Schweiz. Rev. Suisse Zool., 8, 207-243.

Keller, A. A., 1952. Geology of the Mink Creek region, Idaho. Unpublished M. S. Thesis, University of Idaho.

Kellogg, H. F., 1964. Cenozoic stratigraphy and structure of the southern Egan Range Nevada. Geol. Soc. Amer. Bull, 75, 685-708.

Kesling, R. V. and Ploch, R. A., 1960. New Upper Devonian ostracod from southern Indiana. Contr. Mus. Paleont., Univ. Michigan 15 (2), 281-292.

Kietzke, K. K., 1987. Calcareous microfossils from the Upper Triassic of northeastern New Mexico. New Mexico Geol. Soc. Guidebook, 38th Field Conference, 119-126.

__________, 1989a. Calcareous microfossils from the Triassic of the southwestern U. S. (In) S. G. Lucas and A. P. Hunt, eds. Dawn of the age of dinosaurs in the American southwest. New Mexico Mus.. Nat. Hist., Albuquerque, p. 222-232.

__________, 1989b. Calcareous microfossils of the Moenkopi Formation (Lower Triassic, Sythian or Anisian) of central New Mexico. New Mexico Geol. Soc. Guidebook, 40th Field Conference, southeastern Colorado Plateau, 181-190.

__________, 1992. Reassignment of the Todilto Limestone ostracod Metacypris todiltensis Swain to Cytheridella Daday, 1905, with notes on the phylogeny and environmental implications of this ostracod. New Mexico Geol. Soc. Guidebook, 43rd Field Conference, San Juan Basin IV, 173-183.

__________, and Lucas, S. G., 1995. Ostracoda and Gastropoda from the Kayenta Formation (Lower Jurassic) of Arizona, U. S. A. Arizona-Nevada Acad. Sci.

__________, 1991. Ostracoda from the Upper Triassic (Carnian) Tecovas Formation near Kalgary, Crosby County Texas. Texas Jour. Sci. 43, 191-197.

King, J. E., 1973. Late Pleistocene palynology and biogeography of the western Missouri Ozarks. Ecol. Monographs, 43, (4), 539-565.

Kllie, W., 1931. Compagne speleologique de C. Bolivar et R. Jeannel dan l'Amerique du Nord (1928). 3. Crustaces ostracodes. Arch Zool. exped., Gen. 71, 333-344.

Knox, L. W., 1990. Comment in Cypris, newsletter for ostracodologists. Denver, Colorado, p. 33.

Koch, C. L., 1837. Deutschlands Crustaceen, Myriapoden und Arachniden, Ein Beitrag zur deutschen Fauna, Regensburg, Pustet, Heft 11, species 1-6.

__________, 1838. Deutschlands Crustaceen, Myriapoden und Arachniden, Ein Beitrag zur deutschen Fauna, No. 21, Pustet, Regensburg.

Lilljeborg, W., 1853. De Crustaceis ex ordinibus tribus: Cladocera, Ostracoda et Copepoda in Scania occurrentibus, Lund, Berlingske Botryk, i-xv, 1-223.

Lister, K. H., 1975. Quaternary freshwater Ostracoda from the Great Salt Lake Basin, Utah. University of Kansas Paleontological Contr. 78, 33p.

Liubimova (Lyubimova, Lubimova), P. S., 1956. Limnocypridea, Pyrocytheridea and Mandelstamia. (In) Keparisova, L. D., Markovsky, D. P. and Radchenko, G. P., eds.

Materialy po paleontologii, novye semiestva i rody, Vses. Nauchno-Issled. Geol. Inst. (VSEGEI) Materialy, n. ser. no. 12, Paleontology, 106-109, 139-142.

Longwell, C. R., 1921. Geology of the Muddy Mountains, Nevada with a section to the Grand Wash Cliffs. Amer. Jour. Sci., 5th ser., 1, 39-62.

Loranger, D. M., 1951. Userful Blairmore microfossil zone in central and southern Alberta, Canada. Amer. Assoc. Pet. Geol. Bull., 35, 2348-2367.

Lucas, S. G., 1991. Triassic stratigraphy, paleontology and correlation in south-central New Mexico. New Mexico Geol. Soc. Guidebook 42, 243-259.

__________, and Kietzke, K. K., 1993. Calcareous microfossils from the Upper Triassic of Petrified Forest National Park, Arizona. Jour. Arizona-Nevada Acad. Sci., 27, 55-68.

Lundin, R. F., 1997. Note in Cypris, newsletter for ostracodologists, No. 15, Houston, Amoco Prod. Co., p. 44.

Malde, H. E. and Powers, H. A., 1962. Upper Cenozoic stratigraphy of western Snake River Plain, Idaho. Geol. Soc. Amer. Bull., 73, 1197-1220.

Mandelstam, M. F., 1947. Ostracoda from Middle Jurassic deposits of the Magislaka Peninsula. (In) Microfauna, Petroleum Occurrence, Caucasus, Emba and Central Asia. All Union Petroleum Scientific Research, Geol. Exploration Inst. (VNIGRI), State Sci. Tech. Pub. House, Leningrad-Moscow, p. 239-262.

Mansfield, G. R. and Roundy, P. V., 1916. Revisions of the Beckwith and Bear River Formations of southeastern Idaho. U. S. Geol. Survey Prof. Paper 98, 75-84.

Mapel, W. J. and Hail, W. J., 1959. Tertiary geology of the Goose Creek district, Cassia County, Idaho, Box Elder County, Utah and Elko County, Nevada, U. S. Geological Survey Bull., 1035, 217-254.

Masi, L., 1906. Contributo allo sistematica delle Ilyocyprinae. Soc. Ital. Boll. ser. 2, 6, 133-146, 249-268.

Miller, B. H., Graham, R. W., Morgan, H. V., Miller, N. G., McCoy, W. D., Palmer, D. F., Smith, A. J. and Pliny, J. J., 1994. A biota associated with Matayuma-age sediments in west-central Illinois. Quat. Res. 11, 350-365.

__________, Palmer, D. F., McCoy, W. D., Smith, A. J., and Colburn, N., 1993. A pre-Illinoian Pleistocene fossil assemblage from near Connersville, southeastern Indiana. Ibid., 40, 254-261.

Moiola, R. J., 1969. Late Cenozoic of the northern Silver Peak region, Esmeralda County, Nevada. University of California Doctoral Dissertation.

Moore, R. C. and others, 1944. Correlation of the Pennsylvanian Formations of North America. Geol. Soc. Amer. Bull. 55, 657-756.

__________, and others, 1961. Treatise on Invertebrate Paleontology, Part Q, Arthropoda 1, Crustacea, Ostracoda, 442 p.

Moores, E. M., 1969. Mociene-Pliocene sediments, gravity slides and their tectonic significance. Jour. Geol., 76, 88-98.

__________, Scott, R. B. and Lumsden, 1968. Tertiary tectonics of the White Pine-Grant Range region of east-central Nevada. Some regional implications. Geol. Soc. Amer. Bull., 79, 1703-1725.

Morkhoven, F. P. C., van, 1963. Post-Paleozoic Ostracoda, v. 2, Amsterdam, Elsevier Publ. Co., 478 p.

Müller, G. W., 1898. Ostracoden. (In) Voeltzkow, Erg. Zool. Forschungreise im Madagaskar und Ostafrika. Abh. Senckenb. Nat. Ges., XXI

__________, 1990. Deutschlands Süsswasser-Ostracoden. Zoologica, 30, 1-112.

__________, 1912. Ostracoda: Das Tierreich, 31, 1-434.

Müller, O. F., 1776. Zool. Danicae Prodromus, Havniae, 8, xxxii + 282 p.

__________, 1785. Entomostraca seu insecta testacea, quae in aquis Daniae et Norvegiae reperit, descriptit et inconibus illustravit. Lipisae et Havinae, 135 p.

Nick, K. E., Clausen, B. L. and Buchheim, H. P., 1982. Depositional environments and petroleum potential of Miocene lacustrine deposits, west-central Nevada (abst.). Amer. Assoc. Petrol. Geol. Bull., 66, 613.

Nolan, T. B., Merriam, C. W. and Williams, J. S., 1956. The stratigraphic section in the vicinity of Eureka, Nevada. U. S. Geol. Survey, Prof. Paper 276, 77 p.

Oviatt., C. G., Sack, D. and Currey, D. R., 1994. The Bonneville Basin Quaternary, western United States. (In) Global Geological Record of Lake Basins, v. 1, E. Gierlowski Kordesch and Kelts, D. eds. Cambridge University Press, 371-375.

Peale, A. C., 1879. Report on the geology of the Green River District, Wyoming. U. S. Geol. and Geog. Survey Territories, 11th Ann. Rept., 603-606, 635, 640, 641.

Peck, R. E., 1941. Lower Cretaceous Rocky Mountain non-marine microfossils. Jour. Paleontology, 15, 285-304.

__________, 1951a. Non-marine ostracodes-the Subfamily Cyprideinae in the Rocky Mountain area. Ibid., 25, 307-320.

__________, 1951b. A new ostracode genus from the Cretaceous Bear River Formation. Ibid., 25, 575-577.

Peffercorn, H. W. and Gillespie, W. H., 1981. Biostratigraphic significance of plant megafossils near the Mississippian-Pennsylvanian boundary in southern West Virginia and southeastern Virginia. (In) T. G. Roberts, ed. Geol. Soc. Amer. Cincinnati 1981 Field Trip Guidebooks, v. 1, Falls Church, Va., 159-164.

Pickard, M. D., 1955. Subsurface stratigraphy and lithology of Green River Formation in Uinta Basin, Utah. Amer. Assoc. Petrol. Geol., 39, 75-102.

Pierce, H. G., 1988. Chlamydotheca llanoensis, n. sp.; a large ostracode from the latest Pleistocene of the southern High Plains, useful in correlation of the streamside Paleo-Indian sites. Jour. Paleontology, 62, 445-447.

Ramdohr, K. A., 1808. Über die Gattung Cypris Muller und drei zu derselben Gehorige neue Arten. Mag. Gesell. Fr., Berlin, 2, 83-93.

Regnier, J., 1960. Cenozoic geology in the vicinity of Carlin, Nevada. Geol. Soc. Amer. Bull., 71, 1189-1210.

Repetski, J. E. and Henry, T. W., 1983. A late Mississippian conodont faunule from area of proposed Pennsylvanian System stratotype, eastern Appalachians. Fossils and Strata, an System stratotype, eastern Appalachians. Fossils and Strata, 15, 169-170.

Richards, R. W. and Pardee, J. T., 1925. The Melrose phosphate field, Montana. U. S. Geol. Survey Bull. 780a.

Robinson, F. T., McKee, E. H., and Moiola, R. J., 1968. Cenozoic volcanism and sedimentation, Silver Peak region, western Nevada and adjacent California. (In) Studies of Volcanology, R. R. Coates et. al., eds., Geol. Soc. Amer., Memoir 116, 577-601.

Ross, C. P. and Forester, J. D., 1958. Outline of the geology of Idaho. Idaho Bur. of Mines and Geology, Bull. 15, 74 p.

Roth, R., 1933. Some Morrison Ostracoda. Jour. Paleontology, 7, 398-405.

Sars, G. O., 1863. Beretning om en i sommen 1862 foretagen zoologisk reise i Christianias og Trongjemstrifter. Nyt. Mag. Naturv., 12, 218-223, 249-252.

__________, 1890. Oversigt af Norges Crustacees. Norske Videnskabe-Akademi Oslo, Forhandlinger 1880, 1-80.

__________, 1922-1928. An account of the Crustacea of Norway, 9, parts 1, 2, p. 1-32, issued in 1922; parts 3, 4, p. 33-72 (1923); parts 5-12 p. 73-208 (1925); parts 12, 14, p. 209-240 (1926); parts 15, 16, p. 241-277 (1928).

Schafer, H., 1943. Über zwei neue deutsche Arten Süsswasser-Ostracoda. Zool. Anzeiger, 143, (9/10), 210-216.

Schudack, M., Turner, G. and Peterson, F., 1998? Biostratigraphy, paleoecology and biogeography of charophytes and ostracodes from the late Jurassic Morrison Formation, Western interior U. S. A. Geol. Soc. Amer. Special Paper, in press.

Schwalb, A., Lock, S. M. and Dean, W. E., 1995. Ostracode delta 18 O and delta 13 C evidence of Holocene environmental changes in the sediments of two Minnesota lakes. Jour. Paleontology, 14, 281-296.

Scott, H. W., 1944a. Muscle scar patterns on some Upper Paleozoic Ostracoda. Jour. Paleontology, 18, 162-171.

__________, 1944b. Permian and Pennsylvanian fresh-water ostracodes. Ibid., 18, 141-147.

__________, and Smith, W. H., 1951. Molt stages of an Eocene ostracode. Ibid., 25, 327-335.

__________, and Summerson, C. H., 1943. Non-marine Ostracoda from the Lower Pennsylvanian of the southern Appalachian and their bearing on intercontinental correlations. Amer. Jour. Sci., 241, 653-675.

Scourfield, D. J., 1904. Synopsis of the known species of British freshwater Entomostraca, part 3, Ostracoda, Phyllopoda and Branchiura. Quekett Micro. Club, Jour., ser. 2, 9, no. 54, 29-44.

Sharp, R. P., 1039. The Miocene Humboldt Formation in northeastern Nevada. Jour. Geology, 47, 133-160.

Sharpe, R. W., 1893. Report of the freshwater Ostracoda of the United States National Museum, including a revision of the subfamilies and genera of the Family Cyprididae. Proc. U. S. National Mus., 26, 969-1001.

_______, 1897. Contribution to a knowledge of the North American fresh-water Ostracoda included in the Families Cytheridae and Cyprididae. Bull. I11. Lab. Nat. Hist., 4, 414-482.

_______, 1898. Further report on the Ostracoda of the United States National Museum. Proc. U. S. Nat. Mus., 35, 399-430.

_______, 1918. The Ostracoda. (In) Ward and Whipple, Freshwater Biology, New York, Wiley and sons, 790-827.

Shi, C.G., 1964. The Middle and Upper Devonian Ostracoda from Dushan and Douyun, S., Kueichow. Acta Pal. Sinica, Peiping, 12, (1), p. 53.

Silitonga, P. H., 1976. Geology of part of Kittridge Spring Quadrangle, Elko County, Nevada. unpublished M. S. Thesis, Colorado School of Mines 88 p.

Smith, A. J., 1987. The taxonomy and paleoecology of the Holocene freshwater Ostracoda of Pickerel Lake, South Dakota. M. Sc. Thesis, University of Delaware.

__________, 1991. Lacustrine Ostracoda as paleohydrochemical indicators in Holocene lake records of the north-central United States. Doctoral dissertation , Brown University.

__________, 1993. Lacustrine ostracodes as hydrochemical indicators in lakes of the north-central United States. Jour. of Paleolimnology, 8, 121-134.

__________, Delorme, L. D. and Forester, R. M., 1992. A lake's solute history from ostracodes: a comparison of methods. (In) Y. Karaka and A. Maest (eds.), Water-Rock Interaction, 1, Balkema, Rotterdam, 677-680.

Smith, J. F. and Ketner, K. A., 1976. Stratigraphy of post-Paleozoic rocks and summary of resources in the Carlin-Pinon Range area, Nevada. U. S. Geological Survey, Prof. Paper 867-B, 48 p.

Smith, N., 1953. Tertiary stratigraphy of northern Utah and southeastern Idaho. Intermountain Assoc. Petrol. Geol., Guidebook, 4th Ann. Field Trip, 73-77.

__________, 1985. Written Communication.

Sohn, I. G., 1956. Pliocene ostracodes from Jackson Hole, Wyoming. Wyo. Geol. Assoc. Guidebook, 11th Ann. Field Conf., 120-122.

Sohn, I. G., 1957. Upper Jurassic-Lower Cretaceous Cyprideinae (Ostracoda) in the Black Hills (abst.). Geol. Soc. Amer. Bull., 68, 1798.

__________, 1958. Middle Mesozoic nonmarine ostracodes of the Black Hills, Wyoming Geol. Assoc., 13th Ann. Field Conf. Guidebook, 120-126.

__________, 1964. Significance of Triassic ostracodes from Alaska and Nevada. U. S. Geol. Survey. Prof. Paper 501-D, D40-D42.

__________, 1969. Nonmarine ostracodes of early Cretaceous age from Pine Valley Quadrangle, Nevada. U. S. Geol. Survey Prof. Paper 643-3, 9 p.

__________, 1972. Dunkard Ostracoda-an evaluation (abst.). (In) First I. C. White Memorial Symposium, the age of the Dunkard, Abstracts and reference papers, Morgantown, West Virginia Geol. Survey, p. 30, 31.

__________, 1975. Dunkard Ostracoda-an Evaluation. (In) Barlow, J. A., ed., Proceedings of the First I. C. White Memorial Symposium, Ibid., 265-280.

__________, 1976. Antiquity of the muscle scar attachment in Darwinula Brady and Robertson, 1885. (In) Hartmann, G., ed., International Symposium on Evolution of Post-Paleozoic Ostracoda. Abh. Verh. naturwiss. Ver. Hamburg, (NF) 18/19 Suppl., 305-308.

__________, 1977a. Muscle scars of late Paleozoic ostracodes from West Virginia, U. S. Geol. Survey, Jour. Res., 5, 135-141.

__________, 1977b. Age of the Lakota Formation (Lower Cretaceous) in South Dakota and Wyoming as determined by nonmarine ostracodes. Amer. Assoc. Petrol. Geol. Bull, 61, 832.

_____, 1977c. Radiate shell structures in Paleozoic Myodocopid and Palaeocopid ostracodes are epigenetic. Jour. Res. U. S. Geological Survey 5, 125-133.

__________, 1979. Nonmarine ostracodes in the Lakota Formation (Lower Cretaceous) from South Dakota and Wyoming. U. S. Geol. Survey Prof. Paper 1069, 24 p.

__________, 1982. Dryelbidae, n. fam. from continental Upper Jurassic and Lower Cretaceous rocks. (In) Bate, R. H., E. Robinson and L. M. Sheppard, eds. Fossil and Recent Ostracodes, Chichester, Ellis Horwood Ltd., p. 305-325.

__________, 1985. Latest Mississippian (Namurian A) nonmarine ostracodes from West Virginia and Virginia. Jour. Paleontology 59, 446-460.

__________, 1988. Darwinulicopina (Crustacea:Podocopa), a new suborder proposed for nonmarine Paleozoic to Holocene Ostracoda. Proc. Biol. Soc. Washington, 101 (4) , 817-824.

__________, and Morris, R. W., 1963. Cheikella, a new fresh-water ostracode genus, and Telekia, a new name for a homonym. Micropaleontology, 9, 317-331.

Solomon, B. J., McKee, E. H. and Anderson, D. W., 1979. Stratigraphy of depositional environments of Paleogene rocks near Elko, Nevada. (In) Armentrout, J. M. and others, eds. Cenozoic stratigraphy of the western United States, Soc. Econ. Paleont. Mineral., Pacific Section, p. 75-85.

Staplin, F. L., 1963a. Pleistocene Ostracoda of Illinois, Part 1, Subfamilies Candoninae, Cyprinae, general ecology, morphology. Jour. Paleontology, 37, 758-797.

__________, 1963b. Pleistocene Ostracoda of Illinois, Part 2, Subfamilies Cyclocyprinae, Cypridopsinae, Ilyocyprinae; Families Darwinulidae and Cytheridae, stratigraphic ranges and assembly patterns. Jour. Paleontology, 37, 1167-1203.

Stout, L. N., 1975. Polytypic species and hybridization in Quaternary freshwater ostracodes. Paleogeog., Paleoclimat., Paleoecol. 17, 257-266.

Swain, F. M., 1946. Middle Mesozoic Ostracoda from Brazil and New Mexico. Jour. Paleontology, 20, 543-555.

__________, 1947. Tertiary nonmarine Ostracoda of the Salt Lake Formation, northern Utah. Ibid., 21, 518-528.

__________, 1949. Early Tertiary Ostracoda from the western interior United States. Ibid., 23., 172-181.

__________, 1951. Corrections: recent papers on Mesozoic and Cenozoic Ostracoda. Ibid., 25, 236.

__________, 1952. Ostracoda from wells in North Carolina, Part 2, Mesozoic Ostracoda. U. S. Geol. Survey. Prof. Paper 234B, 59-93.

__________, 1955. Ostracoda from San Antonio Bay, Texas. Jour. Paleontology, 29, 561-646.

__________, 1956. Early Tertiary ostracode zones of Unita Basin. Intermountain Assoc. Petrol Geol., Field Conference, Geology and Economic Deposits of east central Utah, 125-138.

__________, 1961. Superfamily Cypridacea. (In) R. C. Moore and others, Treatise on Invertebrate Paleontology, Part Q, Arthropoda 3, Crustacea, Ostracoda. Geol. Soc. Amer., Lawrence, Kansas, 208-254.

__________, 1963. Pleistocene Ostracoda from the Gubik Formation, Arctic coastal plan, Alaska. Jour. Paleontology, 37, 798-834.

__________, 1964a. Early Tertiary Ostracoda from Colorado, Nevada and Utah and their stratigraphic distribution. Ibid., 38, 256-280.

__________, 1964b. Tertiary fresh-water Ostracoda of the Uinta Basin and related forms from southern Wyoming, Idaho and Nevada. Intermountain Assoc. Petrol. Geol. Guidebook, 13th Ann. Field Conf., Salt Lake City, Utah 173-180.

__________, 1977. Paleoecological implications of Holocene and late Pleistocene Ostracoda, Lake Lahonton Basin, Nevada. (In) H. Loffler and D. Danielopol, eds. Aspects of Ecology and Zoogeography of Recent and Fossil Ostracoda. Dr. W. Junk, The Hague, 309-320.

__________, 1982. Marine and brackish water Cretaceous Ostracoda from wells in central and southern Florida. Jour. Micropalaeontology, 1, 115-128.

__________, 1985. Some middle Cenozoic freshwater Ostracoda from western Nevada. Rev. Española de Micropaleontologia, 17, 413-446.

__________, 1986. Late Cenozoic non-marine Ostracoda from southern Idaho. Ibid., 18, 85-114.

__________, 1986b. Middle Cenozoic freshwater Ostracoda from northeastern Nevada. Ibid., 18, 181-227.

__________, 1986c. Freshwater Ostracoda from the Cache Valley Formation (Pliocene) of southeastern Idaho. Ibid, 18, 363-385.

__________,1987a. Some Ostracoda from the Salt Lake Group and Humboldt Formation (Neogene) of northern Utah and southeastern Idaho. Ibid., 19, 113-131.

__________, 1987b. Late Cretaceous? and Paleogene freshwater Ostracoda from central and eastern Nevada. Ibid., 19, 215-238.

__________, 1987c. Some Oligocene and Miocene freshwater Ostracoda from central and southern Nevada and southwestern Texas. Ibid., 19, 413-449.

__________, 1988. Pre-Pleistocene Tertiary freshwater ostracode assemblages in Nevada, northwestern Utah and southern Idaho. (In) Kopp, R. S. and Coheneur, R. E., eds. Cenozoic geology of western Utah, Utah Geol. Assoc., Salt Lake City, 315-317.

__________, 1990a. Miocene and Pliocene freshwater Ostracoda from California, Nevada and Oregon. Rev. Española de Micropaleontologia, 22, 143-159.

__________, 1990b. Species duration and extinction patterns in Cenozoic nonmarine Ostracoda, western United States. (In) Extinction Events in Earth History, E. G. Kaufmann and O. H. Walliser, eds., Lecture Notes in Earth Sciences, Springer-Verlag, Berlin, 422-424.

__________, 1995. Ostracoda from Holocene and Pleistocene lake sediments of Minnesota. Rev. Espanola de Micropaleontologia, 27, 15-67.

Swain, F. M. and Anderson E. G., 1993. Stratigraphy and Ostracoda of the Cotton Valley Group, northern Gulf coastal region. Louisiana Geol. Survey, Geol. Bull., 45, 241 p.

__________, Becker, J. and Dickinson, K. A., 1971. Paleoecology of Tertiary and fossil Quaternary non-marine Ostracoda from the western interior United States. (In) Paleoecologie Ostracodes, H. J. Oertli ed., Bull., Centre Rech., Pau, SNPA, 5 (suppl), 461-478.

__________, and Brown, P. M., 1964. Cretaceous Ostracoda from wells in the southeastern United States. North Carolina Dept. Cons. and Dev., Bull. 78, 55 p.

__________________________, 1972. Lower Cretaceous, Jurassic? and Triassic Ostracoda from the Atlantic coastal region. U. S. Geol. Survey Prof. Paper 795, 55 p.

__________, and Wagoner, J. C., 1979. Biostratigraphy and paleoecology of freshwater Ostracoda from Esmeralda Formation (Neogene) of western Nevada. (In) Taxonomy, Biostratigraphy and Distribution of Ostracoda, N. Krstic ed., Beograd, 191-196.

__________, 1985. Some freshwater Ostracoda from Esmeralda Formation (Neogene) of western Nevada. Rev. Española de Micropaleontologia, 17, 123-140.

Swartz, F. M. and Swain, F. M., 1946. Ostracoda from the Upper Jurassic Cotton Valley Group of Louisiana and Arkansas. Jour. Paleontology, 20, 362-373.

Swirydczuk, K., Wilkinson, B. H. and Smith, G. R., 1979. The Pliocene Glenns Ferry oolite: lake margin carbonate deposition in the southwestern Snake River Plain. Jour. Sed. Petrol. 49, 995-1004.

Tasch, P., 1963. Paleolimnology: Part 3, Marion and Dickinson Counties, Kansas, with additional sections in Harvey and Sedgwick Counties: stratigraphy and biota. Jour. Paleontology, 37, 1233-1251.

Taylor, R. S., 1972. Paleoecology of ostracodes from the Luman Tongue and Tipton Member (early Eocene) of the Green River formation, Wyoming. University of Kansas, Ph.D. Dissertation, 96 p.

Teeter, J. W., 1970. Paleoecology of a Pleistocene microfossil assemblage at the Fairlawn, Ohio mastodon site. Am. Mid. Nat. 83, (2), 583-594.

Tressler, W. L., 1954. Freshwater Ostracoda from Texas and New Mexico. Wash. Acad. Sci., Jour. 44, 138-149.

__________, 1957. The Ostracoda of Great Slave Lake. Ibid., 47, 415-423.

Turner, C. H., 1894. Notes on American Ostracoda with descriptions of new species. Bull. Sci. Lab. Denison Univ., 6, 57-74.

__________, 1895. Freshwater Ostracoda of the United States. Second Rept. State Zool., Minnesota, 277-337.

Turner, H. W., 1900. The Esmeralda Formation, a freshwater lake deposit. U. S. Geol. Survey., Ann. Rept. (21st), part 2, 191-208.

Ulrich, E. O. and Bassler, R. S., 1908. New American Paleozoic Ostracoda. Preliminary revision of the Beyrichiidae with descriptions of new genera. U. S. National Mus. Proc., 35, 277-340.

Vanderpool, H. C., 1928. Fossils from the Trinity Group (Lower Comanchean). Ibid., 2, 95-107.

__________, 1930. Cretaceous section of Maverick County, Texas. Ibid., 4, 252-258.

Vanderpool, H. C., 1933. Upper Trinity microfossils from southern Oklahoma. Ibid. 7, 406-411.

Van Frausen, A. and Wouters, K., 1990. Ostracoda from Holocene calcareous tufa deposits in southern Belgium; a paleoenvironmental analysis. (In) Ostracoda and Global Events, Whatley, R. and Maybury, C., eds., London, Chapman and Hall, 505-511.

Van Houten, F. B., 1956. Reconnaissance of Cenozoic sedimentary rocks of Nevada. Amer. Assoc. Petrol. Geol., Bull, 40, 2801-2805.

Van Nieuwenhuise, D. and He Jun-de, 1986. Implications of Arikareean (Oligocene-Miocene) Ostracoda and foraminiferi asemblages from lacustrine deposits in southwestern Montana. Proc Fourth North American Palentological Convention, Boulder, Colo.

Vavra, V., 1891. Monographie der Ostracoden Böhems. Arch. naturw., Landesf. Bohmens, 8, i-iv, 1-116.

__________, 1898, Süsswassern-Ostracoden, in Ergebn. Hamburg Magalhaensiche Sammelreise, 2, 1026.

Vernet, H., 1878. On an ostracode crustacean of a new genus (Acanthopus) met in deep water of Lake Geneva. Ann. Mag. Nat. Hist., ser. 7, 1,

Weedman, S. D., 1994. Upper Allegheny Group (Middle Pennsylvanian) lacustrine limestones of the Appalachian Basin, U. S.A. (In) Global Geologic Record of Lake Basins, vol. 1, E. Gierlowski-Kordesch and K. Kelts, eds., Cambridge Univ. Press, 127-134.

White, C. A., 1895. The Bear River Formation and its characteristic fauna. U. S. Geol. Survey Bull. 128, 108 p.

Williams, J. S., 1948. Geology of the Paleozoic rocks, Logan Quadrangle, Utah. Geol. Soc. Amer. Bull. 59, 1130-1147.

Winckler, E. M., 1960. Post-Pleistocene ostracodes of Lake Nipissing age. Jour. Paleontology, 34, 923-932.

Winfrey, W. M., 1958. Stratigraphy, correlation and oil potential of the Sheep Pass Formation, east central Nevada. Amer. Assoc. Petrol. Geol. Rocky Mountain Section of Geophysical Records, 77-82.

Wise, C. D., 1962. Unpublished Doctoral Dissertation, University of New Mexico.

Xu, M., 1983. Ostracods from the Mesozoic coal-bearing strata of South China. (In) Maddocks, R. F., ed., Applications of Ostracoda, Univ. Houston, Texas, 352-371.

__________, 1988. Ostracods from the Mesozoic coal-bearing strata of northern Shanxi, China. (In) Hanai, T., Ikeya, N. and Ishizaki, K., eds. Evolutionary Biology of Ostracoda, Amsterdam, Elsevier, 1283-1291.

Young, J. C., 1960. Structure and stratigraphy in north-central Shell Creek Range. (In) Guidebook to the geology of east-central Nevada. J. W. Boettcher and W. W. Sloan, Jr., eds. Intermountain Assoc. Petrol Geol. 11th Ann. Field Conf. Guidebook, 158-172.

Zaddach, E. G., 1844. Synopseos crustaceorum Prussicorum prodromus. Dissertatio Zoologica. Regiomonti (Kohnigsberg).

Zanina, I., Ye., 1960. New Upper Devonian Ostracoda from the central regions of the Russian Platform. Pal. Zhurnal, Moscow, No. 3, p. 87.

Zenker, W., 1854. Monographie der Ostracoden. Arch. Naturg., 20, 1-87.

Plates 1- 34

Permission to reproduce illustrations has been obtained from the following copyright holders or authors and is gratefully acknowledged. The number in parentheses following each figure caption corresponds to the listing below.

1. Paleontological Society
2. Revista Espãnola de Micropaleontologia
3. University of Kansas, Department of Geology
4. Kansas Geological Society
5. I. G. Sohn, 1982
6. North Carolina Geological Society
7. New Mexico Geological Society
8. U. S. Government Printing Office
9. I. G. Sohn, 1956
10. Verlag Paul Parey
11. American Journal of Science
12. Carnegie Museum of Natural History
13. Geological Society of America
14. Journal of Micropaleontology
15. Geological Magazine, Cambridge University Press
16. University of California, Berkeley
17. Arizona Nevada Academy of Science, S. G. Lucas
18. Louisiana Geological Society
19. Robert F. Lundin
20. Illinois Geological Survey

Fossil Non-marine Ostracoda, United States

Plate 1

Figures 1-3. Advenocypris? concinnus Swain, 1985. 1, R side holotype; 2, R side, L, 0.97; 3, dorsal view. Miocene, Esmeralda Formation, Black Spring road section, loc. 3, Nye County, Nevada (after Swain, 1985). (2)

Figure 4. Advenocypris umbonata Swain, 1964. R side, holotype, L, 1.00. Eocene, 315 feet (96 m) above base of Colton-Green River transition beds, east end of Raven Ridge, Rio Blanco County, Colorado (after Swain, 1964). (1)

Figure 5. Amplocypris? sp., Swain, 1987. R side of fractured shell, L, 1.30. Eocene or early Oligocene, Elko Formation, Catlin oil shale works, Elko Nevada (after Swain, 1987). (2)

Figures 6, 7. Candona subovata jones, 1893. R side and dorsal views, L, 1.20. Upper Cretaceous, Montana Formation, Coalville, Utah (after Jones, 1893). (15)

Figures 8, 9. Candona gracilis Chapman, 1896. R side and dorsal view, L, 0.71. Pliocene, Merced Formation, Wildcat Canyon, east of Berkeley, California (after Chapman, 1896). (15)

Figures 10-12. Bradycypris? medicinecreekensis Swain, 1986b. 10, R side, holotype; 11, dorsal view, L, 0.76; 12, dorsal view, L, 1.01; Miocene, Humboldt Formation; 10, 11, "Bobcat" limestone, near base of Medicine Creek section; 12, 161-169 feet (49-51.5 m) above base of Medicine Creek section, Winecup Ranch, Elko County, Nevada (after Swain, 1986b). (2)

Figure 13. Bradycypris? cf. medicinecreekensis Swain, 1990. R side, L, 1.30. Pliocene, Beartrap Quadrangle, Los Angeles County, California (after Swain, 1990). (2)

Figure 14. Bradycypris? sp., Swain and Wagoner, 1985. R side, L, 1.03. Miocene, Esmeralda Formation, Blair Junction section, sample C1J, Esmeralda County, Nevada (after Swain and Wagoner, 1985). (2)

Figures 15, 16. Candona (Pseudocandona?) albicans (Brady), Staplin, 1963. 15, RV female; 16, LV female, L, 0.80. Pleistocene, Mankato or late Cary Substage, Wisconsin Stage, 1.5-2 miles (2.4-3.2 km) east of Mt. Vernon, Indiana (after Staplin, 1963). (1)

Figures 17, 18. Candona acuta Hoff, Staplin, 1963. 17, RV female; 18, LV male, L, 1.12. Pleistocene, 17, Tazewell Substage, Wisconsin Stage, SW1/4 NW1/4 sec. 24, T. 2 N., R. 1 E., Fulton County, Illinois; Mankato or Cary Substage, Wisconsin Stage, 1.5-2 miles (2.4-3.2 km) east of Mt. Vernon, Indiana (after Staplin, 1963). (1)

Figures 19, 20. *Candona cachensis* Swain, 1947. 19, LV interior, holotype; 20, dorsal view, L, 1.23. Pliocene, Salt Lake Group, sec. 15, T. 13 N., R. 2 W., Cache County, Utah (after Swain, 1947). (2)

Figure 21. *Candona bretzi* Staplin, 1963. LV, female, L, 1.15. Pleistocene, Illinoian Stage, SW1/4 NW1/4 sec. 9, T. 8 S., R. 2 W., Calhoun County, Illinois (after Staplin, 1963). (1)

Figures 22a, b. *Candona cachensis* Swain, Swain, 1987. 22a, RV interior, L, 1.03; 22b, adductor muscle scar area, approx. X 310. Pliocene, Cache Valley Formation, Hoopes Creek loc., Franklin County, Idaho (after Swain, 1987). (2)

Figures 23, 24. *Candona* cf. *cachensis* Swain, Sohn, 1956. 23, LV male; 24, dorsal view, L, 1.46. Pliocene, Teewinot Formation, Jackson Hole, Wyoming (after Sohn, 1956). (9)

Figure 25. *Candona candida* (O.F. Müller), Staplin, 1963. RV female, L, 1.15. Pleistocene, Illinoian Stage, SW1/4 SE1/4 NE1/4 sec. 12, just north of Brussels, Johnson County, Illinois (after Staplin, 1963). (1)

Figure 26. *Candona* sp. aff. *C. candida* (O.F. Müller), Dickinson and Swain, 1967. LV, L, 1.25. Pliocene, Hay Ranch Formation ("Upper Humboldt"), Gulf Oil Co. No. 1 Wilkins Ranch well, 1194-1224 feet (364-373 m), Elko County, Nevada (after Dickinson and Swain, 1967). (1)

Figures 27-30. *Candona catherinecreekensis* Swain, 1986. 27, R side, female; 28, dorsal view, female, X 38; 29, LV male, X 36; 30, R side holotype, male, L, 1.69. Pliocene, Glenns Ferry Formation, sec. 14, T. 4 S., R. 1 W., Owyhee County, Idaho (after Swain, 1986). (2)

Figures 31, 32. *Candona caudata* Kaufmann, Staplin, 1963. 31, LV female; 32, LV immature female, L, 1.32. Pleistocene, 31, Tazewell Substage, Wisconsin Stage, Bloomington silts, NW1/4 NE1/4 sec. 11, T. 4 N., R. 3 E., Fulton County, Illinois; 32, Cary or Mankato Substage, Blue Island, sec. 5, T. 37 N., R. 14 E., Chicago, Illinois (after Staplin, 1963). (1)

Plate 1

1
3
4
7
5
8
9
6
2
10
16
15
12
11
13
17
20
19
14
18
21
22a
24
23
25
22b
26
31
30
27
32
29
28

Plate 2

Figures 1a, b. Candona acuta Hoff, Benson, 1967. 1a, RV, adult, L, 1.16; 1b, adductor and frontal muscle scars of RV, approx. X 160. Pleistocene, Sappa Formation, Osborne County, Kansas (after Benson, 1967). (3)

Figures 2, 3. Candona artesensis Swain, 1964. 2, R side holotype shell, L, 0.84; 3, L side of poorly preserved shell, L, 0.85. 2, Eocene, 59 feet (18 m) above base of Green River Formation, Dripping Rock Gulch section, east end of Raven Ridge, Rio Blanco County, Colorado; 3; Eocene, 65 feet (19 m) above base of Colton Formation, same loc. as preceding (after Swain, 1964).

Figures 4, 5. Candona candida (O.F. Müller), Swain, 1963. 4, LV, X 40; 5, RV interior, L, 1.10. Pleistocene, Gubik Formation, Arctic coastal plain, Alaska; 4, Point Barrow core test no. 1, 457-465 feet (139-142 m); 4, Ikpikpuk shot holes, line 8, shot point 10, 45-50 feet (14-15 m) (after Swain, 1963). (1)

Figure 6. Candona caudata Kaufmann, Benson, 1967. LV, L, 1.30. Pleistocene, Sappa Formation, Osborne County, Kansas, U.S.N.M. 649614, loc. U. (after Benson, 1967). (3)

Figures 7-9. Candona chasei Dickinson and Swain, 1967. 7, R side, 1.22; 8, L side of holotype; 9, dorsal view. Pliocene, Hay Ranch Formation (="Upper Humboldt") Gulf Oil Co. no. 1 Wilkins Ranch well 1648-1680 feet (502-512 m), Elko County, Nevada (after Dickinson and Swain, 1967). (1)

Figures 10, 11. Candona clauseni Swain, 1987. 10, R side, L, 1.06, and dorsal view of holotype. Miocene, Esmeralda Formation, Black Spring Road, loc. 1, Nye County, Nevada (after Swain, 1987c). (2)

Figures 12, 13. Candona cf. compressa (Koch), Swain, 1947. 12, LV; 13, interior LV, L, 1.27. Pliocene, Salt Lake Group, sec. 15, T. 13 N., R. 2 W., Cache County, Utah (after Swain, 1947). (1)

Figures 14, 15. Candona crogmaniana Turner, Staplin, 1963. 14, RV male; 15, RV female, L, 1.32. Pleistocene, Illinois (after Staplin, 1963). (1)

Figures 16, 17. Candona devexoidea Dickinson and Swain, 1967. 16, RV female; 17, LV female, L, 1.17. Pliocene, Hay Ranch Formation, 1 mile (1.6 km) south of Elko, Nevada (after Dickinson and Swain, 1967).

Figure 18. Candona cf. devexoidea Dickinson and Swain, Swain, 1986. R side, L, 1.27. Miocene, Humboldt formation, Gulf Oil Co. no. 2 Wilkins Ranch well 3610-3640 feet (1100-1109 m), Elko County, Nevada (after Swain, 1986). (2)

Figures 19-21. Candona diffusoidea Swain, 1987. 19, L side, holotype, L, 1.15; 20, dorsal view; LV interior. Pliocene, Cache Valley Formation, Hoopes Creek loc., Franklin County, Idaho (after Swain, 1987). (2)

Figures 22, 23. Candona dipunta Swain, 1987. 22, R side holotype, L, 1.03; 23, dorsal view, L, 0.77. Pliocene, Cache Valley Formation, Hoopes Creek loc., Franklin County, Idaho (after Swain, 1987). (2)

Figures 24, 25. Candona distincta Furtos, Staplin, 1963. 24, LV female, X 25; 25, RV male, L, 1.02. Pleistocene, Illinois (after Staplin, 1963). (1)

Figures 26, 27. Candona distincta Furtos, Benson, 1967. 26, LV male, X 29; 27, RV female, L, 1.19. Pleistocene, Sappa Formation, Kansas (after Benson, 1967). (3)

Figures 28a, b, 29. Candona sp. aff. C. elliptica Furtos, Swain, 1987. 28a, LV interior, L, 0.72; 28b, adductor muscle scar, approx. X 375; 29, L side, X 64. Pliocene, Cache Valley Formation, Hoopes Creek loc., Franklin County, Idaho (after Swain, 1987). (2)

Plate 2

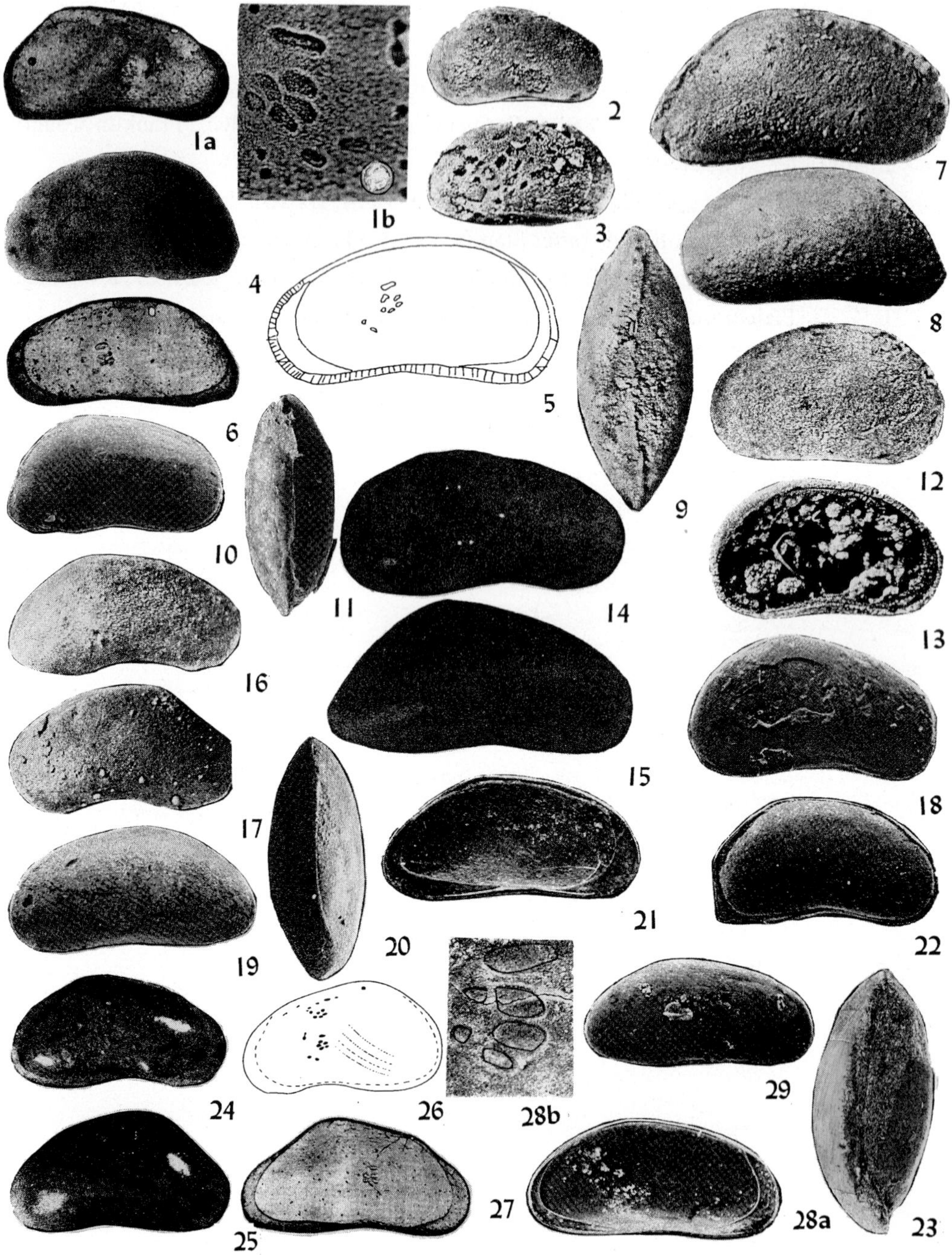
1a
1b
2
3
4
5
6
7
8
9
10
11
12
13
14
15
16
17
18
19
20
21
22
23
24
25
26
27
28a
28b
29

Plate 3

Figures 1, 2. Candona elliptica alta Staplin, 1963. 1, LV female, X 40; 2, RV male, L, 0.92. Pleistocene, Mankato Substage, Wisconsin Stage, northern Illinois (after Staplin, 1963). (1)

Figures 3-5. Candoniella elongatolactoides Dickinson and Swain, 1967. 3, RV interior, L, 0.65; 4, LV?, L, 0.54; 5, dorsal view, L, 0.67. Pliocene, Hay Ranch Formation, 1 mile (1.6 km) south of Elko Nevada (after Dickinson and Swain, 1967). (1)

Figure 6. Candoniella cf. elongatolactoides (Dickinson and Swain), Swain, 1987. RV, L, 0.70. Pliocene, Cache Valley Formation, Hoopes Creek loc., Franklin County, Idaho (after Swain, 1987). (2)

Figure 7. Candoniella cf. elongatolactoides (Dickinson and Swain), Swain, 1986. R side, L, 0.93. Pliocene, Glenns Ferry Formation, sec. 14, T. 4 S., R. 1 W., Owyhee County, Idaho (after Swain, 1986). (2)

Figures 8, 9. Candoniella elongatolactoides (Dickinson and Swain), Swain, 1990. 8, LV, L, 1.32; 9, dorsal view, L, 1.39. Pliocene, Idaho Group, 15 miles (24 km) west of Burns, Oregon (after Swain, 1990). (2)

Figures 10, 11. Candoniella sp. aff. C. elongatolactoides Dickinson and Swain, Swain, 1990. 10, R side of male? shell, L, 1.10; 11, R side of male? shell, L, 1.34. Pliocene, Idaho Group, 15 miles (24 km) west of Burns, Oregon, just below lava flow (after Swain, 1990). (2)

Figure 12. Candona eriensis? Furtos, Staplin, 1963. LV female, L, 1.22. Pleistocene, Tazewell Stage, Bloomington silts, Fulton County, Illinois (after Staplin, 1963). (1)

Figure 13. Candona fluviatilis Hoff, Staplin, 1933. RV female, L, 0.77. Recent, Villa Park, Du Page County, Illinois; also occurs in Pleistocene, Tazewell Stage of Illinois and Wisconsin Stage of Ohio (after Staplin, 1963). (1)

Figures 14, 15. Candona havanaensis Staplin, 1963. 14, RV female, L, 0.77; 15, RV male, L, 1.00. Pleistocene, Tazewell Stage, Illinois River Valley, Illinois (after Staplin, 1963). (1)

Figures 16-18. Candona houghi Staplin, 1963. 16, RV of 8th instar, L, 1.20; 17, LV of 9th instar, L, 1.21; 18, LV of 4th instar, L, 0.40. Pleistocene and Recent, Lake Michigan Basin cores, Illinois (after Staplin, 1963). (1)

Figures 19-21. Candona hyalina Brady and Robertson, Swain, 1963. 19, LV; 20, LV interior; 21, LV interior, L, 1.10. Pleistocene, Gubik Formation, Ikpikpuk shot holes, line 10, shot point 31, 40-50 feet (after Swain, 1963). (1)

Figures 22, 23. Candona illinoiensis Staplin, 1963. 22, RV male; 23, RV female (holotype), L, 0.85. Pleistocene, Illinoian Stage, Brussels Quadrangle, Johnson County, Illinois (after Staplin, 1963). (1)

Figure 24. Candona kentuckyensis Scott and Summerson, 1943. R side, L, 1.07. Lower Pennsylvanian, Kanawha Series, Hance Formation, Molus, Harlan County, Kentucky (after Scott and Summerson, 1943).

Figures 25a, b, 26. Candona kingscreekensis Swain, 1987. 25a, R side, L, 1.43; 25b, enlargement of posteroventral margin, approx. 310; 26, dorsal view, L, 1.31. Pliocene, Cache Valley Formation, Hoopes Creek loc., Franklin County, Idaho (after Swain, 1987). (2)

Figures 27, 28. Candona kittlemani Swain, 1990. 27, R side; 28, LV interior, L, 0.97. Pliocene, Idaho Group, sec. 9, T. 22 S., R. 46 E., Malheur County, Oregon (after Swain, 1990). (2)

Figure 29. Candona cf. lactea Baird, Staplin, 1963. RV female, L, 0.80. Pleistocene, Cary or Mankato Substage, Wisconsin Stage, Chicago Illinois (after Staplin, 1963). (1)

Figure 30. Candona cf. lactea Baird, Benson, 1967. LV, L, 0.70. Pleistocene, Sappa Formation, Osborne County, Kansas (after Benson, 1967). (3)

Figure 31. Candona cf. lactea? Baird, Swain, 1963. LV, L, 0.77. Pleistocene, Gubik Formation, Ikpikpuk shot holes, line 7, shot point 3, 60-70 feet, Arctic coastal plain, northern Alaska (after Swain, 1963). (1)

Figures 32, 33. Candona leightoni Staplin, 1963. 32, LV male; 33, LV female, L, 0.97. Pleistocene, Illinoian, SW$^{1}/_{4}$, NW$^{1}/_{4}$ sec. 9, T. 8 S., R. 2 W., Calhoun County, Illinois (after Staplin, 1963). (1)

Figures 34, 35. Candona magnitata (Holland), Scott, 1944. 34, LV, L, 1.04; 35, LV, X 23. Permian, Nineveh Limestone, Nineveh, Pennsylvania (after Scott, 1944). (1)

Figures 36, 37. Candona bairdioides (Jones and Kirkby, 1879). R side and dorsal view. Pennsyvlanian, Virgilian Series, Cohn Cyclothem, loc. 95. Central Illinois (after Cooper, 1946). (20)

Figures 38, 39. Candona planidorsata Cooper, 1946. R side and dorsal view. Pennsylvanian, Des Moinesian Series, Hanover Limstone, loc. 31, western Illinois (after Cooper, 1946). (20)

Plate 3

1
2
3
4
5
6
7
8
9
10
11
12
13
14
15
16
17
18
19
20
21
22
23
24
25
25a
26
27
28
29
30
31
32
33
34
35
36
37
38
39

Plate 4

Figures 1, 2. Candona? magnitata (Holland), Scott, 1944. 1, RV; 2, dorsal view, L, 1.86. Permian, Nineveh Limestone, Nineveh, Pennsylvania (after Scott, 1944). (1)

Figure 3. Candona (Pseudocandona?) marengoensis (Klie), Staplin, 1963. LV female, L, 0.67. Pleistocene, Mankato or late Cary Substage, Wisconsin Stage, 2 miles E. of Mt. Vernon, Illinois on Route 62 (after Staplin, 1963). (1)

Figure 4. Candona (Pseudocandona?) marengoensis (Klie), Benson, 1967. RV, L, 0.80. Pleistocene, Sappa Formation, Osborne County, Kansas (after Benson, 1967). (3)

Figure 5. Candona minuticandida Staplin, 1963. LV female, L, 0.67. Pleistocene, Tazewell Substage, Wisconsin Stage, SW cor. SE1/4 NE1/4 sec. 4, T. 8 S., R. 2 W., Jackson County, Illinois (after Staplin, 1963). (1)

Figures 6, 7. Candona rawsoni Tressler, Gutentag and Benson 1962 (=C. nyensis of Gutentag and Benson). 6, RV female (holotype of C. nyensis); 7, RV male, L, 0.78. Pleistocene, Illinoian-Wisconsin, Nye sink, NW1/4 SW1/4 sec. 15, T. 6 N., R. 25 E., Beaver County, Oklahoma (after Gutentag and Benson, 1962). (4)

Figures 8, 9. Candona rawsoni Tressler, Gutentag and Benson, 1962. 8, LV female; 9, LV male (see scale bar for magnification); same loc. and horizon as 6 and 7 (after Gutentag and Benson, 1962). (4)

Figures 10, 11. Candona ohioensis Furtos, Staplin, 1963. 10, RV female; 11, LV male, L, 1.78. Pleistocene, pro-Illinoian, SE1/4 SW1/4 sec. 5, T. 24 N., R. 4 W., Tazewell County, Illinois (after Staplin, 1963). (1)

Figure 12. Candona cf. ohioensis Furtos, 1933. LV female, L, 1.24. Lower Oligocene, Chadron Formation, Helmville, Nebraska. (coll. by R. Konizeski).

Figures 13-15. "Candona" paleostraca Tasch, 1963. 13, RV; 14, LV; 15, dorsal view of holotype, L, 0.85. Permian, Wellington Formation, Elmo section VIII B, bed 10, NE1/4 NW1/4 sec. 21, T. 16 S., R. 2 E., Dickinson County, Kansas (after Tasch, 1963). (1)

Figures 16-18. Candona parabretzi Swain and Wagoner, 1985. 16, R side holotype male?, L, 0.91; 17a, b, R side female, L, 0.94, and enlargement of posterior surface, approx. X 650 showing minute reticulation; 18, dorsal view, immature shell, L, 0.47. Miocene, Esmeralda Formation, Blair Junction section, Esmeralda County, Nevada (after Swain and Wagoner, 1985). (2)

Figures 19-21. Candona paracaudata Swain, 1947. 19, R side; 20, RV interior; 21, LV holotype, interior, L, 1.18. Pliocene, Salt Lake Group, sec. 15, T. 13 N., R. 2 W., Cache County, Utah (after Swain, 1947). (1)

Figures 22, 23. Candona cf. paracaudata Swain, Swain, 1987. 22, R side, L, 1.18; 23, dorsal view. Pliocene, Salt Lake Group, South Clarkston section, T. 13 N., R. 12 W., Cache County, Utah (after Swain, 1987). (2)

Figures 24, 25. Candona cf. paracaudata Swain, Swain, 1990. 24, R side, L, 0.81; 25, dorsal view, X 48. Pliocene, Beartrap Quadrangle, Los Angeles County, California (after Swain, 1990). (2)

Figures 26, 27. Candona cf. paraohioensis Staplin, 1963. 26, RV female; 27, LV male, L, 1.03. Pleistocene, pro-Illinoian?, SE1/4 SW1/4 sec. 5, T. 24 N., R. 4 W., Tazewell County, Illinois (after Staplin, 1963). (1)

Figure 28. Candona cf. paraohioensis Staplin, Swain and Wagoner, 1985. R side, L, 0.75. Miocene, Esmeralda Formation, Blair Junction section, Esmeralda County, Nevada (after Swain and Wagoner, 1985). (2)

Figure 29. Candona cf. paraohioensis Staplin, Swain, 1985. R side, L, 1.46. Miocene, Esmeralda Formation, Black Spring road section, Nye County, Nevada (after Swain, 1985). (2)

Figures 30-32. Candona cf. patzcuaro Tressler, Swain, 1987. 30, R side, L, 1.14; dorsal view immature shell, L, 0.65; 32, RV interior, X 35. Pliocene, Cache Valley Formation, Hoopes Creek loc., Franklin County, Idaho (after Swain, 1987). (2)

Figure 33. Candona pearlensis Staplin, 1963. LV, intermediate instar, L, 0.71. Pleistocene, Illinoian Stage, Pearl Quadrangle, NW1/4SW1/4SW1/4 sec. 22, T. 9 S., R. 2 W., Calhoun County, Illinois (after Staplin, 1963). (1)

Plate 4

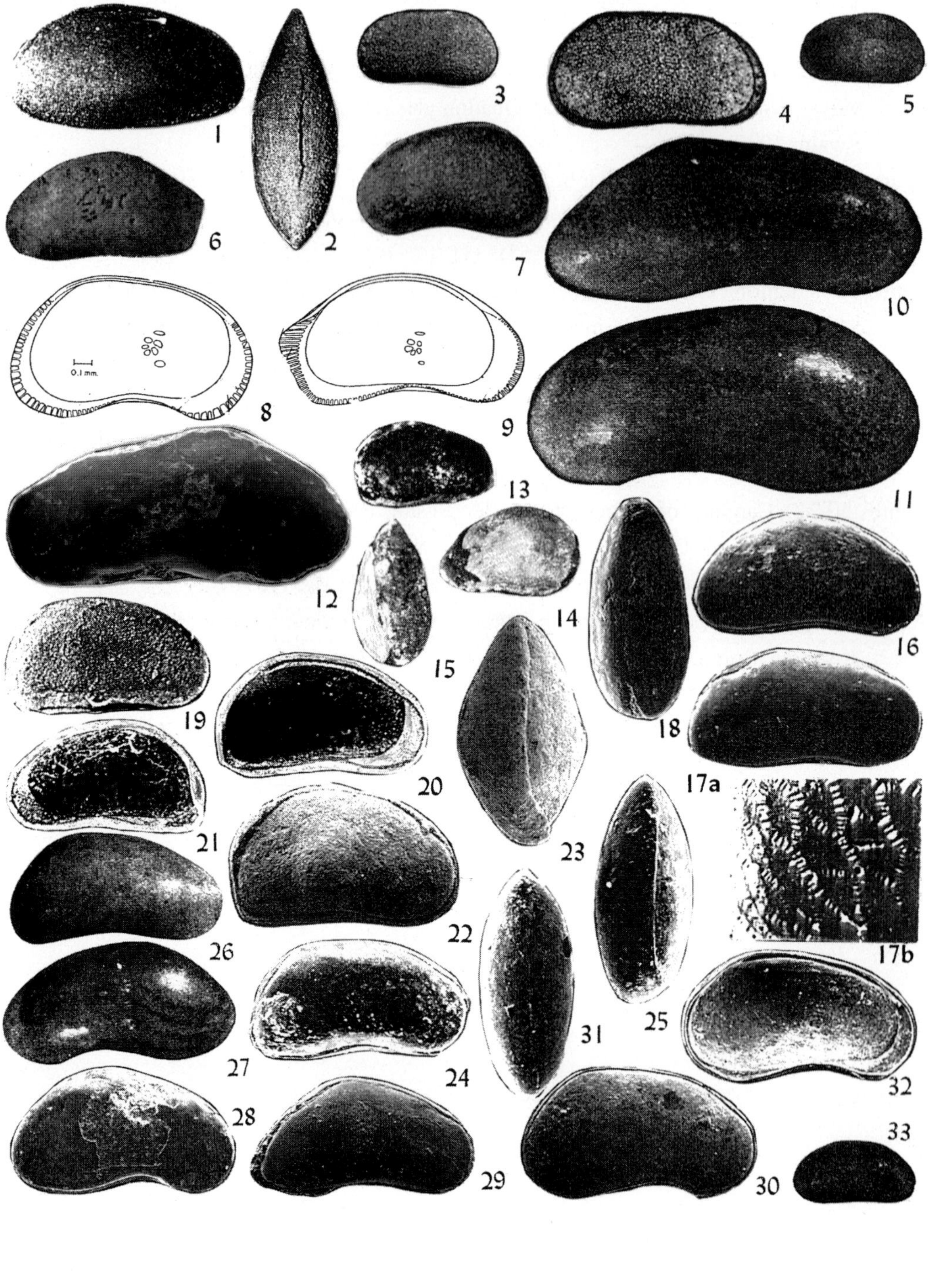

1
2
3
4
5
6
7
8
0.1 mm.
9
10
11
12
13
14
15
16
17a
17b
18
19
20
21
22
23
24
25
26
27
28
29
30
31
32
33

Plate 5

Figures 1, 2. Candona pearlensis Staplin, 1963. 1, LV female, holotype, L, 0.96; 2, LV male, L, 1.01. Pleistocene, Illinoian Stage, Pearl Quadrangle NW1/4SW1/4SW1/4 sec. 22, T. 9 S., R. 2 W., Calhoun County, Illinois (after Staplin, 1963). (1)

Figures 3, 4. Candona poseyensis Staplin, 1963. 3, LV female, holotype, L, 0.95; 4, RV male, L, 1.00. Pleistocene, Mankato or late Cary Substage, Wisconsin Stage, 1.5-2 miles (2.4-3.2 km) east of Mt. Vernon, Indiana on Route 62 (after Staplin, 1963). (1)

Figures 5-7. Candona cf. protzi Hartwig, Swain, 1990. 5, RV exterior, L, 1.26; 6, RV dorsal view, L, 1.25; 7, R side of shell on which adductor muscle scar inpression visible from exterior, X 33. Pliocene, Idaho Group, 15 miles (24 km) west of Burns, Oregon, just below lava flow (after Swain, 1990). (2)

Figure 8. Candona cf. poseyensis Staplin, 1963. RV, L, 0.72. Oligocene, Clarkson Formation, Helmville, Nebraska. (coll. by R. Konizeski).

Figure 9. Candona (Pseudocandona?) punctata (Furtos), Staplin, 1963. RV female, L, 0.83. Pleistocene, late or post-Mankato lake deposit, Sag Bridge, SE cor. sec. 17, T. 37 N., R. 12 E., Chicago, Illinois (after Staplin, 1963). (1)

Figure 10. Candona (Pseudocandona?) pustulosa (Staplin) 1963. LV, holotype, L, 0.87. Pleistocene, Mankato or late Cary Substage, Wisconsin Stage, sec. 26, T. 9 S., R. 12 E., 2.2 miles (3.5 km) west of Shawneetown, Gallatin County, Illinois (after Staplin, 1963). (1)

Figures 11, 12. Candona rangliensis Swain, 1964. 11, L side of shell; 12, R side of holotype shell, L, 1.00. Eocene, 130 feet (39 m) above base of Colton-Green River transition beds, east end of Raven Ridge, Rio Blanco County, Colorado (after Swain, 1964). (1)

Figures 13, 14. Candona sp. aff. C. rangliensis Swain, Swain, 1987c. 13, R side of shell, L, 0.39; 14, dorsal view of shell. Oligocene Pruett Formation, Buck Hill volcanic series, Brewster County, Texas (after Swain, 1987c). (2)

Figure 15. Candona renoensis Gutentag and Benson, 1962. LV, female holotype, L, 0.65. Pleistocene substage, Big Springs Ranch, SW sec. 17, T. 32 S., R. 28 W., Meade County, Kansas (after Gutentag and Benson, 1962). (4)

Figures 16a, b. Candona renoensis Gutentag and Benson, Benson, 1967. 16, LV male, L, 1.44. 16b, adductor muscle scar, approx. X 190. Pleistocene Yarmouth Substage, Dickinson County, Kansas (after Benson, 1967). (3)

Figures 17, 18. Candona scopulosa Furtos, Staplin, 1963. 17, RV male, L, 1.07; 18, RV female, L, 0.99. 17, Pleistocene, Tazewell substage, Wisconsin Stage, NE1/4 NW1/4 NW1/4 sec. 24, T. 42 N., R. 7 E., Kane County, Illinois. 18, Illinoian Stage, SW1/4

NW1/4 SW1/4 sec. 9, T. 8 S., R. 2 W., Calhoun County, Illinois (after Staplin, 1963). (1)

Figures 19, 20. Candona shawneensis Staplin, 1963. 19, RV female, L, 0.87; 20, LV male, L, 1.06. Pleistocene, Mankato or late Cary Substage, Wisconsin Stage, sec. 26, T. 9 S., R. 9 E., Gallatin County, Illinois (after Staplin, 1963). (1)

Figure 21. Candona sigmoides Sharpe, Staplin, 1963. LV female, L, 1.06. Pleistocene, Illinoian Stage, SW1/4 NW1/4 SW1/4 sec. 9, T. 8 S., R. 2 W., Calhoun County, Illinois (after Staplin, 1963). (1)

Figures 22, 23. Candona simpsoni Sharpe, Staplin, 1963. 22, LV female, 23, RV female, L, 0.82. Pleistocene, Tazewell? Substage, Wisconsin Stage, SE1/4 NW1/4 sec. 4, T. 9 S., R. 1 W,. Jackson County, Illinois (after Staplin, 1963). (1)

Figure 24. Candona simpsoni Sharpe, Benson, 1967. RV, L, 0.87. Pleistocene, Sappa Formation, Osburne County, Kansas (after Benson, 1967). (3)

Figures 25, 26. Candona subtriangularis Swain, 1947. 25, RV interior, holotype; 26, RV exterior, partly exfoliated, L, 0.90. Pliocene, Salt Lake Group, sec. 15, T. 13 N., R. 2 W., Cache County, Utah (after Swain, 1947). (1)

Figures 27, 28. Candona cf. subtriangularis Swain, 1985. 27, Right side; 28, dorsal view, L, 1.16. Miocene, Esmeralda Formation, Simon Mine area, 3.3 miles (5.3 km) east of Finger Rock Wash, Mineral County, Nevada (after Swain, 1985). (2)

Figures 29, 30. Candona cf. subtriangularis Swain, Swain, 1990. 29, LV, L, 1.08; 30, RV interior, L, 1.41. Pliocene, Idaho Group NE1/4 SW1/4 sec. 9, T. 22 S., R. 46 E., Malheur County, Oregon (after Swain, 1990). (2)

Figures 31-33. Candona rawsoni Tessler [=Candona swaini Staplin, 1963], Staplin, 1963. 31, LV female, holotype of C. swaini; 32, LV male; 33, dorsal view, L, 1.05. Pleistocene, 31, 32, Cary or Mankato Substage, Glenwood or Calumet Stage of Lake Chicago clay pit, sec. 5, T. 37 N., R. 14 E., Chicago; 33, Lake Mankato? Substage, Lake Michigan core (after Staplin, 1963). (1)

Figures 34, 35. Candona thomasi Staplin, 1963. 34, LV female holotype; 35, LV male, L, 0.76. Pleistocene, Cary or Mankato Substage, Wisconsin Stage, sec. 29, T. 48 N., R. 24 W., Storey County, Iowa (after Staplin, 1963). (1)

Figures 36, 37. Candona salteriana Jones, Cooper, 1946. L side and ventral view, L, 1.32. Pennsylvanian, Virgilian Series, Shumway Cyclothem, central Illinois (after Cooper, 1946). (20)

Plate 5

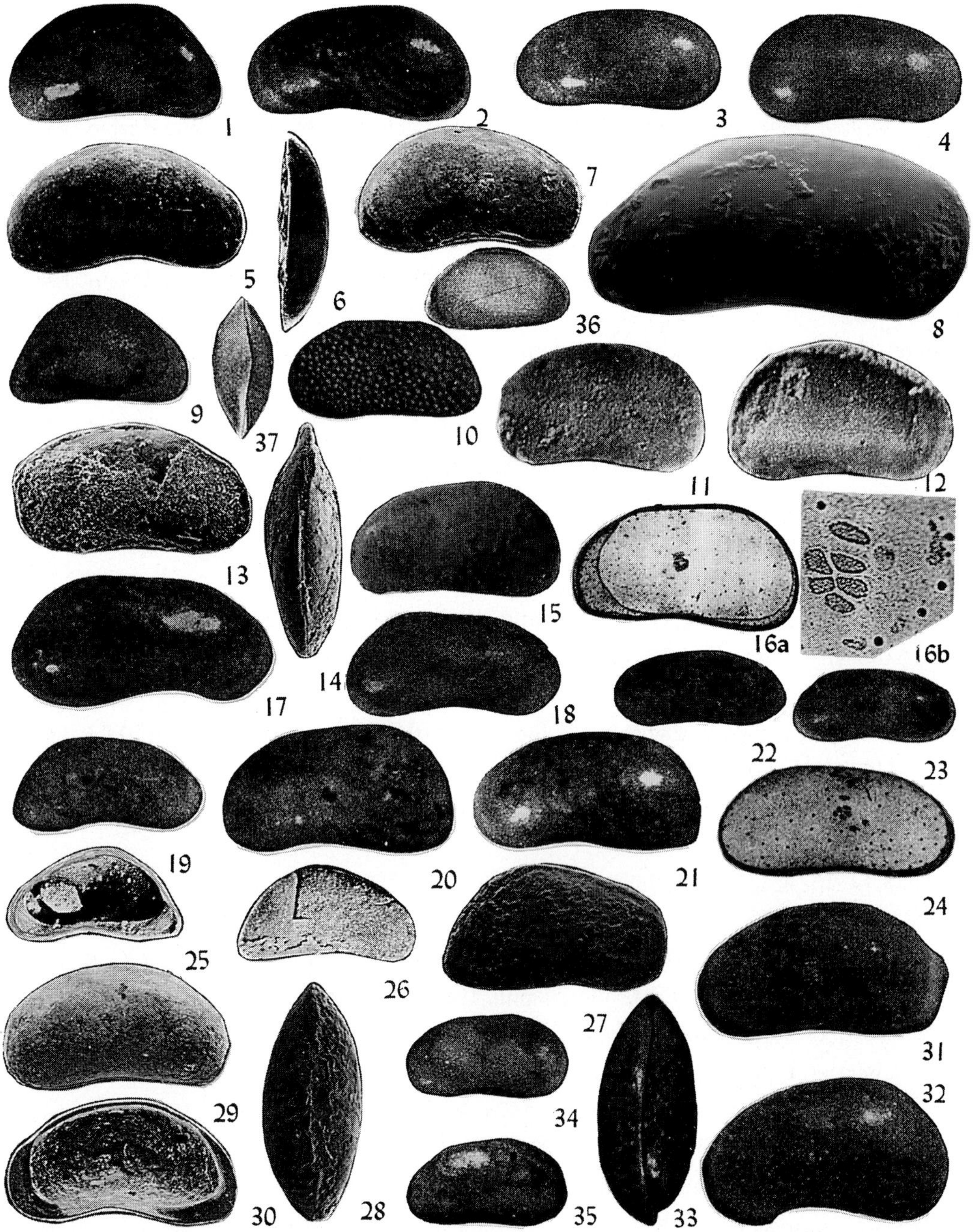
1
2
3
4
5
6
7
36
8
9
37
10
11
12
13
15
16a
16b
17
14
18
22
23
19
20
21
24
25
26
27
31
29
34
32
30
28
35
33

Plate 6

Figures 1, 2. Candona sappaensis (Staplin) [=Candona swaini sappaensis Staplin, 1963], 1963. 1, LV female, holotype, L, 1.12; 2, LV male, L, 1.37. Pleistocene, Kansan Stage, Sappa Formation, SW1/4 NE1/4 sec. 11, T. 2 N., R. 20 W., Harlan County, Nebraska (after Staplin, 1963). (1)

Figures 3, 4. Candona sp. aff. C. symmetrica Hou et al., Swain, 1987. 3, Right side, L, 0.77; 4, dorsal view, L, 0.69. Pliocene, Cache Valley Formation, Hoopes Creek loc., NE1/4 SE1/4 NE1/4 sec. 10, T. 12 S., R. 40 E., Franklin County, Idaho (after Swain, 1987a). (2)

Figures 5, 6. Candona truncata Furtos, Staplin, 1963. 5, RV male, L, 1.11; 6, LV female, L, 1.02. Pleistocene. 5, Tazewell Substage, Wisconsin Stage, NW1/4 NE1/4, sec. 11, T. 4 N., R. 3 E., Fulton County, Illinois; 6, Cary Substage, Wisconsin Stage, 2 miles west of Plumville, Madison County, Ohio (after Staplin, 1963). (1)

Figures 7, 7a. Candona truncata Furtos, Benson, 1967. 7, RV, L, 1.17; 7a, adductor muscle scar, approx. X 135. Pleistocene, Kansan Stage, Sappa Formation, Osborne County, Kansas (after Benson, 1967). (3)

Figures 8, 9. Candona wabashensis Staplin, 1963. 8, LV, female holotype, L, 0.97; 9, LV male, L, 0.85. Pleistocene, Mankato or late Cary Substage, Wisconsin Stage, SW1/4 sec. 24, T. 7 S., R. 14 W., near Mt. Vernon, Indiana (after Staplin, 1963). (1)

Figures 10, 11. Candona wanlessi Staplin, 1963. 10, RV female, holotype, L, 0.83; 11, LV female, L, 0.85. Pleistocene, Tazewell Substage, Wisconsin Stage, SW1/4 NW1/4 sec. 24, T. 2 N., R. 1 E., Fulton County, Illinois, 1 mile (1.6 km) north of Browning (after Staplin, 1963). (1)

Figures 12, 13. Candona willmani Staplin, 1963. 12, RV female, holotype, L, 0.75; 13, LV male, L, 0.95. Pleistocene, Mankato or late Cary Substage, Wisconsin Stage. 12, 1.5-2 miles (2.4-3.2 km) east of Mt. Vernon, Indiana; 13, SW1/4 sec. 24, T. 7 S., R. 14 W., near Mt. Vernon, Indiana (after Staplin, 1963). (1)

Figures 14, 15. Candona sp. Swain, 1949. 14, RV exterior, L, 0.98; 15, immature RV interior, L, 0.70. Paleocene, Ft. Union Formation, Tongue River Member, NW1/4 sec. 31, T. 9 S., R. 49 E., Powder River County, Montana (after Swain, 1949). (1)

Figure 16. Candona sp. Benson, 1969. RV immature interior, L, 0.62. Early Pleistocene, Rita Blanca Lake deposits, Hartley County, Texas (after Benson, 1969). (3)

Figure 17. Candona sp. Swain, 1986. R side of shell, L, 0.53. Pliocene, Glenns Ferry Formation, sec. 14, T. 4 S., R. 1 N., Owyhee County, Idaho (after Swain, 1986). (2)

Figures 18, 19. "Candona" sp., Peck, 1941. 18, "RV"; 19, "LV", L, 1.56. Lower Cretaceous, Deraney Formation, Freedom Quadrangle, Idaho, Wyoming (after Peck, 1941). (1)

Figures 20, 21. "Candona" sp. 2, Sohn, 1956. 20, LV exterior; 21, LV interior, L, 0.66. Pliocene Teewinot Formation, Jackson Hole, Wyoming (after Sohn, 1956). (9)

Figures 22, 23. "Candona" sp. 1, Sohn, 1956. 22, R side of shell, X 30; 23, LV interior, L, 1.00. Pliocene, Teewinot Formation, Jackson Hole, Wyoming (after Sohn, 1956). (9)

Figures 24, 25. "Candona" sp. 3, Sohn, 1956. 24, RV female exterior, L, 1.30; 25, RV female interior, X 30. Pliocene, Teewinot Formation, Jackson Hole, Wyoming (after Sohn, 1956). Compare with Candona truncata. (9)

Figure 26. "Candona" sp. 4, Sohn, 1956. LV female, L, 1.10. Pliocene, Teewinot Formation, Jackson Hole, Wyoming (after Sohn, 1956). (9)

Figures 27, 28. "Candona" sp. 5, Sohn, 1956. 27, LV female; 28, dorsal view LV, L, 1.08. Pliocene, Teewinot formation, Jackson Hole, Wyoming (after Sohn, 1956). (9)

Figures 29, 30. "Candona" sp. 6, Sohn, 1956. 29, R side of female shell; 30, dorsal view of female shell, L, 1.28. Pliocene, Teewinot Formation, Jackson Hole, Wyoming (after Sohn, 1956). (9)

Figure 31. Candona sp. indet., Staplin, 1963. RV, L, 0.96. Pleistocene, Tazewell Substage, Wisconsin Stage, Bloomington slackwater deposits, Fulton County, Illinois (after Staplin, 1963). (1)

Figure 32. Candona sp., R side of poorly preserved shell, L, 1.15. Oligocene Chadron Formation, Helmsville, Nebraska (R. Konizeski, unpublished).

Figures 33, 34. Candoniella bergstromi Swain, 1987. 33, R side; 34, dorsal view, holotype, L, 0.39. Miocene Esmeralda Formation, 2.85 mi (4.5 km) west of Nye County, Nevada (after Swain, 1987). (2)

Figures 35, 36. Candoniella sp. Swain, Becker and Dickinson, Swain, 1985. 35, R side, L, 1.38; 36, ventral view, L, 0.97. Miocene, Esmeralda Formation, 7.6 miles (12.1 km) east of Finger Rock Wash on road to Pactolus Mine, Nye County, Nevada (after Swain, 1985). (2)

Figure 37. Cyclocypris forbesi Sharpe, Staplin, 1963. RV, L, 0.66. Pleistocene Illinoian Stage. SE1/4 SW1/4 sec. 5, T. 24 N., R. 2 W., Tazewell County, Illinois, pro-Illinois, 2 1/2 miles east of Pekin (after Staplin, 1963). (1)

Plate 6

1
2
3
4
5
6
7a
7b
8
9
10
11
12
13
14
15
16
17
18
19
20
21
22
23
24
25
26
27
28
29
30
31
32
33
34
35
36
37

Plate 7

Figures 1-4. Candona (Pseudocandona) ikpikpukensis Swain, 1963. 1, RV, holotype; 2, LV; 3, LV interior, L, 1.10; 4, RV interior, L 1.00. Pleistocene, Gubik Formation, Arctic coastal plain, Alaska, Ikpikpuk shot holes, line 6, shot point 8, at 30 40 feet (9 12m) (after Swain, 1963). (1)

Figures 5, 6. Candona subreniformis Jones, 1893. Exterior of RV and dorsal view, L, 1.20. Upper Cretaceous, Montana Formation, Coalville, Wyoming (after Jones, 1893). (15)

Figure 7. Candona candida depressa Chapman, 1896. R side, L, 1.69. Pliocene, Merced Formation, Wildcat Canyon, east of Berkeley, California (after Chapman, 1896). (16)

Figures 8, 9. Candona lactea acuminata Chapman, 1896. R side and dorsal views, L, 1.02. Pliocene, Merced Formation, Wildcat Canyon, east of Berkeley, California (after Chapman, 1896). (16)

Figures 10, 11. Cypris purbeckensis Forbes, Jones, 1893. R side and dorsal view, L, 1.10. Upper Cretaceous, Montana Formation, Coalville, Utah (after Jones, 1893). (15)

Figure 12. Candona (Pseudocandona) sp. aff. C. (P.) maragoensis (Klie). LV, L, 0.61. Pleistocene, bone bed at Boney Spring, Benton County, Missouri (J.E. King, unpublished).

Figures 13-15. Candona (Pseudocandona) n. sp. 13, LV; 14, RV, L,0.77; 15, RV interior, L,0.81. Pleistocene, bone bed at Boney Spring, Benton County, Missouri (J.E. King, unpublished).

Figures 16, 17. Cypridea (Pseudocypridina) inornata (Peck), Peck, 1951. RV. Lakota Formation, L, 1.40; ventral view, Kootenai Formation, L, 1.12. Lower Cretaceous, near Hot Springs, South Dakota and central Montana; respectively (after Peck, 1951). (1)

Figures 18-22. Cypridea (Bisulcocypridea) arvadensis (Swain, 1949) R side, holotype, L, 1.15; ventral view, L side, L valve interior, dorsal view, L, 1.34. Paleocene, Wasatch Formation, basal part, sec. 6, T. 54 N., 76 W., Sheridan County, Wyoming (after Swain, 1949). (1)

Figures 23, 24. Cypridea (Pseudocypridina?), n. sp., Sohn, 1958. L side, ventral side, L, 1.23. Lower Cretaceous, Lakota Formation, NM 1/4 sec. 29, T. 51 N., R. 65 W., Crook County, Wyoming (after Sohn, 1958). (9)

Figures 25-28. Cypridea (Pseudocypridina) laevicula Peck, 1951. R side, dorsal view, ventral view, L side, L, 0.84. Lower Cretaceous, Lower Bear River Formation, near Evanston, Wyoming (after Peck, 1951). (1)

Figures 29, 30. Cypridea (Pseudocypridina) piedmonti Roth, 1933. R side, ventral view, L, 1.07. Upper Jurassic or Lower Cretaceous, Morrison Formation or Lakota Formation, sec. 28, T. 4 N., R. 6 E., Meade County, South Dakota (after Peck, 1951). (1)

Figures 31-33. Cytheridella todiltensis (Swain, 1946), Kietzke, 1992. R side, R side, L side, holotype, L, 0.65. Upper Jurassic, Morrison Formation, Todilto Member, six miles (9.6 km) north of Thoreau, New Mexico (after Swain, 1946). (1)

Figures 34, 35. Cypria subangulata Chapman, 1896. R side, dorsal view, L, 0.63. Pliocene, Merced Formation, Wildcat Canyon and San Pablo Ridge, east of Berkeley, California (after Chapman, 1896). (16)

Figures 36, 37. Cyclocypris californica Chapman, 1896. L valve dorsal view, L, 1.02. Pliocene, Merced Formation, Wildcat Canyon, east of Berkeley, California (after Chapman, 1896). (16)

Figures 38, 39. Cypris procera Chapman, 1896. L side and dorsal view, L, 1.27. Pliocene, Merced Formation, Wildcat Canyon and Bollinger Canyon, east of Berkeley, California (after Chapman, 1896). (16)

Figures 40, 41. Cypridopsis pliocaenica Chapman, 1896. L side and dorsal view, L, 0.98. Pliocene, Merced Formation, San Pablo Valley, east of Berkeley, California (after Chapman, 1896). (16)

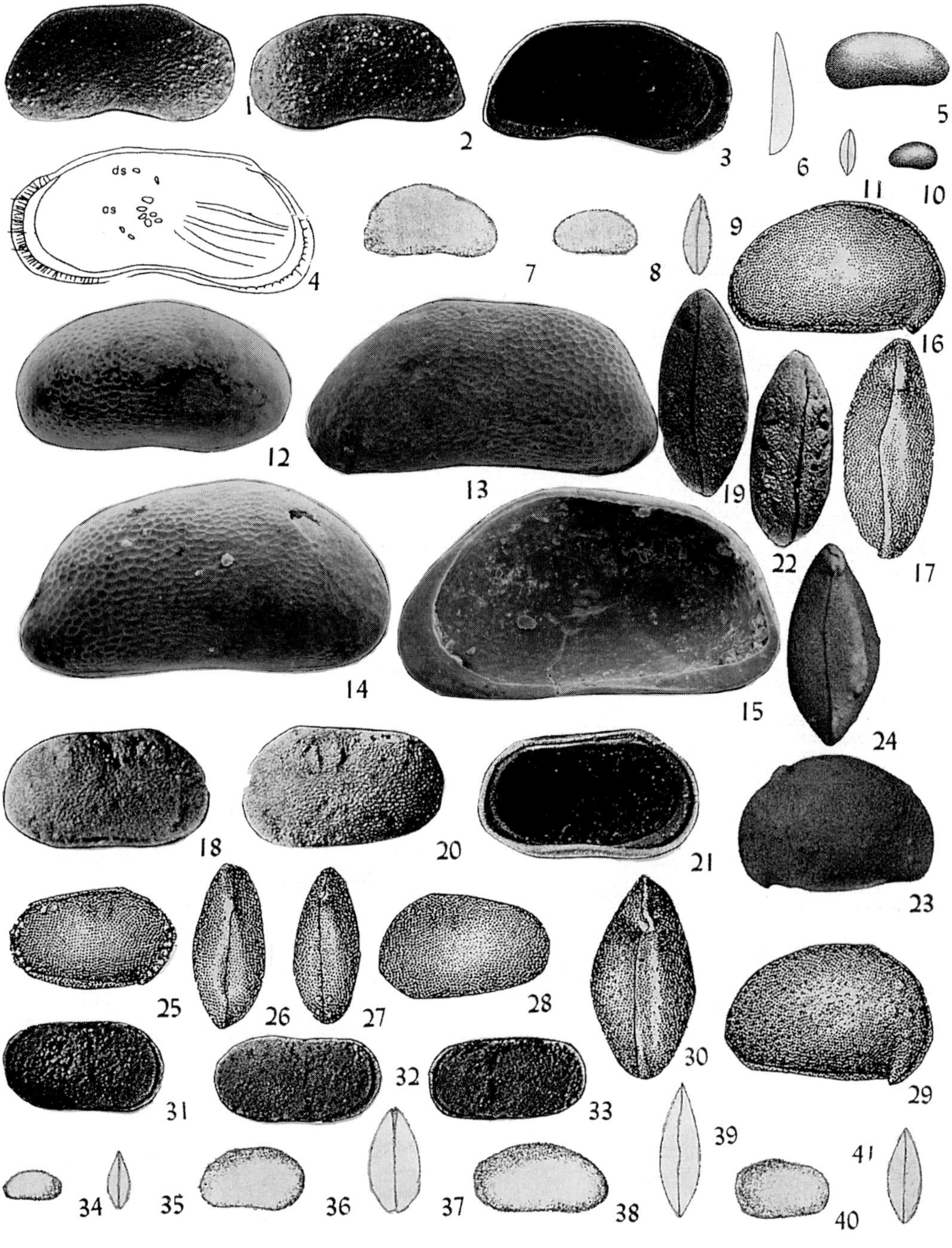
1
2
3
4
ds
as
5
6
7
8
9
10
11
12
13
14
15
16
17
18
19
20
21
22
23
24
25
26
27
28
29
30
31
32
33
34
35
36
37
38
39
40
41

Plate 8

Figures 1, 2. "Cypris" purbeckensis Forbes, Jones, 1886. L side and dorsal view, L, 0.76. Upper Jurassic, "Atlantosaurus beds", near Canyon City, Colorado (after Jones, 1886). (15)

Figures 3, 4. "Cytherideis" marshii Jones, 1886. L? side and dorsal view, L, 0.86. Upper Jurassic, "Atlantosaurus beds", near Canyon City, Colorado (after Jones, 1886). (15)

Figures 5, 6. "Cytheridea" atlantosaurica Jones, 1886. R? side and dorsal view, L, 0.80. Upper Jurassic, Morrison Formation, "Atlantosaurus beds", near Canyon City, Colorado (after Jones, 1886). (15)

Figures 7, 8. "Cytheridea" truncata Jones, 1883. L side and dorsal view, L, 0.60. Lower Cretaceous, Lower Bear River Formation or Gannett Group, Cokeville, Wyoming (after Jones, 1893). (15)

Figures 9, 10. "Cytheridea" tenuis Jones, 1893. L side and dorsal view, L, 0.60. Lower Cretaceous, Lower Bear River Formation or Gannett Group, Cokeville, Wyoming (after Jones, 1893). (15)

Figures 11, 12. "Cytheridea" aequalis Jones, 1893. L side and dorsal view, L, 0.60. Lower Cretaceous, Lower Bear River Formation or Gannett Group, Cokeville, Wyoming (after Jones, 1893). (15)

Figures 13, 14. "Cytherideis" impressa Jones, 1893. R side and dorsal view, L, 0.60. Lower Cretaceous, Lower Bear River Formation or Gannett Group, Cokeville, Wyoming (after Jones, 1893). (15)

Figures 15, 16. Darwinula leguminella (Forbes), Jones, 1886. R side and dorsal view, L, 0.93. Upper Jurassic, "Atlantosaurus beds", Morrison Formation, near Canyon City, Colorado (after Jones, 1886). (15)

Figures 17, 18. Cypridea (Bisulcocypridea) arvadensis (Swain, 1964). R side female? L, 1.00, R side, male? L, 1.12. Eocene, Wasatch Formation, 4.5 miles (7.2 km) east of Arvada, Wyoming, between Arvada and Roland Coals (after Swain, 1964a). (1)

Figures 19. Cypridea (Pseudocypridina) piedmonti Roth, Sohn, 1958. L side, holotype, L, 1.30. Lower Cretaceous, Lakota Formation, sec. 18, T. 4 N., R. 6 E., Meade County, South Dakota (after Sohn, 1958). (9)

Figures 20-24. Dryelba persulcata (Peck), Sohn, 1982. 20, R side, holotype, L, 1.10; 21, dorsal view, female; 22, ventral view, female; 23, R side, male; 24, dorsal view, male, L, 1.06. Lower Cretaceous, Lower Bear River Formation, Lincoln County, Wyoming (after Peck, 1951). (1)

Figures 25, 26. Candona acutula Delorme, Lister, 1975. 25, RV, female, L, 0.96; 26, LV interior, male, L, 1.03. Pleistocene, Lake Bonneville Group, Saltair core hole SE$^{1}/_{4}$ sec. 25, T. 1 N., R. 3 W., Salt Lake County, Utah (after Lister, 1975). (3)

Figures 27, 28. Candona adunca Lister, 1975. 27, LV, holotype: 28, RV interior, L, 1.43. Pleistocene, Lake Bonneville Group, Saltair core hole, SE$^{1}/_{4}$ sec. 25, T. 1 N., R. 3 W., Salt Lake County, Utah (after Lister, 1975). (3)

Figures 29, 30. Candona camuramarginis Lister, 1975. 29, LV, holotype; 30, RV interior, L, 0.93. Pleistocene, Lake Bonneville Group, Saltair core hole, SE$^{1}/_{4}$sec. 25, T. 1 N., R. 3 W., Salt Lake County, Utah (after Lister, 1975). (3)

Figures 31-33. Candona caudata Kaufmann, Lister, 1975. 31, LV, female, L, 1.05; 32, RV, female interior; 32, LV male, L, 1.03. Pleistocene, Lake Bonneville Group, Saltair core hole SE$^{1}/_{4}$ sec. 25, T. 1 N., R. 3 W., Salt Lake County, Utah (after Lister, 1975). (3)

Figure 34. Candona cf. crogmaniana Turner, Lister, 1975. LV, L, 0.80. Pleistocene, Lake Bonneville Group, Saltair core hole, SE$^{1}/_{4}$ sec. 25, T. 1 N., R. 3 W., Salt Lake County, Utah (after Lister, 1975). (3)

Figure 35. Candona pronopa Lister, 1975. RV, holotype, L, 0.85. Pleistocene, Lake Bonneville Group, Saltair core hole SE$^{1}/_{4}$ sec. 25, T. 1 N., R. 3 W., Salt Lake County, Utah (after Lister, 1975). (3)

Figures 36-38. Candona rawsoni Tressler, Lister, 1975. 36, RV, female, L, 1.07; 37, RV, female interior; 38, RV male, L, 1.06. Pleistocene, Lake Bonneville Group, Saltair core hole SE$^{1}/_{4}$ sec. 25, T. 1 N., R. 3 W., Salt Lake County, Utah (after Lister, 1975). (3)

Figures 39, 40. Cyclocypris ampla Furtos, Lister, 1975. 39, R side; 40, RV interior, L, 0.60. Pleistocene, Lake Bonneville Group, Saltair core hole SE$^{1}/_{4}$ sec. 25, T. 1 N., R. 3 W., Salt Lake County, Utah (after Lister, 1975). (3)

Figures 41, 42. Cyclocypris serena (Koch), Lister, 1975. 41, R side; 42, RV interior, L, 0.59. Pleistocene, Lake Bonneville Group, Saltair core hole SE$^{1}/_{4}$ sec. 25, T. 1 N., R. 3 W., Salt Lake County, Utah (after Lister, 1975). (3)

Figures 43-45. Cyprideis salebrosa van den Bold, Lister, 1975. 43, LV, female, L, 0.88; 45, LV interior, female; RV male, L, 0.88. Pleistocene, Lake Bonneville Group, Saltair core hole SE$^{1}/_{4}$ sec. 25, T. 1 N., R. 3 W., Salt Lake County, Utah (after Lister, 1975). (3)

Figures 46, 47. Cypridopsis vidua (Muller), Lister, 2975. 46, RV; 47, RV interior, L, 0.65. Pleistocene, Lake Bonneville Group, Saltair core hole SE$^{1}/_{4}$ sec. 25, T. 1 N., R. 3 W., Salt Lake County, Utah (after Lister, 1975). (3)

Figures 48, 49. Paracypridopsis sp. (Lister, 1975). 48, R side; 49, LV interior, L, 0.57. Pleistocene, Lake Bonneville Group, Saltair core hole SE1/4 sec. 25, T. 1 N., R. 3 W., Salt Lake County, Utah (after Lister, 1975). (3)

Figures 50, 51. Carbonita agnes (Jones, 1870). R side and dorsal views, L, 0.64. Pennsylvanian, Des Moinesian Series, Lonsdale Limestone, western Illinois (after Cooper, 1946). (20)

Figures 52, 53. Carbonita orbiculata Cooper, 1946. R side and dorsal views, L, 0.95. Pennsylvanian, Atoka Series, Seville Limstone, western Illinois (Cooper, 1946).

Figures 54, 55. Carbonita? tenuis Cooper, 1946. R side and ventral views, L, 0.74. Pennsylvanian, Virgilian Series, Cohn Cyclothem, central Illinois (after Cooper, 1946). (20)

Plate 8

Plate 9

Figures 1, 1a, 2. Candoniella? sp. aff. C. suzini Schneider, Swain, 1987c. R side, L, 1.16; dorsal view and surface detail. Miocene, Horse Camp formation, Loc. 6, sec. 12, T. 10 N., R. 58 E., Nye County, Nevada (after Swain, 1987c). (2)

Figures 3, 4. Candoniella bergstromi Swain [=Candoniella sp. 2 of Swain, 1985[, 1986. 3, R side, L, 0.68; 4, dorsal view. Miocene, Esmeralda Formation, loc. 4, 7.6 miles (12.1km) east of Finger Rock Wash, road to Pactolus Mine, Nye County, Nevada (after Swain, 1986). (2)

Figures 5, 6. Candoniella sp. aff. C. parca Hou et al., Swain, 1986. 5, R side; 6, dorsal view, L, 0.88. Miocene, Humboldt Formation, section at Cobre, Elko County, Nevada (after Swain, 1986). (2)

Figure 7. Candoniella pinecreekensis Swain, 1986. L side of holotype, L, 0.83. Miocene or late Oligocene, Humboldt or Indian Well Formation, Palisade section, SE1/4 secs. 1, 12, T. 31 N., R. 51, E., Eureka County, Nevada (after Swain, 1986). (2)

Figures 8, 9. Candoniella cf. pinecreekensis Swain, Swain, 1987c. 8, L side; 9, dorsal view, L, 0.75. Eocene, Twin Bridges Limestone, center sec, 36, T. 32 N., R. 55 E., Twin Bridges, Elko County, Nevada (after Swain, 1987c). (2)

Figures 10, 11. Candoniella whitei (Scott and Smith), Swain, 1964 [=Candona whitei Scott and Smith, 1951]. 10, L side; 11, dorsal view of shell, L, 0.93. Eocene, Colton-Green River transition beds, 2 feet above base, Dripping Rock Gulch section, Rio Blanco County, Colorado (after Swain, 1964). (1)

Figures 12, 13. Candonopsis? cf. kingsleyi (Brady and Robertson), Swain, 1963. 12, RV, L, 0.76; 13, LV interior, L, 0.82. Pleistocene, Gubik Formation, Fish Creek shot holes, line 1-48, shot point 1, at 80 feet, Arctic coastal plain, Alaska (after Swain, 1963). (1)

Figure 14. Candonopsis sp. aff. C. shahejiensis Hou et al., Swain, 1985. R side of shell, L, 093. Miocene, Esmeralda Formation, loc. 4, road to Pactolus Mine, 7.6 miles (12.1 km) east of Finger Rock Wash, Nye County, Nevada (after Swain, 1985). (2)

Figure 15. Candonopsis sp. aff. C. shahejiensis Hou et al., Swain, 1987c. RV, L, 0.58. Oligocene, Edwards Creek Tuff, locality 15, T. 19 N., R. 38 E., Churchill County, Nevada (after Swain, 1987c). (2)

Figures 16-18. Carbonita n. sp., Sohn, 1985. 16, R side of steinkern, L, 0.91; 17, dorsal view of internal mold; 18, multiple adductor muscle scar, approx. X 280, of internal mold. Upper Mississippian, Blue Stone Formation, Bramwell Member, Flat Top Quadrangle, Mercer County, West Virginia (after Sohn, 1985). (1)

Figures 19-21. Carbonita? sp., Sohn, 1985. 19, R side of internal mold, L, 1.20; 20, dorsal view of internal mold. Upper Mississippian, Blue Stone Formation, Bramwell Member, Flat Top Quadrangle, Mercer County, West Virginia (after Sohn, 1985). (1)

Figure 22. Caspiollina microstriata Dickinson and Swain, 1967. LV (broken) of holotype, L, 0.98. Pliocene, Hay Ranch Formation, Gulf Oil Company, Wilkins Ranch No. 1 well, Elko County, Nevada, 1017-1047 feet (310-319 m) (after Dickinson and Swain, 1967). (1)

Figures 23-26. Chlamydotheca llanoensis Pierce, 1988. 23. RV exterior, holotype, L, 2.84; 24, RV interior, L, 2.77; 25, dorsal view, L, 2.63; 26, adductor muscle scar field. Late Pleistocene Blackwater Draw Formation, Lubbock Lakes Archeological Rite, Lubbock County, Texas (after Pierce, 1988). (1)

Figure 27. Clinocpyris? sp., Swain, 1986b. R side, L, 1.11. Miocene, Humboldt Formation, Cobre section, 1011 feet above above (after Swain, 1986b). (2)

Figures 28, 29. Cyclocypris fragilis Staplin, 1963. 28, LV; 29, RV, L, 0.51. Pleistocene, Mankato or late Cary Substage, Wisconsin Stage, creek 1.5-2 miles east of Mt. Vernon, Indiana on Route 62 (after Staplin , 1963). (1)

Plate 9

1
1a
2
3
4
5
6
7
8
9
10
11
12
13
14
15
16
17
18
19
20
21
22
23
24
25
26
27
28
29

Plate 10

Figures 1, 2. Chlamydotheca? dixieflatensis Swain, 1987. 1, L side of holotype shell, broken medially, but retaining part of original shell, L, 1.10; 2, R side of partly exfoliated shell, L, 0.98. Eocene, Twin Bridges Limestone, Twin Bridges, near center sec. 36, T. 32 N., R 55 E., Elko County, Nevada (after Swain, 1987b). (2)

Figures 3, 4. Candoniella whitei (Scott and Smith), Swain, 1987. 3, L side of shell, L, 0.39; 4, dorsal view of shell. Late Paleocene or Eocene, Sheep Pass Formation, Grant Range, Johnson Canyon, Nye County, Nevada (after Swain, 1987b). (2)

Figures 5, 6. Clinocypris sp. Swain, 1987. 5, RV exterior, L, 1.90; 6, LV exterior, L, 2.07. Late Cretaceous or Paleocene, Sheep Pass Formation, Egan Range, Nevada (after Swain, 1987b). (2)

Figures 7, 8. Cyclocypris cf. forbesi Sharpe, Swain, 1987c. 7, L side of shell, L, 0.29; 8, dorsal view of badly weathered shell, L, 0.29, both are silicified. Oligocene, Pruett? Formation, Buck Hill volcanic series, loc. 10, 1 mile (1.6 km) SW of old Neville Ranch, Brewster County, Texas (after Swain, 1987c). (2)

Figures 9-11. Cyclopyris? ketneri Swain, 1987. 9, R side of holotype shell; 10, dorsal view, L, 0.42; 11, adductor muscle scar of exfoliated shell, approx. X 540. Eocene, Twin Bridges Limestone, Twin Bridges, center sec. 36, T. 32 N., R. 55 E., Elko County, Nevada (after Swain, 1987). (2)

Figures 12, 13. Cyclocypris laevis (O.F. Müller) Staplin, 1963. 12, RV; 13, LV, L, 0.49. Pleistocene, Cary or Mankato Substage, Wisconsin Stage, sec. 5, T. 37 N., R. 14 E., Chicago Lake clays (after Staplin, 1963). (1)

Figures 14, 15. Cyclocypris ovum (Jurine), Staplin, 1963. 14, RV; 15, dorsal view, L, 0.45. Pleistocene, 14, Illinoian Stage, SW1/4 NW1/4 SW1/4 sec. 9, T. 8 S., R. 2 W., Calhoun County, Illinois. 15, Late or post-Mankato Substage, Wisconsin Stage, Sag Bridge, SE cor. sec. 17, T. 37 N., R. 12 E., Chicago, Illinois (after Staplin, 1963). (1)

Figures 16, 17. Cyclocypris sp. aff. C. serena (Koch), Swain, 1987. 16, R side of shell; 17, dorsal view, L, 0.50. Pliocene, Salt Lake Group, Cache Valley Formation, NW1/4 SE1/4 NE1/4 sec. 10, T. 12 S., R. 40 E., Franklin County, Idaho (after Swain, 1987a). (2)

Figures 18, 19. Cyclocypris sharpei Furtos, Staplin, 1963. 18, LV; 19, RV, L, 0.51. Pleistocene, Illinoian Stage, Pearl Quadrangle, SW1/4 NW1/4 SW1/4 sec. 9, T. 8S., R. 2 W., Calhoun County, Illinois (after Staplin, 1963). (1)

Figures 20, 21. Cypria anterosinuata Staplin, 1963. 20, RV, L, 0.35; 21, LV, holotype, L, 0.45. 20, Pleistocene, Mankato or late Cary Substage, Wisconsin Stage, 1.5-2 miles east of Mt. Vernon, Indiana, on Route 62; 21, same age, sec. 13, T. 9 S., R. 10 E., Gallatin County, Illinois (after Staplin, 1963). (1)

Figures 22-25. Cypria hoopescreekensis Swain, 1987. 22, LV exterior holotype, L, 0.56; 23, LV interior; 24, dorsal view; 25, adductor muscle scar of fig. 23, approx. X 420. Pliocene, Cache Valley Formation, Hoopes Creek loc., NW$^{1}/_{4}$ SE$^{1}/_{4}$ NE$^{1}/_{4}$ sec. 10, T. 12 S., R. 40 E., Franklin County, Idaho (after Swain, 1987a). (2)

Figures 26, 27. Cypria maculata Hoff, Staplin, 1963. 26, LV; 27: RV, L, 0.47. Pleistocene, late Cary or Mankato Substage, Wisconsin Stage, 1.5-2 miles E of Mt. Vernon, Indiana on Route 62 (after Staplin, 1963). (1)

Figures 28, 29. Cypria mediana Hoff, Staplin, 1963. 28, RV, juvenile?; 29, LV juvenile, L, 0.67. Pleistocene, late Cary or Mankato Subtage, Wisconsin Stage, 1.5-2 miles (2.4-3.2 km) E of Mt. Vernon, Indiana on route 62 (after Staplin, 1963). (1)

Figure 30. Cypria nevadensis Dickinson and Swain, 1967. LV, holotype, a poorly preserved specimen, L, 1.15. Pliocene, Hay Ranch Formation ("Upper Humboldt"), Gulf Oil Company, Wilkins Ranch no. 1 well, 1536-1588 feet (465-484 m) (after Dickinson and Swain, 1967). (1)

Figure 31. Cypria obesa Sharpe, Staplin, 1963. RV, L, 0.72. Pleistocene, Illinoian Stage, SW$^{1}/_{4}$ NW$^{1}/_{4}$ SW$^{1}/_{4}$ sec. 9, T. 8 S., R. 2 W., Calhoun County, Illinois (after Staplin, 1963). (1)

Figure 32. Cypria turneri Hoff, Staplin, 1963. LV, L, 0.61. Pleistocene, Illinoian Stage, SW$^{1}/_{4}$ NW$^{1}/_{4}$ SW$^{1}/_{4}$ sec. 9, T. 8 S., R. 2 W., Calhoun County, Illinois (after Staplin, 1963). (1)

Figures 33, 34. Cypricercus elegantulus Swain, 1986b. 33, R side of shell, L, 1.32; 34, dorsal view exfoliated anteriorly, L, 1.48. Miocene, Humboldt Formation, conglomeratic limestones near base, Medicine Creek section, T. 39 N., R. 65 E., Winecup Ranch, Elko County, Nevada (after Swain, 1987b). (2)

Figures 35, 36. Cypicercus cf. elegantulus Swain, 1987c. 35, L side; 36, dorsal view, L, 0.45. Miocene, Esmeralda Formation, 2.85 miles (4.5 km) west of Black Spring, loc. 3, Nye County, Nevada (after Swain, 1987c). (2)

Figures 37-39. Cypricercus reticulatus Zaddach, Staplin, 1963. 37, RV female, L, 0.93; 38, LV female; 39, RV juvenile, L, 0.72. Pleistocene, Mankato or late Cary Substage. Wisconsin Stage, SW$^{1}/_{4}$ sec. 24, T. 7 S., R. 14 W., near Mt. Vernon, Indiana (after Staplin, 1963). (1)

Figure 40. Cypricercus reticulatus wrenshallensis Swain, 1995. R side holotype, L, 0.44. Pleistocene, clay pit at Wrenshall, Minnesota (after Swain, 1995). (2)

Figure 41. Cypricercus minensis Swain, 1985. Ventral view, L, 1.14. Miocene, Esmeralda Formation, loc. 2, Finger Rock Wash section, lower shale, 0.5 mile (0.8 km) south of Rawhide Ranch, Mineral County, Nevada (after Swain, 1985). (2)

Plate 10

1
2
5
6
3
4
7
8
11
12
13
14
15
16
17
18
19
10
9
20
21
22
23
24
26
27
28
29
25
30
31
32
33
34
35
36
37
38
39
40
41

Plate 11

Figure 1. Cypricercus minensis Swain, 1985. R side, holotype L, 1.07. Miocene, Esmeralda Formation, Finger Rock Wash section, 0.5 mile (0.8 km) south of Rawhide Ranch, Mineral County, Nevada (after Swain, 1985). (2)

Figures 2, 3. Cypricercus sp. aff. minensis Swain, Swain, 1985. 2, L side, L, 0.96; 3, dorsal view, L, 1.01. Miocene, Esmeralda Formation, 2.85 miles (4.5 km) west of Black Spring, Nye County, Nevada (after Swain, 1985). (2)

Figure 4. Cypricercus hunterensis Swain, 1986b. R side of holotype, L, 0.95. Miocene, Humboldt Formation, Gulf Oil Company, no. 1 Thousand Springs well, SE1/4 SW1/4 sec. 8, T. 40 N., R. 66 E., Elko County, Nevada, 1960-2000 feet (595-610 m) (after Swain, 1986b). (2)

Figures 5, 6. Cypricercus mineralensis Swain, 1985. 5, R side, holotype, L, 1.16; 6, dorsal view, L, 0.77. Micoene, Esmeralda Formation, 2.85 miles (4.5 km) west of Black Spring, Nye County, Nevada (after Swain, 1985). (2)

Figures 7, 8. Cypricercus cf. mineralensis Swain, Swain, 1987. 7, L side of shell, L, 0.60; 8, dorsal view, L, 0.57. Miocene, Horse Spring Formation, approx. NW1/4 sec. 31, T. 13 S., R. 16 E., Fossil Ridge, Clark County, Nevada (after Swain, 1987c). (2)

Figures 9-11. Cypricercus oblongatus Swain and Wagoner, 1985. 9, RV exterior, L, 0.61; 10, LV interior, L, 0.61; 11, ventral view, holotype, L, 0.60. Miocene, Esmeralda Formation, Blair Junction, Esmeralda County, Nevada (after Swain and Wagoner, 1985). (2)

Figures 12, 13. Cypricercus palisadensis Swain, 1986b. 12, R side holotype, L, 1.07; 13, dorsal view, L, 0.75. Miocene or late Oligocene, Humboldt or Indian Well Formation, 2 miles (3.2 km) SE of Palisade along slope above dam on Pine Creek, Eureka County, Nevada (after Swain, 1986b). (2)

Figure 14. Cypricercus? sp. aff. C. palisadensis Swain, Swain, 1987. R side, L, 0.63. Eocene, Twin Bridges Limestone, Twin Bridges, Elko County, Nevada (after Swain, 1987b). (2)

Figures 15, 16. Cypericercus cf. palisadensis Swain, Swain, 1986b. 15, R side; 16, dorsal view, L, 0.80. Miocene, Humboldt Formation, Medicine Creek section, T. 39 N., R. 65 E., Winecup Ranch, Elko County, Nevada (after Swain, 1986b). (2)

Figures 17-19. Cypricercus parafuscatus Dickinson and Swain, 1967. 17, R side holotype; 18, R side; 19 dorsal view, L, 0.63. Pliocene, Hay Ranch Formation (= Upper Humboldt), Gulf Oil Company, no. 1 Wilkins Ranch well, sec. 21, T. 38 N., R. 61 E.,

Elko County Nevada; 17, depth 1334-1374 (407-419 m); 18, depth 957-987 feet (292-301 m); 19, same as 17 (after Dickinson and Swain, 1967). (1)

Figures 20, 21. Cypricercus sp. aff. C. reticulatus (Zaddach) Swain, 1987. 20, R side, L, 0.56; 21, L side, L, 0.62. Pliocene, Cache Valley Formation, Hoopes Creek loc., NW1/4 SE1/4 NE1/4 sec. 10, T. 12 S., R. 40 E., Franklin County, Idaho (after Swain, 1987a). (2)

Figures 22, 23. Cypricercus cf. sacramentoensis Swain, Swain, 1987c. 22, R side; 23, dorsal view, L, 0.62. Miocene or late Oligocene, Humboldt or Indian Well Formation, center N line sec. 29, T. 31 N., R. 55 E., Dixie Flats Quadrangle, Elko County, Nevada (after Swain, 1987c). (2)

Figure 24. Cypricercus? sacramentoensis Swain, Swain, 1987. L side, L, 0.87. Miocene North Creek Formation, sec. 3, T. 14 N., R. 69 E., 5.85 miles (9.4 km) east of Sacramento Pass, White Pine County, Nevada (after Swain, 1987c). (2)

Figures 25. Cypricercus sacramentoensis Swain, 1986b. R side of holotype, L, 0.68. Miocene, North Creek Formation, sec. 3, T. 14 N., R. 69 E., 5.85 miles (9.4 km) east of Sacramento Pass, White Pine County, Nevada (after Swain, 1986b). (2)

Figures 26, 27. Cypricercus stami Dickinson and Swain, 1967. 26, L side; 27, dorsal view, holotype, L, 1.00. Pliocene, Hay Ranch Formation (=Upper Humboldt), Gulf Oil Company, no. 1 Wilkins Ranch well, sec. 21, T. 38 N., R. 61 E., Elko County, Nevada; 1466-1496 feet (447-456 m) (after Dickinson and Swain, 1967). (1)

Figures 28, 29. Cypricercus sp. aff. C. stami Dickinson and Swain, Swain, 1986. 28, L side, L.0.93; 29, dorsal view, L, 0.98. Pliocene, Glenns Ferry Formation, sec. 14, T. 4 S., R. 1 W., Owyhee County, Idaho (after Swain, 1986a). (2)

Figures 30, 31. Cypricercus toanoensis Swain, 1986. 30, R side holotype, L, 1.01; 31, dorsal view. Miocene Humboldt Formation, Medicine Creek section, T. 39 N., R. 65 E., Winecup Ranch, Elko County, Nevada; 111-139 feet (34-42 m) above base of exposure (after Swain, 1986b). (2)

Figure 32. Cypricercus tuberculatus (Sharpe), Staplin, 1963. LV, L, 0.91. Pleistocene, Cary or Mankato Substage, Wisconsin Stage, sec. 29, T. 48 N., R. 24 W., Story County, Iowa (Staplin, 1963). (1)

Figures 33a, b. Cypricercus tuberculatus (Sharpe), Benson, 1967. 33a, LV, L,1.18; 33b, adductor muscle scar field, approx. X 105. Pleistocene, Yarmouth Stage, Dickinson County, Kansas (Benson, 1967). (3)

Plate 11

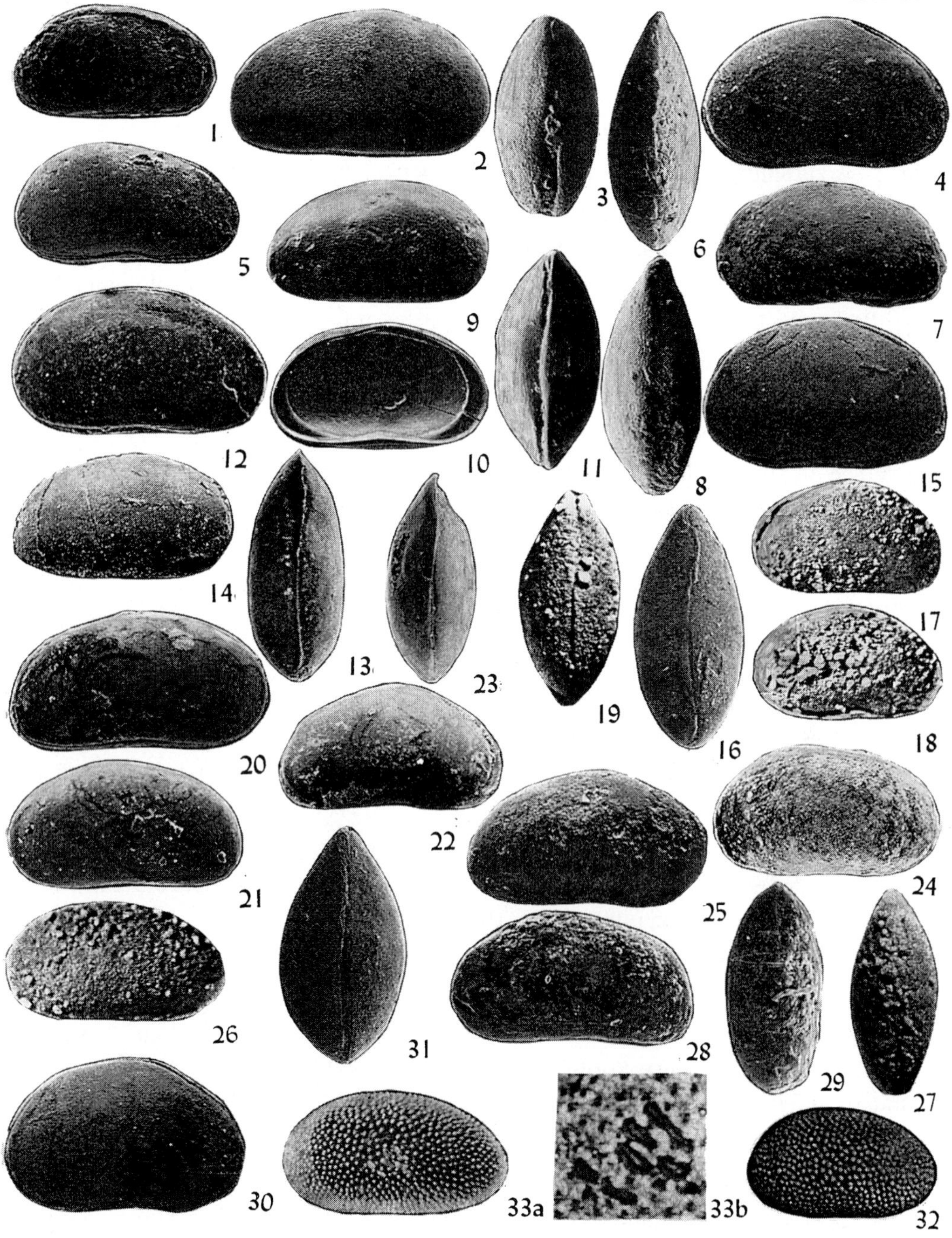
1
2
3
6
4
5
9
6
7
12
10
11
8
15
14
13
23
19
16
17
18
20
22
24
21
25
26
31
28
29
27
30
33a
33b
32

Plate 12

Figure 1. Cypricercus sp. 1, Swain, 1987. L side of poorly preserved silicified shell, L, 0.83. Oligocene, Buck Hill volcanic series, Pruett? Formation, 2 miles (3.2 km) NNE of Burnt House Camp, Brewster County, Texas (after Swain, 1987c). (2)

Figure 2. Cypriconcha? sp., Swain, 1987. LV interior, imperfect, L, 0.77. Eocene, Twin Bridges Limestone, Twin Bridges, Elko County, Nevada (after Swain, 1987b). (2)

Figures 3-5. Cypridea anomala Peck, 1941. 3, RV "holotype"; 4, dorsal view; 5, LV, L, 0.92. Lower Cretaceous, Draney Formation, Freedom Quadrangle, Idaho-Wyoming (after Peck, 1941). (1)

Figure 6. Cypridea anomala Peck, Peck , 1951. Ventral view of "holotype", L, 0.89. Lower Cretaceous, Cloverly Formation, 5 miles (8 km) north of Gardner, Montana (after Peck, 1951). (Note discrepancy in holotype designations with Fig. 3). (1)

Figures 7, 8. Cypridea (Bisulcocypridea) arvadensis (Swain), (Peck, 1951). 7, R side; 8, ventral view, L, 1.15. Eocene, Green River Formation, 10 miles (16 km) east of Kemmerer, Wyoming (after Peck, 1951). (1)

Figures 9-12. Cypridea (Bisulcocypridea) bicostata Sohn, 1969. 9, R side; 10, L side; 11, dorsal view; 12, ventral view, L, 1.24. Lower? Cretaceous, Newark Canyon Formation, U.S.G.S. loc. 27754, Elko County, Nevada (after Sohn, 1969). (see discussion of Sheep Pass Formation). (8)

Figures 13, 14. Cypridea (Bisulcocypridea) bicostata Sohn, Swain, 1987. 13, L side, X 38; 14, R side, L, 0.84. Upper Cretaceous, Sheep Pass Formation, lower part, Egan Range, Nevada, Sheep Pass Canyon loc. (after Swain, 1987b). (2)

Figures 15, 16. Cypridea (Bisulcocypridea) bisulcata (Swain) 1949. 15, R side holotype; 16, ventral view, L, 1.00. Eocene, Flagstaff Limestone, in lower 50 feet (15.2 m), 1 mile west of Ephraim, Utah (after Swain, 1949). (1)

Figures 17, 18. Cypridea (Bisulcocypridea) bisulcata (Swain), Swain, 1964. 17, R side; 18, ventral view, L, 1.16. Eocene, Colton-Green River transition beds, 55 feet (16.7m) above base, east end of Raven Ridge, Dripping Rock Gulch section, Rio Blanco County, Colorado (after Swain, 1964a). (1)

Figures 19-21. Cypridea brevicornis Peck, 1941. 19, R side; 20, L side; 21, dorsal view, L, 0.80. Lower Cretaceous, Draney Formation, Freedom Quadrangle, Idaho-Wyoming (after Peck, 1941). (1)

Figures 22-24. Cypridea compta Peck, 1941. 22, R side "holotype"; 23, L side; 24 dorsal view, L, 0.80. Lower Cretaceous, Draney Formation, Freedom Quadrangle, Idaho-Wyoming; T. 5 S., R. 45 E. (after Peck, 1941). (1)

Figures 25, 26. Cypridea compta Peck, Peck, 1951. 25, R side "holotype", Lower Cretaceous Kootenai Formation, Univ. Missouri loc. 0-976-5; 26, R side L, 0.73, Lower Cretaceous, Cloverly Formation, 4 miles (6.4 km) west of Lander, Wyoming (after Peck, 1951). (Note discrepancy in designation of holotype with Fig. 22). (1)

Figures 27-29. Cypridea dequeenensis Swain and Brown, 1964. 27, LV holotype; 28, R side; 29, RV interior, L, 0.97. Lower Cretaceous, Trinity Group, De Queen Limestone, Gypsum Bluff, Pike County, Arkansas (after Swain and Brown, 1964) (6)

Figure 30. Cypridea (Cypridea) dequeenensis Swain and Brown, Swain and Brown, 1972. L side, L, 0.67. Upper Jurassic or Lower Cretaceous (Barremian?), Unit H, Dare County, North Carolina, well NC-DA-OT-12; depth 7306-7316 feet (2227-2230 m) (after Swain and Brown, 1972). (8)

Figures 31-33. Cypridea diminuta Vanderpool, Swain and Brown 1964. 31, R side holotype (refigured); 32, LV interior, L, 0.70; 33, ventral view. Lower Cretaceous, De Queen Limestone, Gypsum Bluff, Pike County, Arkansas (after Swain and Brown, 1964). (6)

Figure 34. Cypridea sp. aff. C. wyomingensis Jones, 1893. R side, L, 0.95. Upper Jurassic, Shongaloo Formation, Standard Oil Company, no. 1 Zimmerman well, Union County, Arkansas; 7030 feet (2143 m) (after Swain and Anderson, 1993). (18)

Figures 35-37. Cypridea inornata Peck, 1941. 35, LV holotype; 36, RV; 37, dorsal view, L, 1.40. Lower Cretaceous, Kootenai Formation, near Great Falls, Montana (after Peck, 1941). (1)

Figures 38-40. Cypridea (Pseudocypridina) henrybelli Sohn, 1979. 38, L side; 39, R side; 40, dorsal view, L, 1.16. Lower Cretaceous, Lakota Formation, sec. 15, T. 8 S., R 4 E., Fall River County, South Dakota (after Sohn, 1979). (8)

Figures 41-44. Cypridea (Pseudocypridina) laeli Sohn, 1979. 41, R side; 42, L side; 43 dorsal view; 44, ventral view, L, 1.12. Lower Cretaceous, Lakota Formation, U.S.G.S. col. 31001, Crook County, Wyoming (after Sohn, 1979). (8)

Plate 12

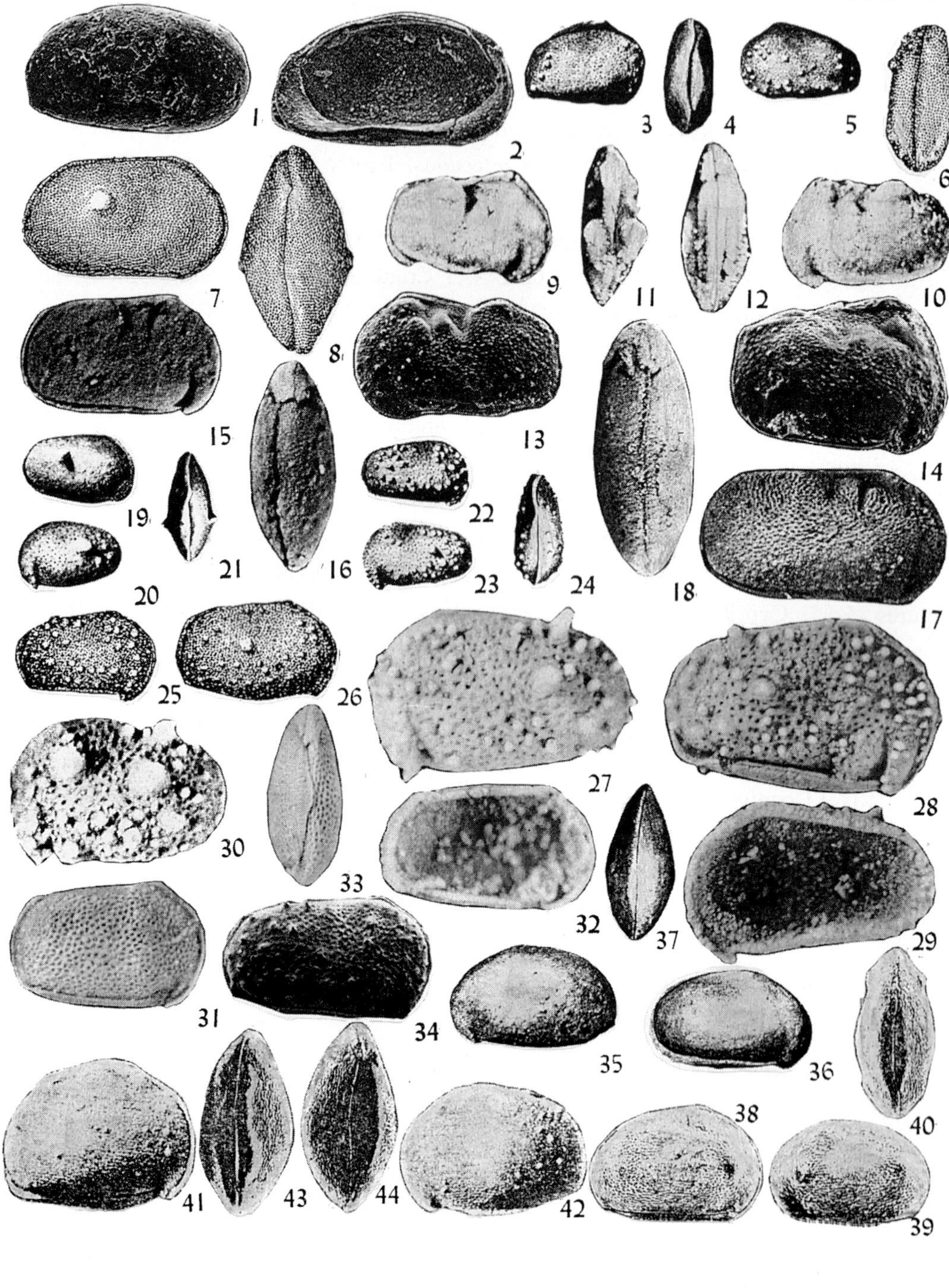
1
2
3
4
5
6
7
8
9
10
11
12
13
14
15
16
17
18
19
20
21
22
23
24
25
26
27
28
29
30
31
32
33
34
35
36
37
38
39
40
41
42
43
44

Plate 13

Figures 1, 2. Cypridea wyomingensis Jones 1893. R side and ventral view, L, 0.92. Lower Cretaceous, Lower Bear River Formation or Gannett Group, Cokeville, Wyoming (after Jones, 1893). (15)

Figures 3, 4. Cypridea wyomingensis Jones, 1893. L side and dorsal view, L, 1.11. Lower Cretaceous, Lower Bear River Formation or Gannett Group, Cokeville, Wyoming (after Jones, 1893). (15)

Figure 5. Cyridea (Pseudocypridina) piedmonti Roth, 1933. R side, holotype, L, 0.87. Upper Jurassic or Lower Cretaceous, Morrison or Lakota Formation, lower shale, sec. 28, T. 4 N., R. 6 E., Meade County, South Dakota (after Roth, 1933). (1)

Figures 6, 7. Longispinella longispina (Peck), Sohn, 1958. 6, R side; 7, dorsal view, L, 0.90. Lower Cretaceous, Lakota Formation, NW1/4 sec. 29, T. 51 N., R. 65 W., Crook County, Wyoming (after Sohn, 1958). (Introduced here for comparison with other cyprideaids). (9)

Figure 8. Cypridea (Ulwellia) menevoides Swain and Brown, 1972. L side, holotype showing RV overlap of LV, L, 0.98. Upper Jurassic or Lower Cretaceous (probably the latter), Unit H, NC-Cur-OT-12 well, Currituck County, North Carolina; 4308-4318 feet (1313-1316 m) (after Swain and Brown, 1972). (8)

Figures 9-11. Cypridea nitidula Peck, 1941. 9, R side, holotype; 10, LV interior; 11, ventral view, L, 1.02. Lower Cretaceous, Draney Formation, Freedom Quadrangle, Idaho-Wyoming (after Peck, 1941). (1)

Figure 12. Cypridea (Bisulcocypridea) nyensis (Swain, 1964). LV, holotype, L, 1.17. Paleocene?, Sheep Pass Formation (part), Grant Range, Nye County, Nevada (after Swain, 1964a). (1)

Figure 13. Cypridea (Bisulcocypridea) nyensis (Swain), Swain, 1987. L side of abraded, weakly nodose shell, L, 0.99. Paleocene?, Sheep Pass Formation (part), Grant Range, Johnson Canyon locality, Nye County, Nevada (after Swain, 1987b). (2)

Figures 14, 15. Cypridea obesa Peck, 1951. 14, R side; 15, ventral view, holotype, L, 0.92. Lower Cretaceous, Cloverly Formation, near Lander Wyoming (after Peck, 1951). (1)

Figures 16-18. Cypridea (Cypridea) pecki Sohn, 1969. 16, R side; 17, L side; 18, ventral views of holotype, L, 0.78. Lower Cretaceous? (Aptian?), Newark Canyon Formation, center W1/2 NW1/4 sec. 21, T. 29 N., R. 53 E., Elko County, Nevada (after Sohn, 1969). (8)

Figures 19-22. Cypridea (Pseudocypridina) piedmonti Roth, Sohn, 1979. 19, L side; 20, R side; 21, RV interior; 22, dorsal view, L, 1.08. Lower Cretaceous, Lakota Formation,

Chilson? Member, Meade County, South Dakota, U.S.G.S. coll. 26089 (after Sohn, 1979). (8)

Figures 23, 24. Cypridea quadrata Peck, 1951. 23, R side, holotype; 24, ventral view, L, 0.92. Lower Cretaceous, Cloverly Formation, near Lander, Wyoming (after Peck, 1951). (1)

Figures 25-27. Cypridea (Bisulcocypridea) skeeteri (Peck) (Peck, 1951) (as Cypridea skeeteri Peck). 25, R side, holotype; 26, L side; 27, dorsal view, L, 0.89. Upper Cretaceous, Bear River Formation, 9 miles (14.5 km) SE of Evanston, Wyoming (after Peck, 1951). (1)

Figures 28, 29. Cypridea wyomingensis Jones, 1894 (Vanderpool, 1928). (= C. tuberculcata gypsumensis Vanderpool, 1928). 28, LV; 29, dorsal view, L, 0.48. Lower Cretaceous, Trinity Group, De Queen Clay, Gypsum Bluff, near Murphreesboro, Pike County, Arkansas (after Vanderpool, 1928). (1)

Figures 30, 31. Cypridea dequeenensis Swain and Brown, 1964. (= Cypridea ventrosa bispinosa Vanderpool, 1928). 30, RV: 31, LV interior, L, 0.44. Lower Cretaceous, Trinity Group, De Queen Formation, Gypsum Bluff, near Murphreesboro, Pike County, Arkansas (after Vanderpool, 1928) (=Cypridea wyomingensis Jones, 1893 by Peck, 1941). (1)

Figures 32-34. Cypridea wyomingensis (Jones), Peck, 1941. 32, R side; 33, L side; 33, ventral view, L, 1.04. Lower Cretaceous, Draney Limestone, Freedom Quadrangle, Idaho-Wyoming (after Peck, 1941). (1)

Figure 35. Cypridea wyomingensis (Jones), Swain and Brown, 1964. RV, 1.09, Lower Cretaceous, Trinity Group, De Queen Formation, Gypsum Bluff, Pike County, Arkansas (after Swain and Brown, 1964). (6)

Figure 36. Cypridea (Cypridea) wyomingensis (Jones), Swain and Brown, 1972. R side, L, 0.98. Lower Cretaceous, Unity G, NC-DA-OT-11 well, Dare County, North Carolina; 4446-4456 feet (1355-1358 m) (after Swain and Brown). (8)

Figures 37, 38. Cypridea? sp., Sohn, 1958. 37, R side; 38, ventral view, L, 1.17. Lower Cretaceous, Lakota Formation, SW1/4 sec. 15, T. 8 S., R. 4 E., Fall River County, South Dakota (after Sohn, 1958). (9)

Figure 39. Cypridea (C.) sp., Swain and Brown, 1972. R side, L, 0.53. Upper Jurassic or Lower Cretaceous, Unit H, NC-DA- OT-12 well, Dare County, North Carolina; 7145-7156 feet (2178-2181 m) (after Swain and Brown, 1972). (8)

Figures 40, 41. "Cypridea" sp. 1, Sohn, 1979. 40, R side; 41, dorsal view, L, 0.40. Lower Cretaceous, Lakota Formation, Chilson Member, Unit 2, U.S.G.S. coll. 26098, Fall River County, South Dakota (after Sohn, 1979). (8)

e 42. Cypridea sp. 1, Nadeau, 1945, unpublished. R side, magnification unknown. Upper Jurassic or Lower Cretaceous, Morrison Formation?, Canyon City, Colorado.

Figure 43. Cypridea sp. 2, Nadeau, 1945, unpublished. R side, magnification unknown. Upper Jurassic or Lower Cretaceous, Morrison Formation?, Canyon City, Colorado?

Figures 44, 45. Cypridacea? incertae sedis, Nadeau, 1945, unpublished. 44, R side; 45, edge view, magnification unknown. Upper Jurassic or Lower Cretaceous, Morrison Formation?, Canyon City, Colorado.?

Plate 13

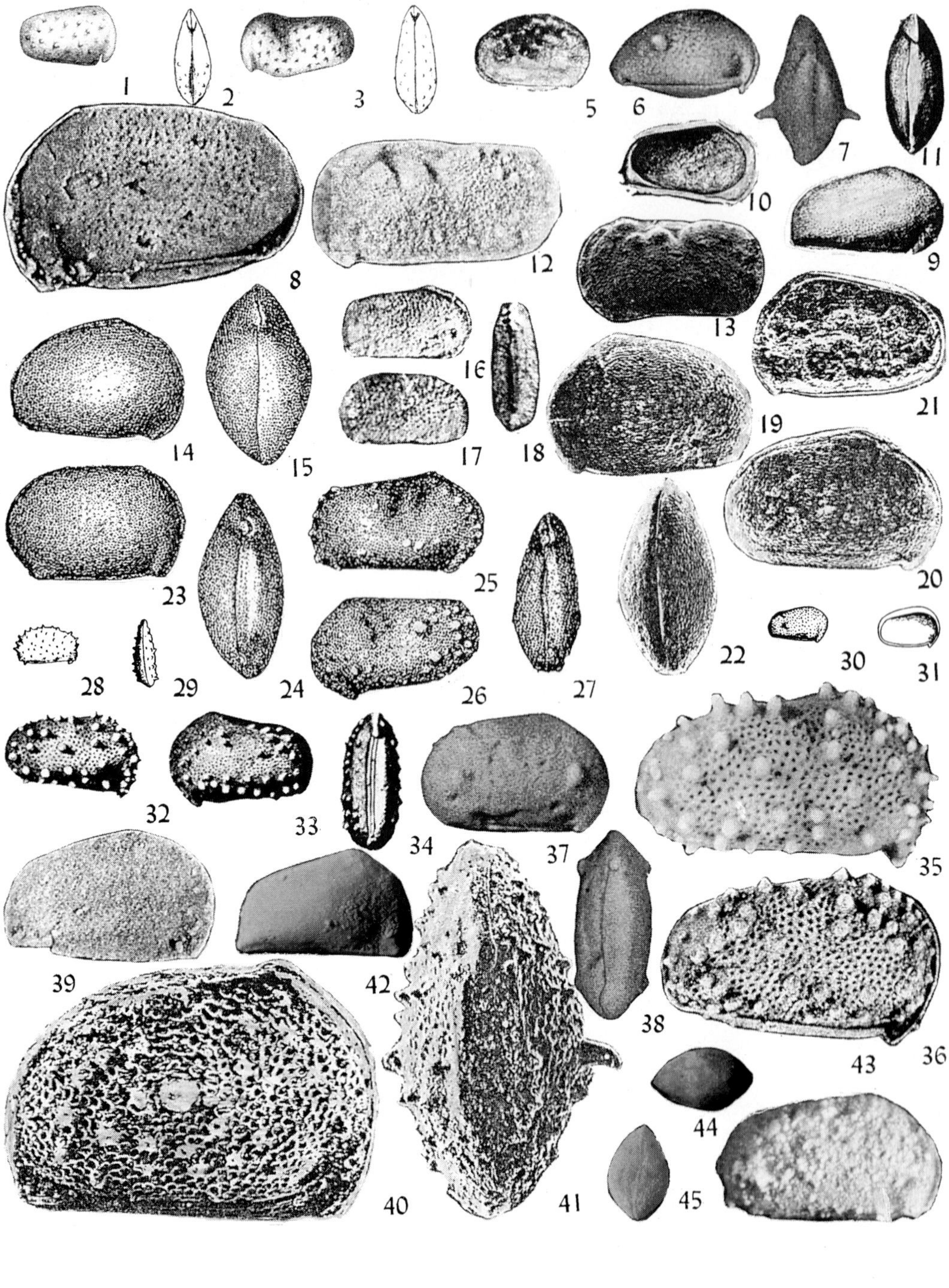
1
2
3
5
6
7
11
8
12
10
9
13
21
14
15
16
17
18
19
23
24
25
27
22
20
28
29
26
30
31
32
33
34
37
35
39
42
38
36
43
40
41
44
45

Plate 14

Figure 1. Cypridacea, incertae sedis. Nadeau, 1945, unpublished. R? side, magnification unknown. Upper Jurassic or Lower Cretaceous, Morrison Formation?, Canyon City, Colorado.? (Resembles Dogelinella).

Figures 2, 3. Cypridacea, incertae sedis, Benson, 1967. 2, LV, L, 1.03; 3, LV, interior. Pleistocene, Yarmouth Stage, Dickinson County, Kansas. USNM 649609, loc. 8 (after Benson, 1967). (Resembles Dogelinella). (13)

Figure 4. Cypridacea, incertae sedis 2, Nadeau, 1945, unpublished. R? side, magnification unknown. Upper Jurassic or Lower Cretaceous, Morrison Formation?, Canyon City, Colorado.?

Figures 5, 6. Cypridacea, incertae sedis 3, Nadeau, 1945, unpublished. R sides of 2 shells, magnification unknown. Upper Jurassic or Lower Cretaceous, Morrison Formation?, Canyon City, Colorado. (Resembles Trapezoidella).

Figure 7. Cypridacea, incertae sedis 4, Nadeau, 1945, unpublished. R? side, magnification unknown. Upper Jurassic or Lower Cretaceous, Morrison Formation?, Canyon City, Colorado.

Figure 8. Cypridacea, incertae sedis 5, Nadeau, 1945, unpublished. R? side, magnification unknown. Upper Jurassic or Lower Cretaceous, Morrison Formation?, Canyon City, Colorado?

Figures 9, 10. Cypridacea, incertae sedis 6, Nadeau, 1945, unpublished. 9, R? side; 10, L? side, magnification unknown. Upper Jurassic or Lower Cretaceous, Morrison Formation?, Canyon City, Colorado?

Figure 11. Cypridacea, incertae sedis 7, Nadeau, 1945, unpublished. R? side, magnification unknown. Upper Jurassic or Lower Cretaceous, Morrison Formation?, Canyon City, Colorado.

Figure 12. Cypridacea, incertae sedis 7, Nadeau, 1945, unpublished. R? side, magnification unknown. Upper Jurassic or Lower Cretaceous, Morrison Formation?, Canyon City, Colorado?

Figure 13. Cypridacea, incertae sedis 8, Nadeau, 1945, unpublished. R side, magnification unknown. Upper Jurassic or Lower Cretaceous, Morrison Formation?, Canyon City, Colorado?

Figure 14. Cypridacea, incertae sedis 9, Nadeau, 1945, unpublished. R? side, magnification unknown. Upper Jurassic or Lower Cretaceous, Morrison Formation?, Canyon City, Colorado?

Figure 15. Cypridacea, incertae sedis 10, Nadeau, 1945, unpublished. R? side, mag unknown. Upper Jurassic or Lower Cretaceous, Morrison Formation?, Canyon City, Colorado?

Figure 16. Ostracoda, incertae sedis, Nadeau, 1945, unpublished. R? side, magnification unknown. Upper Jurassic or Lower Cretaceous, Morrison Formation?, Canyon City, Colorado?

Figure 17. Ostracoda, incertae sedis, Nadeau, 1945, unpublished. L? side, magnification unknown. Upper Jurassic or Lower Cretaceous, Morrison Formation?, Canyon City, Colorado?

Figures 18-20. Cyprideis littoralis Brady, Gutentag and Benson, 1962. 18, RV female, L, 0.99; 19, LV male; 20, RV female, interior, L, 0.81. Pleistocene, Illinoian-Wisconsin Stages, Nye Sink, NW1/4 SW1/4 sec. 15, T. 6 N., R. 25 ECM, Beaver County, Oklahoma (after Gutentag and Benson, 1962). (4)

Figures 21-24. Cyprideis maladensis Swain, 1987. 21, L side, female, holotype, L, 0.97; 22, 22a, 23, R side, enlargement of an elongate sieve plate, approx. X1680, dorsal view, female, L, 0.88; 24, dorsal view male, L, 0.75. Pliocene, Salt Lake Group, secs. 4-26, T. 16 S., R. 36 E., Oneida County, Idaho (after Swain, 1987a). (2)

Figures 25, 25a. Cyprideis salebrosa van den Bold, Benson, 1967. 25, LV, transmitted light, L, 0.99; 25a, adductor muscle scar area. Late Pleistocene, USNM 649597, Beaver County, Oklahoma (after Benson, 1967). (13)

Plate 14

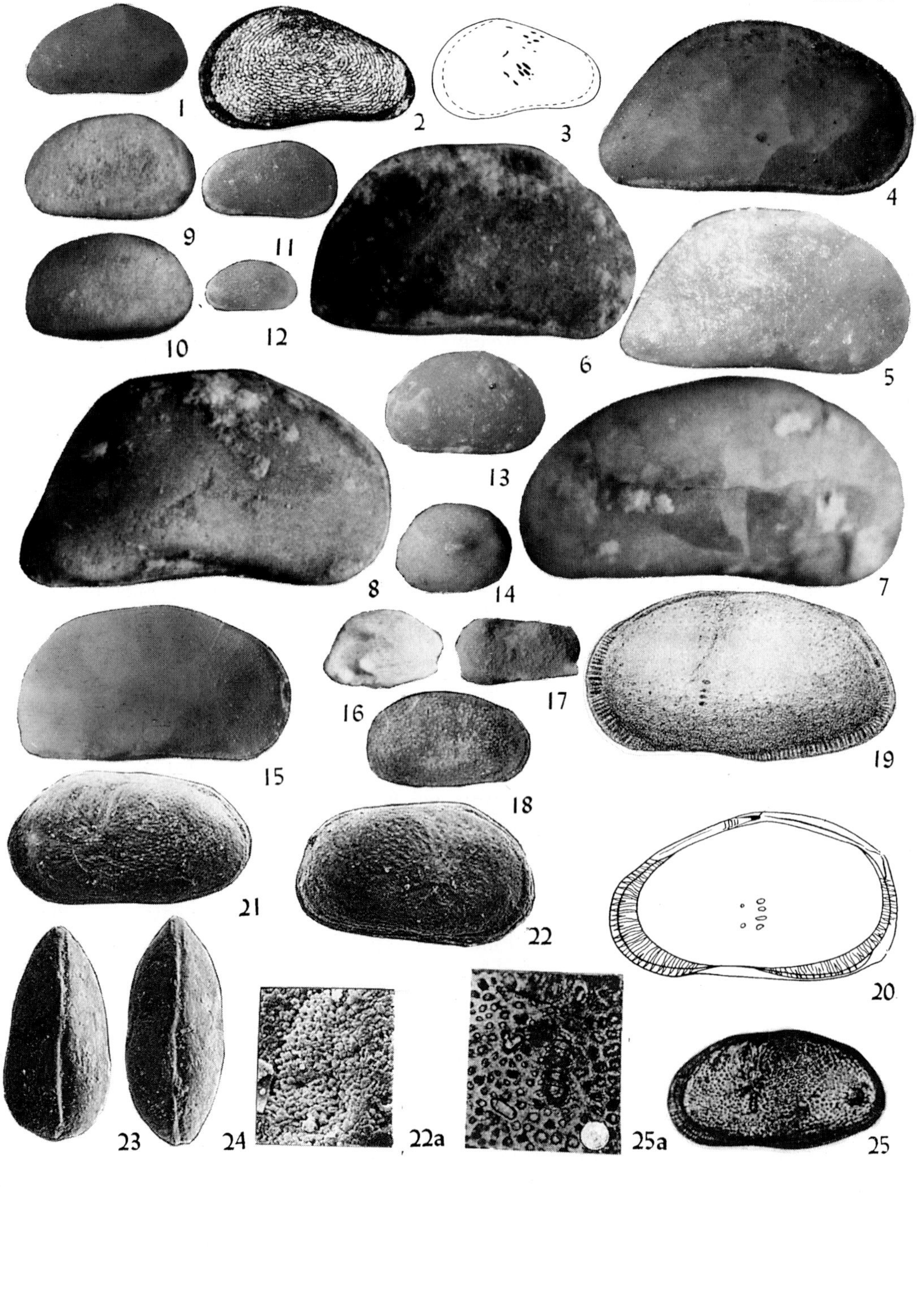

Plate 15

Figures 1-3. Cyprideis sp., group of C. salebrosa van den Bold, Benson, 1969. 1, RV, L, 0.80; 2, LV, L, 0.65; 3, dorsal view RV. Early Pleistocene, Rita Blanca lake beds, Hartley County, Texas (after Benson, 1969). (3)

Figure 4. Cyprideis? sp., Swain, 1986. R side, L, 0.70. Pliocene, Cache Valley Formation, Hoopes Creek loc., NW1/4 SE1/3 NE1/4 sec. 10, T. 12 S., R. 40 E., Franklin County, Idaho (after Swain, 1986a). (2)

Figure 5. Cyprideis n. sp., Swain, 1986. R side, L, 1.10. Miocene, Humboldt Formation, Gulf Oil Company, no. 1 Thousand Springs well, SE1/4 SW1/4 sec. 8, T. 40 N., R 66 E., Elko County, Nevada: 1670-1680 feet (509-512 m) (after Swain, 1986a). (2)

Figures 6, 6a. Eocypridina aciculata (Scott and Summerson, 1943). 6, LV internal mold, L, 5.3; 6a, detail of juncture of three sets of radii, approx. X 50. Lower Pennsylvanian, Kanawha Series, Hance Formation, along U.S. 119, just north of Molus, Harlan County, Kentucky (after Scott and Summerson, 1943). This form may be marginal marine, brackish water in origin. (11)

Figure 7. Eocypridina radiata (Jones, Kirkby, and Brady). Scott and Summerson, 1943. LV interior mold, L, 5.0. Lower Pennsylvanian, Kanawha Series, Hance Formation, along U.S. 119, just north of Molus, Harlan County, Kentucky (after Scott and Summerson, 1943). This form may be marginal marine, brackish water in origin. (11)

Figures 8-10. Cypridopsella? esmeraldensis Swain and Wagoner, 1985. 8, R side, holotype, L, 0.98; 9, L side; 10, ventral view. Miocene, Esmeralda Formation, Blair Junction section, Esmeralda County, Nevada (after Swain and Wagoner, 1985). (2)

Figure 11. Cypridopsella esmeraldensis Swain and Wagoner, Swain, 1985. R side of large shell, L, 1.44. Miocene, middle Esmeralda Formation, 2.85 miles (4.5 km) west of Black Spring, Nye County, Nevada (after Swain, 1985). (2)

Figures 12, 13. Cypridopsis dickinsonia Tasch, 1963. 12, LV; 13, dorsal view, L, 0.74. Permian, Wellington Formation, Elmo Bed, NE1/4 NW1/4 sec. 21, T. 16 S., R. 2 E., Dickinson County, Kansas (after Tasch, 1963). (1)

Figures 14-16. Cypridopsis fabulina Scott and Summerson, 1943. 14, LV; 15, dorsal view, female; 16, dorsal view male, L, 0.80. Lower Pennsylvanian, Briceville Formation, between Coalfield and Christman, Morgan County, Tennessee (after Scott and Summerson, 1943). (11)

Figures 17-18. Cypridopsis ovata (Holland), Scott, 1944. 17, LV; 18, dorsal view, L, 0.78. Permian, Dunkard Series, Nineveh Formation, Nineveh, Pennyslvania (after Scott, 1944). (11)

'igures 19, 20. Cypridopsis parvula (Holland). L side, R side, L. 0.59. Permian, Nineveh Formation, Wood County, West Virginia (after Scott, 1944).

Figures 21, 22. Cypridopsis ovata (Holland) (=Whipplella ovata Holland, 1934, original figures). 21, R side; 22, dorsal view, L, 0.78. Permian, upper Dunkard Series, upper Greene Group, Nineveh Formation, Nineveh, Pennsylvania (after Holland, 1934). (12)

Figures 23, 24. Cypidopsis parvula (Holland) (=Whipplella parvula Holland, 1934, original figures). 23, R side; 24, dorsal view, L, 0.59. Permian, upper Dunkard Series, upper Greene Group, Nineveh Formation, Nineveh, Pennsylvania (after Holland, 1934). (12)

Figures 25, 26. Cypridopsis saltlakensis Swain, 1947. 25, R side; 26, LV? interior, L, 0.65. Pliocene, Salt Lake Group, sec. 15, T. !3 N., R. 2 W., Cache County, Utah (after Swain, 1947). (1)

Figure 27. Cypridopsis vidua (O.F. Müller), Gutentag and Benson, 1962. RV, L, 0.50. Pleistocene, Sangamon-Wisconsin Stage, Odee Formation, sec. 33, T. 34 S., R. 29 W., Meade County, Kansas (after Gutentag and Benson, 1962). (4)

Figure 28. Cypridopsis vidua (O.F. Müller), Staplin, 1963. LV, L, 0.67. Pleistocene, Illinoian Stage, Brussels Quadrangle, SW1/4 SE1/4 NE1/4 sec. 12, just north of Brussels, Johnson County, Illinois (after Staplin, 1963). (1)

Figures 29, 29a. Cypridopsis vidua (O.F. Müller), Benson, 1967. 29, LV transmitted light, L,0.87; 29a, adductor muscle scar area, approx. X 140. Pleistocene, Yarmouthian Substage, Dickinson County, Kansas, USNM no. 649606, loc. P (after Benson, 1967). (3)

Figure 30. Cypridopsis? sp. L side, L, 0.60, Swain, 1986. Miocene, Humboldt Formation, Gulf Oil Company, no. 2 Wilkins Ranch well, sec. 19, T. 39 N., R. 69 E., Elko County, Nevada, 2635-2665 feet (803-813 m) (after Swain, 1986b). (2)

Figure 31. Cypridopsis? sp., King 1970, unpublished. LV, L, 0.61. Pleistocene, bone bed at Boney Spring, Benton County, Missouri (Dissertation area of James E. King, University of Arizona, sample dated 16,500 B.P.)

Figures 32, 33. Cyprinotus sp. aff. C. cingalensis Brady, Swain, 1986b. 32, RV, poorly preserved; 33, LV partly exfoliated, L, 1.01. Miocene, Humboldt Fonation, Cobre section, Elko County, Nevada; 1011 feet (308m) above base (after Swain, 1986b). (2)

Figures 34, 35. Cyprinotus glaucus Furtos, Staplin, 1963. 34, LV female; 35, RV female, L, 1.10 X 40. Pleistocene, Cary or Mankato Substage, Wisconsin Stage, Blue Island, Chicago, Illinois, sec. 5, T. 37 N., R. 14 E., (after Staplin, 1963). (1)

Figure 36. Cyprinotus sp., Staplin, 1963. LV, L, 1.12. Pleistocene, late Kansan stage, Sappa Silt, SE1/4 sec. 15, SW1/4 sec. 14, T. 98 N., R. 48 W., Lyon County, Iowa (after Staplin, 1963). (1)

Plate 15

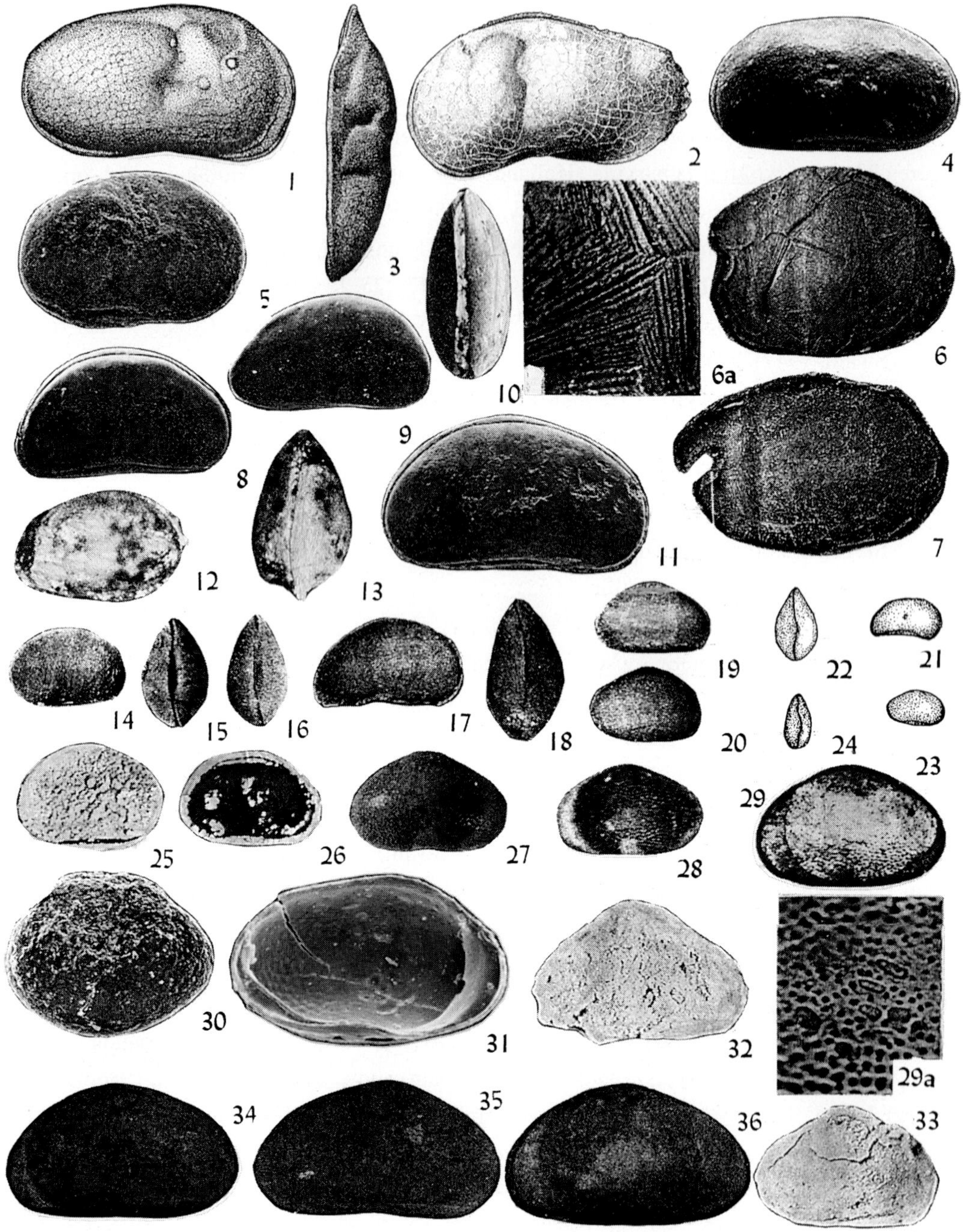
1
2
3
4
5
6
6a
7
8
9
10
11
12
13
14
15
16
17
18
19
20
21
22
23
24
25
26
27
28
29
29a
30
31
32
33
34
35
36

Plate 16

Figures 1, 2. Cyprinotus? sp. aff. C. yuchengensis Hou and Li, Swain, 1985. 1, R side, L, 0.93; 2, dorsal view. Micoene, Esmeralda Formation, 7.6 miles (12.1 km) east of Finger Rock Wash on road to Pactolus Mine, Nye County, Nevada (after Swain, 1985). (2)

Figures 3, 4. Cyprinotus? sp. 1, Swain, 1986b. 3, R side; 4, dorsal view, L, 0.93. Micoene, Humboldt Formation , Gulf Oil Company, no. 1 Thousand Springs well, SE1/4 SW1/4 sec. 8, T. 40 N., R. 66 E., Elko County, Nevada; 1250-1260 feet (381-384 m) (after Swain, 1986b). (2)

Figures 5, 5a. Cyprinotus sp., Benson, 1967. 5, RV by transmitted light, L, 1.10; 5a, adductor muscle scar area, approx. X 140. Pleistocene, Yarmouthian Stage, Dickinson County, Kansas (after Benson, 1967). (3)

Figures 6, 7. Cyprinotus sp. 1, Sohn, 1956. 6, R side; 7, dorsal view; L, 1.36. Pliocene, Teewinot Formation, Jackson Hole, Wyoming, USNM 562 027 (after Sohn, 1956). (9)

Figure 8. Cypris osmondi Dickinson and Swain, 1967. LV, holotype, L, 0.93. Pliocene, Hay Ranch Formation (=upper Humboldt), Gulf Oil Company, no. 1 Wilkins Ranch well sec. 21, T. 38 N., R. 61 E., Elko County, Nevada; 1558-1588 feet (475-484 m) (after Dickinson and Swain, 1967). (1)

Figures 9, 10. Cypris potamides Dickinson and Swain, 1967. 9, RV holotype, L, 0.95; 10, LV. Pliocene, Hay Ranch Formation (=upper Humboldt), Gulf Oil Company, no. 1 Wilkins Ranch well sec. 21, T. 38 N., R. 61 E., Elko County, Nevada; 1496-1526 feet (456-465 m); 10, 1344-1374 feet (410-419 m) (after Dickinson and Swain, 1967). (1)

Figures 11, 12. Cypris potamides Dickinson and Swain, Swain, 1967. 11, R side, L, 0.88; 12, dorsal view. Pliocene or Miocene, Salt Lake Group, hillside SE1/4 NE1/4 sec. 30, T. 11 N., R. 14 W., Box Elder County, Utah (after Swain, 1987a). (2)

Figures 13, 14. Cypris purbeckensis Forbes?, Jones, 1885, Peck, 1941. 13, R side; 14, dorsal view, L, 1.10. Lower Cretaceous, Draney Formation, Freedom Quadrangle, Idaho-Wyoming (after Peck, 1941). (1)

Figures 15, 16. Cyprois ephraimensis Swain, 1964. 15, LV, holotype; 16, RV, L, 1.01. Eocene, Colton Formation, 59 feet (18 m) above base, Dripping Rock Gulch section, Raven Bridge, Rio Blanco County, Colorado (after Swain, 1964a). (1)

Figures 17, 18. Cyprois ephraimensis Swain, 1964 (=C. cf. marginata (Strauss) Swain, 1949). 17, R side; 18, dorsal view, L, 0.96. Eocene, Flagstaff Limestone, 1 mile east of Ephraim, Utah (after Swain, 1949). (1)

Figures 19-21. Cyprois whiteflatensis Swain, 1987. 19, R side, L, 1.50; 20, RV interior; 21, enlargement of adductor muscle scar area, approx. X340. Eocene, Twin Bridges

Limestone, center sec. 36, T. 52 N., R. 55 E., Twin Bridges, Elko County, Nevada (after Swain, 1987b). (2)

Figure 22. Cyprois sp. 1, R. Konizeski, unpublished. R side, L, 1.54. Oligocene, Chadron Formation, Helmville, Nebraska (coll. by R. Konizeski).

Figure 23. Cyprois sp. 2, R. Konizeski, unpublished. R side, L, 1.38. Oligocene, Chadron Formation, Helmville, Nebraska (coll. by R. Konizeski).

Figures 24-27. Cytheridella alpinensis Swain, 1987. 24, L side holotype; 25, dorsal view, L, 0.68, female; 26, R side; 27, dorsal view, L, 0.40, male. Oligocene, Buck Hill Volcanic series, Pruett? Formation, Alpine Quadrangle, 1 mile (1.6 km) SW of old Neville Ranch, Brewster County, Texas (after Swain, 1987c). (2)

Figure 28. Cytheridella sp. aff. C. ilosvayi Daday, Swain, 1987c. L side of shell, L, 0.53. Oligocene, Buck Hill Volcanic series, Pruett Formation, Alpine Quadrangle, 1 mile (1.6 km) SW of old Neville Ranch, Brewster County, Texas (after Swain, 1987c). (2)

Figures 29, 30. Cytherissa lacustris (Sars), Staplin, 1963. RV, female; 30, RV immature, L, 0.87. Pleistocene, late Mankato Substage, Wisconsin Stage, core, northern Lake Michigan (after Staplin, 1963). (1)

Figures 31, 32. Darwinula sarytirmensis Sharapova, Kietzke, and Lucas, 1995. L side and ventral view, L, 1.2 approx. Lower Jurassic, Kayenta Formation, SW1/4 sec. 35, T. 27 N., R 12 E., Coconino County, Arizona (after Kietzke and Lucas, 1995). (7)

Figure 33. Darwinula? sp. B., Kietzke, 1989. L side, L, 0.74. Lower Triassic, Moenkopi Formation, NW1/4 NE1/4 NE1/4 sec. 36, T. 7 N., R. 4 W., Cibola County, New Mexico, at White Ridge (after Kietzke, 1989). (7)

Figure 34. Darwinula sp. A., Kietzke, 1989. L side, L, 0.90. Lower Triassic, Moenkopi Formation, NW1/4 NE1/4 NE1/4 sec. 36, T. 7 N., R. 4 W., Cibola County, New Mexico, at White Ridge (after Kietzke, 1989). (7)

Figure 35. Darwinuloides sp. B., Kietzke, 1989. R side, L, 0.86. Lower Triassic, Moenkopi Formation, NW1/4 NE1/4 NE1/4 sec. 36, T. 7 N., R. 4 W., Cibola County, New Mexico, at White Ridge (after Kietzke, 1989). (7)

Figures 36, 37. Darwinuloides magna Jiang in Xu, Kietzke and Spencer, 1995. L side and dorsal view, L, 1.23. Lower Jurassic, Kayenta Formation, SW1/4 sec. 35, T. 27 N., R. 12 E., Coconino County, Arizona (after Kietzke and Spencer, 1995). (17)

Figure 38. Gerdalia cf. triassica Belousova, Lucas and Kietzke, 1993. L side, L, 0.75. Upper Triassic, Petrified Forest Formation, Petrified Forest National Park, Arizona (after Lucas and Kietzke, 1993). (17)

Plate 16

1
2
3
4
5a
6
7
8
9
5
10
12
13
14
18
16
11
15
17
22
23
21
20
31
32
37
33
25
36
34
35
19
26
38
28
30
24
29
27

Plate 17

Figure 1. Darwinuloides sp. A., Kietzke, 1989. L side, L, 0.63. Lower Triassic, Moenkopi Formation, NE1/4 NE1/4 NW1/4 sec. 36, T. 7 N., R. 4 W., Cibola County, New Mexico, at White Ridge (after Kietzke, 1989). (7)

Figures 2-5. Cytherissa simplissima Swain, 1963. 2, RV; 3, RV interior; 4, LV interior; 5, dorsal view, L, 0.91. Pleistocene, Gubik Formation, Arctic coastal plain, northern Alaska, Ikpikpuk shot holes, line 10, shot points 24, 31, at 40-50 feet (12-15 m) (after Swain, 1963). (1)

Figure 6. Cytherissa simplissima Swain, 1963. LV interior, L, 0.84. Pleistocene, Gubik Formation, Arctic coastal plain, northern Alaska, Ikpikpuk shot holes, line 10, shot point 8, at 30-40 feet (9-12 m) (after Swain, 1963). (1)

Figure 7. Cytheridella? sp., immature, J. E. King, unpublished. R side, L, 0.24. Pleistocene, bone bed at Boney Spring, Benton County, Missouri (coll. by J.E. King, 1970).

Figure 8. Cytheromorpha? sp., Swain, 1985. LV, L, 0.57. Miocene, Esmeralda Formation, 3.3 miles (5.3 km) east of Finger Rock Wash, Simon Mine area, Nye County, Nevada (after Swain, 1985). (2)

Figures 9, 10. Darwinula aurea (Brady and Robertson), Staplin, 1963 (=D. stevensoni (B. & R.)). 9, LV; 10, RV, L, 0.70. Pleistocene, Cary of Mankato Substage, Wisconsin Stage, sec. 5, T. 37 N., R. 14 W., Blue Island, Chicago, Illinois (after Staplin, 1963). (1)

Figures 11, 12. Darwinula hollandi Scott, 1944. 11, RV, holotype; 12, dorsal view, L, 0.42. Permian, Nineveh Formation, Nineveh, Pennsylvania (after Scott, 1944). (Probably *not* Darwinula, may be a paracyprid). (1)

Figure 13. Darwinula rogersii (Jones), Swain and Brown, 1972. LV, L, 0.84. Upper Triassic, Newark Supergroup, Cumnock County, North Carolina (after Swain and Brown, 1972). (8)

Figures 14, 15. Darwinula? simplus (Roth), (=Paracypris simplus Roth, 1933). 14, R side; 15, dorsal view, L, 1.08. Upper Jurassic or Lower Cretaceous, Morrison Formation, sec. 28, T. 4 N., R. 6 E., Lawrence County, South Dakota (after Roth, 1933). (=Lower Cretaceous Lakota Formation by Sohn, 1957, 1958). (Has RV overlapped by LV, and thus is not typical Darwinula). (1)

Figure 16. Darwinula cf. stevensoni (Brady and Robertson), Swain, 1949. RV, L, 0.79, X 40, poorly preserved. Paleocene, Fort Union Formation, Tongue River Member, NW1/4 sec. 31, T. 9 S., R. 49 E., Moorhead Coal District, Powder River County, Montana, 125 feet (38 m) above Anderson Coal (after Swain, 1949). (1)

Figures 17, 18. Darwinula cf. stevensoni (Brady and Robertson), Swain, 1986. 17, LV interior; 18, adductor muscle scar area, showing two frontal scars, approx. X 310. Pliocene, Cache Valley Formation, Hoopes Creek loc., NW1/4 SE1/4 NE1/4 sec. 10, T. 12 S., R. 40 E., Franklin County, Idaho (after Swain, 1986a). (2)

Figures 19, 20. Darwinula cf. stevensoni (Brady and Robertson), Swain, 1987c. 19, L side, L, 0.45; 20, dorsal view, L, 0.47. Miocene, Horse Camp Formation, sec. 12, T. 10 N., R 58 E., 3.8 miles (6.2 km) east of Currant, Nye County, Nevada (after Swain, 1987c). (2)

Figures 21, 22. Darwinula subquadrata Swain and Brown, 1972. 21, LV, holotype, L, 0.83; 22, RV, L, 0.89. Upper Triassic, Newark Supergroup, Cumnock Formation, Bethany Church loc. near Gulf, Chatham County, North Carolina (after Swain and Brown, 1972). (8)

Figures 23-25. Darwinula sp., Sohn, 1977. 23, R side, L, 0.66; 24, ventral view; 25, adductor muscle scar, approx. X 600. Pennsylvanian, Monongahela Formation, Monongalia County, West Virginia, U.S.G.S. coll, 12899-pc (after Sohn, 1977). (8)

Figure 26. Darwinula sp., Swain, 1986. R side, L, 0.80. Miocene, Humboldt Formation, Gulf Oil Company, no. 2 Wilkins Ranch well, sec. 19, T. 39 N., R. 69 E., Elko County, Nevada: 4910-4940 feet (1497-1506 m) (after Swain, 1986b). (2)

Figure 27. Darwinula sp., Swain, 1987. L side, L, 0.69. Paleocene or Eocene, Sheep Pass Formation, Grant Range, Johnson Canyon section, Nye County, Nevada (after Swain, 1987b). (2)

Figures 28, 29. Darwinula sp., Swain, 1987c. 28, L side, L, 0.55, dorsum tilted toward observer; 29, dorsal view, L, 0.33. Oligocene, Buck Hill Volcanic series, Pruett? Formation, 1 mile (1.6 km) SW of old Neville Ranch, Brewster Conuty, Texas (after Swain, 1987c). (2)

Figures 30-33. Darwinula sp. 1, Sohn, 1985. 30, R side of internal mold; 31, dorsal view of mold; 32, posterior view, L, 0.60; 33, adductor muscle scar, approx. X 400. Upper Mississippian, Bluestone Formation, Bramwell Member, Top Flat Quadrangle, Mercer County, West Virginia, U.S.G.S. coll. 12983-PC (after Sohn, 1985). (1)

Figure 34. Darwinula sp. 2 Sohn, 1985. R side of internal mold, L, 1.10. Upper Mississippian, Bluestone Formation, Bramwell Member, Top Flat Quadrangle, Mercer County, West Virginia (after Sohn, 1985). (1)

Figure 35. Darwinula sp. 3, Sohn, 1985. R side of internal mold, L, 1.24. Upper Mississippian, Bluestone Formation, Bramwell Member, Top Flat Quadrangle, Mercer County, West Virginia (after Sohn, 1985). (1)

Figure 36. Darwinula? sp. 1, Swain and Brown, 1972. LV, 1.90. Upper Triassic, Supergroup, Cumnock Formation, Bethany Church loc., 0.5 mile (0.8 km) west of Gulf, Chatham County, North Carolina (after Swain and Brown, 1972). (8)

Figure 37. Darwinula? sp. 1, Swain and Brown, 1972. LV, internal mold, L, 0.88. Upper Triassic, Newark Supergroup, undifferentiated, near Carpenter, Wake County, North Carolina (after Swain and Brown, 1972). (8)

Figure 38. Darwinula? sp. 2, Swain and Brown, 1972. RV, L, 0.87. Upper Triassic, Newark Supergroup, Cumnock Formation, Bethany Church, 0.5 mile (0.8 km) west of Gulf, Chathman County, North Carolina (after Swain and Brown, 1972). (8)

Figure 39. Darwinula? sp. 3, Swain and Brown , 1972. R side, L, 0.65. Upper Triassic, Newark Supergroup, Cumnock Formation, Bethany Church, 0.5 mile (0.8 km) west of Gulf, Chathman County, North Carolina (after Swain and Brown, 1972). (8)

Figure 40. Darwinula? sp. 4, Swain and Brown , 1972. RV, L, 0.65. Upper Triassic, Newark Supergroup, Cumnock Formation, Bethany Church, 0.5 mile (0.8 km) west of Gulf, Chathman County, North Carolina (after Swain and Brown, 1972). (8)

Figure 41. Darwinula sp., Kietzke, 1987. R side, L, 0.50. Upper Triassic, Chinle Formation, upper shale, Bull Canyon, NW1/4 sec. 28, T. 9 N., R. 26 E., Guadalupe County, New Mexico (after Kietzke, 1987). (7)

Figure 42. Darwinula sp aff. D. luilingchuanensis Zhong, Lucas and Kietzke, 1993. R side, L, 0.66. Upper Triassic, Petrified Forest Formation, Petrified Forest National Park, Arizona (after Lucas and Kietzke, 1993). (17)

Figures 43, 44. Gerdalia sp., Kietzke, 1989. R side, L, 1.10; R side. Lower Triassic, Moenkopi Formation, NW1/4 NE1/4 NW1/4 sec. 36, T. 7 N., R. 4 W., Cibola County, New Mexico (after Kietzke, 1989). (7)

Figure 45. Eucypris serrata (Müller), Lister, 1975. RV, L, 1.91. Pleistocene, Lake Bonneville Group, Saltair core hole, SE1/4 sec. 25, T. 1 N., R. 3 W., Salt Lake County, Utah (after Lister, 1975). (3)

Plate 17

1
2
3
5
4
9
10
6
34
42
11
36
41
44
38
43
12
8
45
7
15
13
16
14
17
19
20
21
18
22
23
26
27
24
30
32
39
25
31
37
33
28
29
40
35

Plate 18

Figure 1. *Darwinula* sp., Sohn, 1976. RV, L, 0.63. Pennsylvanian, Monongahela Formation, Waynesburg Limestone Member, just below Little Waynesburg Coal, roadcut, Interstate 79 just north of Pursglove, Monongalia County, West Virginia (after Sohn, 1976). (10)

Figures 2, 3. *Darwinula* sp., Nadeau, 1945, unpublished. 2, LV; 3, RV, magnification unknown. Upper Jurassic or Lower Cretaceous, Morrison Formation?, Canyon County, Colorado? (Nadeau, 1945).

Figures 4, 5. *Disopontocypris? hendersoni* Swain, 1987c. 4, R side, holotype, L, 0.97; 5, dorsal view. Miocene, Esmeralda Formation, 2.85 miles (4.5 km) west of Black Spring, Nye County, Nevada (after Swain, 1987c). (2)

Figures 6-8. *Dogelinella? coaldalensis* Swain and Wagoner, 1985. 6, R side, holotype, L, 0.50; 7, L side, L, 0.34; 8, enlargement of shell surface, approx. X 500. Miocene, Esmeralda Formation, Blair Junction section, sample C1-2, Esmeralda County, Nevada (after Swain and Wagoner, 1985). (2)

Figures 9, 10. *Dogelinella?* sp., Swain, 1987c. 9, R side, L, 0.43; 10, dorsal view. Miocene, North Creek Formation, 5.85 miles (9.4 km) east of Sacramento Pass summit, north of U.S. highway 50, White Pine County, Nevada (after Swain, 1987c). (2)

Figures 11, 12. *Dongyingia? lariversi* Swain, 1985. 11, L side, holotype, L, 0.84; 12, RV interior, L, 1.04. Miocene, Esmeralda Formation, 2.85 miles (4.5 km) west of Black Spring, Nye County, Nevada (after Swain, 1985). (2)

Figures 13-15. *Dongyingia? lariversi* Swain, Swain, 1987c. 13, R side, L, 0.46; 14, R side, L, 0.54; 15, dorsal view. Miocene, Esmeralda Formation, Warrior Mine section, 7650 feet (2332 m) S. 26° E. of NW cor. Simon 1:24000 Quadrangle, Nye County, Nevada (after Swain, 1987c). (2)

Figures 16-19. *Dryelba pustulosa* Sohn, 1982. 16, RV; 17, LV, L, 1.20; 18, dorsal view; 19, ventral view, L, 1.35 X 30. Upper Jurassic, Morrison Formation, Crook County, Wyoming; U.S.G.S. coll. 31746 (after Sohn, 1982). (5)

Figures 20-23. *Elkocythereis bramletti* Dickinson and Swain, 1967. 20, LV, female, L, 0.76; 21, LV interior, L, 0.82; 22, dorsal view; 23, LV male, L, 0.89. Pliocene, Hay Ranch Formation, Gulf Oil Company, no. 1 Wilkins Ranch well sec. 21, T. 38 N., R. 61 E., Elko County Nevada; 957-987 feet (292-301 m) (21, 22); 1283-1314 feet (391-401 m) (20); 1017-1047 feet (310-319 m) (23) (after Dickinson and Swain, 1967). (1)

Figures 24-28. *Elkocythereis minutidontis* Dickinson and Swain, 1967. 24, R side, female, L, 0.73; 25, LV female, L, 0.75; 26, RV interior, L, 0.76, female; 27, LV interior, male, L,

0.76; 28, dorsal view, female, L, 0.82. Pliocene, Hay Ranch Formation, 2 miles north of Twin Bridges, Elko County, Nevada, sec. 24, T. 32 N., R. 55 E., 65-67.5 feet (19.8-20.6 m) above base of exposure (after Dickinson and Swain, 1967). (1)

Figures 29-32. Elkocythereis postdorsoalata Swain, 1986. 29, R side, female, L, 0.56; 30, dorsal view, female, L, 0.77; 31, R side, male, L, 0.61; 32, L side, female, L, 0.73. Pliocene, Glenns Ferry Formation, sec. 14, T. 4 S., R. 1 Q., Owyhee County, Idaho (after Swain, 1986a). (12)

Figure 33. Elkocythereis postdorsoalata Swain, Swain, 1987. R side, L, 0.52. Pliocene, Cache Valley Formation NW1/4 SE1/4 NE1/4 sec. 10, T. 12 S., R. 40 E., Hoopes Creek loc., Franklin County, Idaho (after Swain, 1987a). (2)

Figures 34-36. Eucypris bobcatensis Swain, 1986b. 34, R side, holotype, L, 0.98 (plain light); 35, R side, (SEM); 36, dorsal view, (SEM). Miocene, Humboldt Formation, "Bobcat Limestone," Medicine Creek section, T. 39 N., R. 65 E., 161-199 feet (49-61 m) above base, Winecup Ranch, Elko County, Nevada (after Swain, 1986b). (2)

Figures 37a, b. Eucypris sp. aff. E. bobcatensis Swain, Swain, 1987. R side, L, 1.16 and enlargement of adductor muscle scar area, approx. X 250 of partly exfoliated shell. Eocene, Twin Bridges Limestone, west side of Huntingdon Creek, sec. 36, T. 32 N., R. 55 E., Elko County, Nevada (after Swain, 1987b). (2)

Figure 38. Elkocythereis sp. aff. E. bramletti Dickinson and Swain, 1967. Exterior of LV, L, 0.82. Miocene, Esmeralda Formation, 3.3 miles (5.3 km) east of Finger Rock Wash on road to Simon Mine, Mineral County, Nevada (after Swain, 1985). (2)

Plate 18

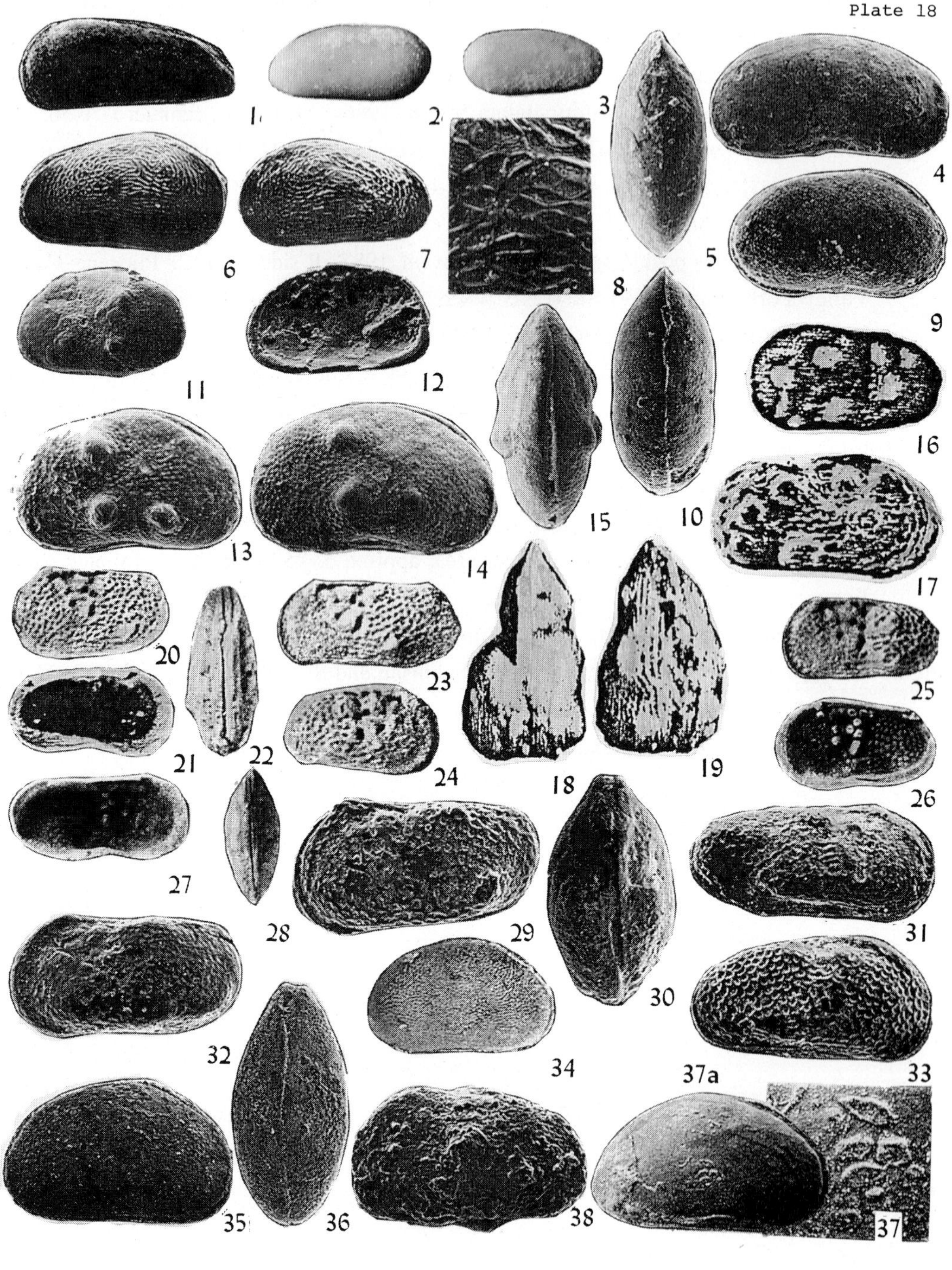
1
2
3
4
5
6
7
8
9
10
11
12
13
14
15
16
17
18
19
20
21
22
23
24
25
26
27
28
29
30
31
32
33
34
35
36
37a
37
38

Plate 19

Figures 1-3. Eucypris fingerrockensis Swain, 1985. 1, L side, holotype, L, 1.07; 2, L side, L, 1.15; 3, ventral view, L, 1.22. Miocene, Esmeralda Formation; 1, 3, Finger Rock Wash, 0.5 mile (0.8 km) south of Rawhide Ranch, Mineral County, Nevada; 2, 2.85 miles (4.5 km) west of Black Spring, Nye County, Nevada (after Swain, 1985). (2)

Figures 4, 5. Eucypris fingerrockensis Swain, Swain, 1987. 4, R side, poorly preserved shell, L, 0.58; 5, dorsal view, L, 0.63. Miocene, Horse Camp Formation, sec. 12, T. 10 N., R 58 E., near Currant, Nye County, Nevada (after Swain, 1987c). (2)

Figures 6-8. Eucypris meadensis Gutentag and Benson, 1962. 6, LV female, holotype; 7, LV holotype, interior, L, 1.20. Recent, spring, Meade County, Kansas State Park; NW1/4 SW1/4 sec. 14, T. 33 S., R. 29 W. (after Gutentag and Benson, 1962). (4)

Figures 9-11. Eucypris microreticulata Swain, 1986. 9, R side, holotype; 10, dorsal view; 11, R side, female?, L, 0.96. Miocene, Humboldt Formation, Gulf Oil Company, no. 1 Thousand Springs well, center SE1/4 SW1/4 sec. 8, T. 40 N., R 66 E., Elko County, Nevada; 9, 10 from 1890-1900 feet (576-579 m), 11, from 1710-1720 feet (521-524 m) (after Swain, 1986b). (2)

Figure 12. Eucypris microreticulata Swain, Swain, 1987. L side, L, 0.51. Miocene, Horse Spring Formation, Fossil Ridge, approx. NW1/4 sec. 3, T. 17 S., R. 16 E., Clark County, Nevada (after Swain, 1987c). (2)

Figure 13. Eucypris cf. microreticulata Swain, Swain, 1990. R side, L, 0.82. Pliocene, Beartrap Quadrangle, Los Angeles County, California (after Swain, 1990). (2)

Figures 14, 15. Eucypris ornatoides Dickinson and Swain, 1967. 14, R side, holotype; 15, dorsal view, L 1.18. Pliocene, Hay Ranch Formation, Gulf Oil Company, no.1 Wilkins Ranch well, sec. 21, T. 38 N., R. 61 E., Elko County, Nevada, 1953-1983 feet (595-604 m) (after Dickinson and Swain, 1967). (1)

Figures 16-18. Eucypris ornatoides Dickinson and Swain, Swain, 1986b. 16, L side, L, 0.88; 17, R side; 18, dorsal view, L, 0.95. Miocene, Humboldt Formation; 16, conglomeratic limestone near base of Medicine Creek section; 17, 18, "Bobcat limestone", 161-199 feet (49-61 m) above base of Medicine Creek section, T. 39 N., R. 65 E., Winecup Ranch, Elko County, Nevada (after Swain, 1986b). (2)

Figures 19, 20. Eucypris ornatoides elongata Swain, 1986b. 19, R side, holotype, L, 0.97; 20, dorsal view. Miocene, Humboldt Formation, Gulf Oil Company no.1 Thousand Springs well, center SE1/4 SW1/4 sec. 8, T. 40 N., R. 66 E., Elko County, Nevada; 1680-1690 feet (512-515 m) (after Swain, 1986b). (2)

Figures 21, 22. Eucypris punctatula Dickinson and Swain, 1967. 21, LV; 22, RV interior, holotype, L, 1.18. Pliocene, Hay Ranch Formation, 1 mile south of Elko, Nevada on highway 46 (after Dickinson and Swain, 1987). (1)

Figure 23. Eucypris trapeziformis Staplin, 1963. RV, holotype, L, 0.94. Pleistocene, Kansan Stage, NW1/4 SE1/4 sec. 27, T. 17 N., R. 11 W., Cass County, Illinois (after Staplin, 1963). (1)

Figure 24. Eucypris? sp., Swain, 1986. R? side, L, 0.39. Pliocene, Glenns Ferry Formation, sec. 14, T. 4 S., R. 1 W., Owyhee County, Idaho (after Swain, 1986a). (12)

Figures 25, 26. Eucypris sp. 1, Swain, 1986b. 25, L side; 26, dorsal view, L, 1.08. Miocene, Humboldt Formation, Gulf Oil Company, no.1 Thousand Springs well, SE1/4 SW1/4 sec. 8, T. 40 N., R. 66 E., Elko County, Nevada; 2550-2600 feet (777-793 m) (after Swain, 1986b). (2)

Figures 27, 28. Eucypris sp. 2, Swain, 1986b. 27, R side, L, 0.98; 28, dorsal view, L, 0.78. Miocene, Humboldt Formation, Medicine Creek sec. T. 39 N., R. 65 E., Winecup Ranch, Elko County, Nevada; 199-210 feet (61-64 m) above base of exposure (after Swain, 1986b). (2)

Figure 29. Eucypris sp. 4., Swain, 1986b. R side of exfoliated shell, L, 0.92. Miocene or Oligocene, Humboldt or Indian Well Formation, Palisade sec., 2 miles SE of Palisade, Elko County, Nevada (after Swain, 1986b). (2)

Figure 30. Fusocandona cachevalleyensis Swain, 1987. Dorsal view, L, 1.20. Pliocene, Cache Valley Formation, Hoopes Creek locality NW1/4 SE1/4 NE1/4 sec. 10, T. 12 S., R. 40 E., Franklin County, Idaho (after Swain, 1987a). (2)

Figures 31, 32. Gutschickia levis Cooper, 1946. R side and dorsal views, L, 0.73. Pennsylvanian, Des Moinesian Series, Hanover Limestone, western Illinois (after Cooper, 1946). (20)

Figures 33, 34. Gutschickia ovata Cooper, 1946. R side and dorsal views, L, 1.15. Pennsylvanian, Des Moinesian Series, Hanover Limestone, western Illinois (after Cooper, 1946). (20)

Plate 19

1
2
3
4
5
6
7
8
9
10
11
12
13
14
15
16
17
18
19
20
21
22
23
24
25
26
27
28
29
30
31
32
33
34

Plate 20

Figures 1, 2a, b. Fusocandona cachevalleyensis Swain, 1987. 1, R side, holotype, L, 1.20; 2a, b, LV interior, L, 0.62, and muscle scar area showing dorsal field as well as adductor and frontal spots, approx. X 215. Pliocene, Cache Valley Formation, Hoopes Creek loc. NW1/4 SE1/4 NE1/4 sec. 10, T. 12 S., R. 40 E., Franklin County, Idaho (after Swain, 1987a). (2)

Figures 3, 4. Fusocandona cachevalleyensis Swain, Swain, 1990. 3, RV interior, L, 0.87; 4, LV dorsal view, L, 1.50. Pliocene Idaho Group, NE1/4 SW1/4 sec. 9, T. 22 S., R. 46 E., Malheur County, Oregon (after Swain, 1990). (2)

Figures 5, 6. Fusocandona exotica Swain, 1987. 5, R side of holotype, L, 1.20; 6, ventral view. Pliocene, Cache Valley Formation, Hoopes Creek loc. NW1/4 SE1/4 NE1/4 sec. 10, T. 12 S., R. 40 E., Franklin County, Idaho (after Swain, 1987a). (2)

Figures 7-11. Fusocandonopsis owyheensis Swain, 1986. 7, R side, holotype, male; l, 2.04; 8, dorsal view, male, L, 1.23; 9, R side of highly ornamented male, L, 1.77; 10, LV female, interior, L, 0.8; 11, dorsal view, female, L, 0.75. Pliocene, Glenns Ferry Formation, sec. 14, T. 4 S., R. 1 W., Owyhee County, Idaho (after Swain, 1986a) (12)

Figures 12, 13. Geisinidae, gen indet., sp. 1, Sohn, 1985. 12, L side; 13, dorsal view, L, 0.80. Upper Mississippian, Bluestone Formation, Bramwell Member, Flat Top Quadrangle, Mercer County, West Virginia, U.S.N.M. 365128 (after Sohn, 1985). (1)

Figures 14, 15. Geisinidae, gen. indet., sp. 2, Sohn, 1985. 14, RV; 15, dorsal view, L, 0.92. Upper Mississippian, Bluestone Formation, Bramwell Member, Flat Top Quadrangle, Mercer County, West Virginia, U.S.N.M. 365130 (after Sohn, 1985). (1)

Figure 16. Gutschickia convexa Tasch, 1963. RV, L, 1.74 "note sharp rise of left valve above right along hinge and absence of overlap around ends" (Tasch, 1963), Permian, Wellington Formation, Marion 1A, bed 6, NE1/4 NE1/4sec. 10, T. 17 S., R. 2 E., Marion County, Kansas. (1)

Figures 17-19. Gutschickia deltoidea (Holland, 1933). 17, L side, holotype; 18, R side; 19, dorsal view, L, 1.48. Permian, upper part of Greene Group, Nineveh Formation, Nineveh, Pennsylvania, about 8 miles (12.8 km) NW of Waynesburg, Greene County, Pennsylvania (after Holland, 1933, as Whipplella deltoidea Holland). (12)

Figure 20. Gutschickia deltoidea (Holland); Scott, 1944. R side, L, 1.34. Permian, Nineveh Formation, Nineveh, Pennsylvania (after Scott, 1944). (1)

Figures 21-23. Gutschickia ninevehensis (Holland, 1933). 21, LV; 22, RV, holotype; 23, dorsal view, L, 1.07. Permian, upper part of Dunkard Series, upper part of Greene Group, Nineveh Formation, Nineveh, Pennsylvania, about 8 miles (12.8 km) NW of

Waynesburg, Greene County, Pennsylvania (after Holland, 1934, as Whipplella ninevehensis Holland). (12)

Figures 24-26. Gutschickia ninevehensis (Holland), Sohn, 1977. 24, R side, holotype; 25, L side; 26, dorsal view, L, 1.07. Permian, Nineveh Limestone Member, Greene Formation, Nineveh, Pennsylvania (after Sohn, 1977). (8)

Figures 27, 28. Gutschickia wellingtonensis Tasch, 1963. 27, L side, holotype; 28, dorsal view, L, 1.52. Permian, Wellington Formation, Marion 1A, bed 6, NE1/4 NE1/4 NE1/4 sec. 10, T. 17 S., R. 2 E., Marion County, Kansas (after Tasch, 1963). (1)

Figures 29-31. Gutschickia sp., Sohn, 1977. 29, L valve, internal mold; 30, dorsal view of L valve mold, L, 1.12; 31, adductor muscle scar area on mold, approx. X 425. Pennsylvanian, Monongahela Formation, Monongalia County, West Virginia, U.S.N.M. 168122 (after Sohn, 1977). (8)

Figures 32, 33. Gutschickia sp., Sohn, 1977. 32, R side; 33, dorsal view, L, 1.05; identified as Gutschickia ninevehensis (Holland) by Cooper, 1946. Pennsylvanian, shales in Cohn cyclothem, Clark County, Illinois (after Sohn, 1977). (8)

Figure 34. Gutschickia sp., Sohn, 1977. L side, L, 1.03. Pennsylvanian, Monongahela Formation, Monongalia County, West Virginia, U.S.G.S. coll. 12899-PC (after Sohn, 1977). (8)

Figures 35, 36. Gutschickia subangulata Cooper, 1946. R side and dorsal views, L, 1.17. Permian, Greene Formation, Nineveh Limestone, western Pennsylvania (after Cooper, 1946). (20)

Plate 20

1
2a
2b
3
4
5
6
7
8
9
10
11
12
13
14
15
16
17
18
19
20
21
22
23
24
25
27
28
29
30
31
32
33
34
35
36

Plate 21

Figures 1, 2. Gutschickia sp., Sohn, 1977. 1, R side of a corroded shell, L, 1.03, showing adductor muscle scar area; 2, dorsal view, L, 1.20. Pennsylvanian, Monongahela Formation, Monongalia County, West Virginia, U.S.N.M. 168124 (after Sohn, 1977). (8)

Figures 3-6. Gutschickia spp., Sohn, 1977. 3, R side of a partly exfoliated shell; 4, dorsal view; 5, L side, L, 0.70; 6, adductor muscle scar area, approx. X 280. Upper Mississippian, Bluestone Formation, Bramwell Member, Crumpler Quadrangle, Wyoming County, West Virginia. 3, 4, U.S.N.M. 365177; 5, 6, U.S.N.M. 365112 (after Sohn, 1985). (1)

Figures 7-9a, b. Hemicyprinotus ionensis Swain, 1985. 7, R side, holotype, L, 0.75; 8, dorsal view; 9, a, b, LV interior, adductor muscle scar area, approx. X 245. Miocene, Esmeralda Formation, 2.85 miles (4.5 km) west of Black Spring, Nye County, Nevada (after Swain, 1985). (2)

Figures 10, 11. Hemicyprinotus sp. aff. H. ionensis Swain, Swain, 1987c. 10, R side of poorly preserved shell, L, 0.87; 11, dorsal view, L, 0.34? Miocene, Horse Spring Formation, approx. NW$^{1}/_{4}$ sec. 31, T. 17 S., R. 16 E., Clark County, Nevada (after Swain, 1987c). (2)

Figure 12. Hemicyprinotus ionensis Swain, Swain, 1990. Internal mold of LV, L, 0.58. Miocene, Truckee Formation, 2 miles (3.2 km) WSW of Hazen, Churchill County, Nevada (after Swain, 1990). (2)

Figure 13. Hemicyprinotus krafti Dickinson and Swain, 1967. LV, holotype, L, 1.00. Pliocene, Hay Ranch Formation, Gulf Oil Company, no.1 Wilkins Ranch well, Elko County, Nevada; 1648-1680 feet (502-512 km) (after Dickinson and Swain, 1967). (1)

Figures 14-16. Hemicyprinotus oneidensis Swain, 1987. 14, R side, holotype, L, 1.06; 15, L side; 16, dorsal view. Pliocene, Salt Lake Group, Cache Valley Formation, secs. 4-26, T. 16 S., R. 36 E., Oneida County, Idaho (after Swain, 1987a). (12)

Figures 17, 18. Hemicyprinotus reniformis Swain, 1964. 17, L side, holotype, L, 0.78; 18, R side, paratype. Eocene, Green River Formation, 804 feet above base, Dripping Rock Gulch, east end of Raven Bridge, Rio Blanco County, Colorado (after Swain, 1964a). (1)

Figures 19, 20. Hemicyprinotus? simplex Swain, 1986b. 19, R side, holotype; 20 dorsal view, L, 0.72. Miocene, Humboldt Formation, section at Cobre, Elko County, Nevada; 204-232 feet (62-71 m) above base of exposure (after Swain, 1986b). (2)

Figures 21, 22. Ussuriocypris subreniformis Swain and Wagoner, 1985. 21, R side, holotype, L, 1.09; 22, dorsal view, L, 0.57? of another shell. Miocene, Esmeralda Formation, Blair Junction section, Esmeralda County, Nevada, sample C1H (after Swain and Wagoner, 1985). (2)

Figures 23, 24. Ussuriocypris subreniformis Swain and Wagoner, Swain, 1985. 23, R side; 24, ventral view. Miocene, Esmeralda Formation, 2.85 miles (4.5 km) west of Black Spring, Nye County, Nevada (after Swain, 1985). (2)

Figures 25, 26. Hemicyprinotus watsonensis Swain, 1964. 25, R side, holotype, L, 0.83; 26, L side of partly exfoliated shell, L, 0.75. Eocene, Colton and Green River Formations, Dripping Rock Gulch, east end of Raven Bridge, Rio Blanco County, Colorado (after Swain, 1964a). (1) 25, 53 feet (16 m) above base of Colton Formation; 26, 130 feet (40 m) above base of Colton-Green River transition beds.

Figures 27, 34. Hastifaba pervulgata Cooper, 1946. 27, L side; 34, end view of holotype, L,1.00. Pennsylvanian, Des Moinesian Series, Jamestown Limestone, loc. 52, central Illinois (after Cooper, 1946). (20)

Figures 28, 29. Hemicyprinotus watsonensis emaciata Swain, 1964. 28, LV, holotype; 29, LV, dorsal view, L, 0.88. Eocene, Colton and Green River Formations, Dripping Rock Gulch section, east end of Raven Ridge, Rio Blanco County, Colorado. 28, 53 feet (16 m) above base of Colton Formation; 29, 64 feet (19.5 m) above base of Green River Formation (after Swain, 1964). (1)

Figures 30, 31. Hemicyprinotus sp. aff. H. watsonensis Swain, Swain, 1987. 30, R side; 31, dorsal view, L, 0.85. Eocene, Twin Bridge Limestone, west side Huntingdon Creek, near center sec. 36, T. 32 N., R. 55 E., Elko County, Nevada (after Swain, 1987b). (2)

Figures 32, 33. Hemicyprinotus? sp., Swain, 1986b. 32, LV; 33, dorsal view, L, 0.90. Miocene, Humboldt Formation, Gulf Oil Company, no.1 Thousand Springs well, center SE1/4 SW1/4 sec. 8, T. 40 N., R. 66 E., Elko County, Nevada; 1700-1710 feet (518-521 m) (after Swain, 1986b). (2)

Plate 21

1
2
3
4
5
6
7
8
9a
9b
10
11
12
13
14
15
16
17
18
19
20
21
22
23
24
25
26
27
28
29
30
31
32
33
34

Plate 22

Figures 1, 2. Hemicyprinotus? sp., Swain, 1985. 1, R side; 2, dorsal view, L, 0.95. Miocene, Esmeralda Formation, 2.85 miles (4.5 km) west of Black Spring, Nye County, Nevada (after Swain, 1985). (2)

Figures 3, 4. Herpetocypris andersoni Swain, 1986. 3, R side, female, holotype; L, 0.74; 4, LV male. Pliocene, Glenns Ferry Formation, sec. 14, T. 4 S., R. 1 W., Owyhee County, Idaho (after Swain, 1986a). (2)

Figures 5-8. Herpetocypris franklinensis Swain, 1987. 5, R side, holotype, L, 1.20; 6, LV interior; 7, RV interior; 8, dorsal view. Pliocene, Cache Valley Formation, Hoopes Creek loc. NW1/4 SE1/4 NE1/4 sec. 10, T. 12 S., R. 40 E., Franklin County, Idaho (after Swain, 1987a). (2)

Figure 9. Candoniella? sp., Swain, 1986b (as Herpetocypris sp. aff. H. reptans [Baird]). LV, L, 0.62. Miocene, Humboldt Formation, Gulf Oil Company, no.1 Thousand Springs well, center SE1/4 SW1/4 sec. 8, T. 40 N., R. 66 E., Elko County, Nevada; 50-55 feet (15-17 m) (after Swain, 1986b). (2)

Figure 10. Herpetocypris stenoformis Dickinson and Swain, 1987. LV, holotype, L, 1.00. Pliocene, Hay Ranch Formation, Gulf Oil Company, no.1 Wilkins Ranch well, sec. 21, T. 38 N., R. 61 E., Elko County, Nevada: 1680-1710 feet (512-521 m) (after Dickinson and Swain, 1967). (1)

Figure 11. Herpetocypris stenoformis Dickinson and Swain, Swain and Wagoner, 1985. L side, L, 0.39. Miocene, Esmeralda Formation, Blair Junction section, Esmeralda County, Nevada, sample C1C2 (after Swain and Wagoner, 1985). (2)

Figure 12. Herpetocypris cf. stenoformis Dickinson and Swain, Swain, 1990. R? side, L, 0.61. Pliocene, Idaho Group, just below lava flow, 15 miles (20 km) west of Burns, Harney County, Oregon (after Swain, 1990). (2)

Figures 13-15. Herpetocypris? winfreyi Swain, 1987. 13, RV, exfoliated, L, 0.52; 14, R side, holotype; 15, dorsal view, L, 0.70. Paleocene or Eocene, Sheep Pass Formation, Grant Range, Johnson Canyon loc., Nye County, Nevada (after Swain, 1987b). (2)

Figures 16, 17. Candoniella whitei (Scott and Smith), Swain, 1949. 16, R side; 17, R side, L, 0.88. Eocene, Green River Formation, secs. 3, 4, 9, T. 14 N., R. 14 E., near Sunnyside, Carbon County, Utah (after Swain, 1949). (1)

Figure 18. Herpetocypris sp., Benson, 1967. LV, L, 0.98. Pliocene, Laverne Formation, Beaver County, Oklahoma, U.S.N.M. 64913 (after Benson, 1967). (3)

Figure 19. Herpetocypris? sp., Swain, 1986b. L side, L, 0.63. Miocene, Humboldt Formation, Gulf Oil Company, no.2 Wilkins Ranch well, sec. 19, T. 39 N., R. 69 E., Elko County, Nevada; 4270-4300 feet (1302-1311 m) (after Swain, 1986b). (2)

Figures 20-24. Heterocypris blairensis Swain and Wagoner, 1985. 20, R side, holotype, L, 0.89, sample C1C; 21, R side, L, 0.63, sample C1K; 22, L valve interior, L, 0.61, sample C1A; 23, ventral view, L, 0.58, sample, C1C2; 24, dorsal view, L, 0.69, sample C1D. Miocene, Esmeralda Formation, Blair Junction section, Esmeralda County, Nevada (after Swain and Wagoner, 1985). (2)

Figure 25. Heterocypris blairensis Swain and Wagoner, Swain, 1985. R side, L, 1.41. Miocene, Esmeralda Formation, Finger Rock Wash section, 0.5 mile (0.8 km) south of Rawhide Ranch, Stewart Valley, Mineral County, Nevada; ash bed at base of exposure (after Swain, 1985). (2)

Figures 26, 27. Hastifaba robusta Cooper, 1946. L side and dorsal views, of holotype. Pennsylvanian, Des Moinesian Series, Hanover Limestone, loc. 28, central Illinois (after Cooper, 1946). (20)

Figures 28, 29. Heterocypris humboldtensis Dickinson and Swain, 1967. 28, LV, holotype; 29, dorsal view, L, 0.88. Pliocene, Hay Ranch Formation, ("Upper Humboldt"), Gulf Oil Company, no.1 Wilkins Ranch well, sec. 21, T. 38 N., E. 61 E., Elko County, Nevada; 1648-1680 feet (505-512 m) (after Dickinson and Swain, 1967). (1)

Figures 30, 31. Heterocypris riflensis Swain, 1964. 30, R side, holotype, L, 0.93; dorsal view, paratype. Eocene, Green River Formation, 550 feet (168 m) above base, Anvil Point section near Rifle, Colorado (after Swain, 1964a). (1)

Figure 32. Heterocypris whiteriverensis Swain, 1864. R side of holotype, a shell damaged dorsally, L, 0.92. Eocene, Colton-Green River transition beds, 130 feet (40 m) above base, 8 miles (13 km) west of Rangeley, east end of Raven Ridge, Rio Blanco County, Colorado (after Swain, 1964a). (1)

Figure 33. Heterocypris wilkinensis Dickinson and Swain, 1967. LV, holotype, L, 1.05. Pliocene, Hay Ranch Formation ("Upper Humboldt"), Gulf Oil Company, no. 1 Wilkins Ranch well, sec. 21, T. 38 N., R. 61 E., Elko County, Nevada; 1526-1588 feet (465-484 m) (after Dickinson and Swain, 1967). (1)

Figures 34-37. Hilboldtina multiplicata Scott and Summerson, 1943. 34, L side; 35, R side; 36, dorsal view; 37, ventral view of holotype, L, 1.07. Lower Pennsylvanian, Kanawha Series, Scott Formation, on old WPA road, NW of Newcomb, Campbell County, Tennessee, 1/4 mile (0.4 km) SE of road (after Scott and Summerson, 1943). (11)

Figures 38, 39. Hastifaba spinosa Cooper, 1946. L side and dorsal view of genoholotype, L, 1.16. Pennsylvanian, Des Moinesian Series, Jamestown Limestone, central Illinois (after Cooper, 1946). (20)

Plate 22

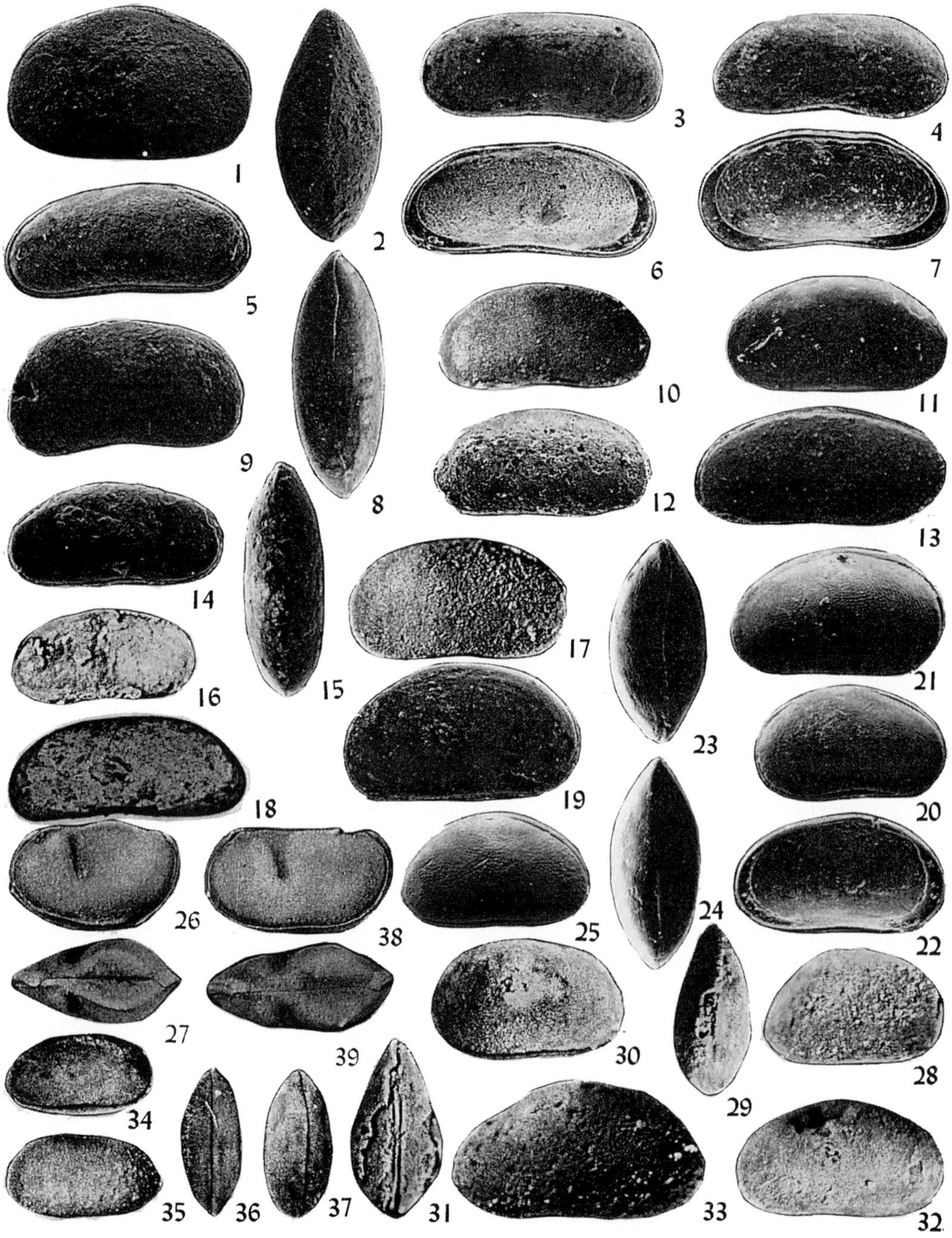
1
2
3
4
5
6
7
8
9
10
11
12
13
14
15
16
17
18
19
20
21
22
23
24
25
26
27
28
29
30
31
32
33
34
35
36
37
38
39

Plate 23

Figure 1. Huabeinia? sp. aff. H. huidongensis Hou and Li, 1978, Swain, 1985. R side, L, 0.63. Miocene, Esmeralda Formation, loc. 3, 2.85 miles (4.5 km) west of Black Spring, Nye County, Nevada (after Swain, 1985). (2)

Figures 2, 3. Huabeinia? sp. aff. H. huidongensis Hou and Li, Swain, 1987c. 2, R side, L, 0.33; 3, dorsal view. Miocene, Esmeralda Formation, 2.85 miles (4.5 km) west of Black Spring, Nye County, Nevada (after Swain, 1987c). (2)

Figure 4. Ilyocyprimorpha linonondosa (Swain, 1949), Swain, 1964. LV, holotype, L, 1.19. Paleocene, Fort Union Formation, roof of Roland Coal, SE1/4 sec. 13, T. 19 S., R. 45 E., Powder River County, Montana (after Swain, 1949). (1)

Figures 5, 6. Herpetocypris lata Chapman, 1896. L side and dorsal view, L, 2.12. Pliocene, Merced Formation, Bollinger Canyon, east of Berkeley, California (after Chapman, 1896). (16)

Figures 7, 8. Herpetocypris merriamiana Chapman, 1896. RV and dorsal view, L, 1.49. Pliocene, Merced Formation, Bollinger Canyon, east of Berkeley, California (after Chapman, 1896). (1)

Figures 9, 10. Huabeinia sp. aff. H. huidongensis Hou and Li, Swain, 1985. (=Pseudocandona? sp., Swain and Wagoner, 1985). R side and dorsal view, L, 0.46. Miocene, Esmeralda Formation, Blair Junction section, Esmeralda County, Nevada, samples C10 and C1C3 (after Swain and Wagoner, 1985). (2)

Figure 11. Kassinina? gasspeakensis Swain, 1987. Dorsal view, L, 0.49. Horse Spring Formation, loc. 9, Clark County, Nevada (after Swain, 1987c). (2)

Figures 12, 13. Ilyocypris bradyi Sars, Swain, 1963. 12, LV; 13, RV, L, 0.83. Pleistocene, Gubik Formation, Arctic coastal plain, Alaska. 12, Teshekpuk Lake, Party 43, line 1-A-48, shot point 27a, depth 60 feet (18 m); 13, Oumalik Test well no.2, 0-10 feet (after Swain, 1963). (1)

Figure 14. Ilyocypris bradyi Sars, Gutentag and Benson, 1962. R side, L, 0.58. Pleistocene Odel Formation, Meade County, Kansas (after Gutentag and Benson, 1962). (4)

Figure 15. Ilyocypris bradyi Sars, Staplin, 1963. LV, female, L, 0.93. Pleistocene, Mankato Substage, Wisconsin Stage, SE1/4 NE1/4 sec. 4, T. 21 N., R. 7 E., Whiteside County, Illinois (after Staplin, 1963). (1)

Figures 16, 17. Ilyocypris gibba (Ramdohr), Swain, 1963. 16, RV, nodose, L, 0.92; 17, RV interior, L, 0.80. Pleistocene, Gubik Formation, Arctic coastal plain, Alaska, Ikpikpuk Shot Holes, line 15-47, shot point 14, at 40-50 feet (12-15 m) (after Swain, 1963). (1)

Figure 18. Ilyocypris gibba (Ramdohr), Gutentag and Benson, 1962. RV, L, 0.92. Pleistocene, late Kansan or Yarmouthian Stage, above Pearlette Ash, SE1/4 NE1/4 sec. 1, T. 25 S., R. 7 W., Reno County, Kansas (after Gutentag and Benson, 1962). (4)

Figure 19. Ilyocypris gibba (Ramdohr), Staplin, 1963. LV female, L, 0.96. Pleistocene, Mankato Substage, Wisconsin Stage, SE1/4 NE1/4 sec. 4, T. 21 N., R. 7 E. Whiteside County, Illinois (after Staplin, 1963). (1)

Figure 20. Ilyocypris gibba (Ramdohr), Benson, 1967. LV interior showing adductor and dorsal muscle field scars, L, 0.92. Pleistocene, Yarmouthian Stage, Dickinson County, Kansas (after Benson, 1967). (3)

Figure 21. Ilyocyprimorpha shafferi (Staplin, 1963). LV female, holotype, L, 1.40. Pleistocene, Tazewell Stage, center E line SW1/4 sec. 4, T. 4 N., R. 3 E., Fulton County, Illinois (after Staplin, 1963). (1)

Figure 22. Ilyocypris shawneetownensis Staplin, 1963. LV female, holotype, L, 0.92. Pleistocene, Mankato Substage, Wisconsin Stage, sec. 13, T. 9 S., R. 10 E., mile 852.7 of Ohio River, Gallatin County, Illinois (after Staplin, 1963). (1)

Figures 23, 24. Ilyocypris thousandspringensis Swain, 1986. 23, L side, holotype; 24, dorsal view, L, 0.67. Miocene, Humboldt Formation, Gulf Oil Company, no.1 Thousand Springs well, center SE1/4 SW1/4 sec. 8, T. 40 N., R. 66 E., Elko County, Nevada; 2600-2650 feet (793-808 m) (after Swain, 1986b). (2)

Figure 25. Ilyocypris? sp., Sohn, 1956. RV, L, 0.50. Pliocene, Teewinot Formation, Jackson Hole, Wyoming, U.S.N.M. L54-94 (after Sohn, 1956). (9)

Figures 26-29. Jonesina? arcuata (Bean), Scott and Summerson, 1943. 26, RV; 27, LV, internal mold; L, 0.92; 29, dorsal view, male, internal mold, L, 0.92. Lower Pennsylvanian, Kanawha Series, Briceville Formation, along highway between Coalfield and Christmas, Morgan County, Tennessee (after Scott and Summerson, 1943). (11)

Figures 30, 31. Limnocypridea morrisonensis (Roth), Sohn, 1979. R side and dorsal view, L, 2.40. Lower Cretaceous, Lakota Formation by Sohn, sec. 28, T. 4 N., R. 6 E., Lawrence County, South Dakota (after Roth, 1933). (1) (=Upper Jurassic Morrison Formation by Roth).

Figures 32, 44. Ilyocypris biplicata (Koch), Lister, 1975. LV, L, 1.25, LV interior, L, 0.83. Pleistocene, Lake Bonneville Group, Saltair core hole, SE1/4 sec. 25, T. 1 N., R. 3 W., Salt Lake County, Utah (after Lister, 1975). (3)

Figures 33-36. "Jonesina" pahasapensis Roth, 1933. 33, RV, holotype?; 34, RV; 35, ventral view; 36, dorsal view, L, 1.00. Lower Cretaceous, Lakota Formation by Sohn, 1979, sec. 28, T. 4 N., R. 6 E., Lawrence County, South Dakota (after Roth, 1933). (1) (=Theriosynoecum? pahasapensis [Roth], van Morkhoven, 1963; ?=Dryelba pahasapensis [Roth], Sohn, 1983) (=Upper Jurassic, Morrison Formation by Roth, 1933).

Figures 37, 38. Jonesina? bispinosa Scott and Summerson, 1943. 37, RV, holotype; 38, RV, L, 1.10. Lower Pennsylvanian, Kanawha Series, Mingo Formation, roof of mine in coal about 250 feet above road pass of U.S. 25 E., between Clear Creek and Little Cannon Creek, Bell County, Kentucky (after Scott and Summerson, 1943). (11) (Note similarity to Hastifaba Cooper, 1946).

Figure 39. Jonesina? sp., Scott and Summerson, 1943. RV, L, 0.43. Lower Pennsylvanian, no locality given (after Scott and Summerson, 1943). (11) (Note similarity to Hastifaba Cooper, 1946).

Figures 40-43. Kassinina? gasspeakensis Swain, 1987c. 40, R side, L, 0.69; 41, dorsal view, L, 0.62; 42, R side, immature shell, L, 0.30; 43, 11, RV, L, 0.82 and dorsal view, holotype. Miocene, Horse Spring Formation, Fossil Bridge, approx. NW1/4 sec. 31, T. 17 S., R. 16 E., Clark County, Nevada, unsurveyed area (after Swain, 1987c). (2)

Plate 23

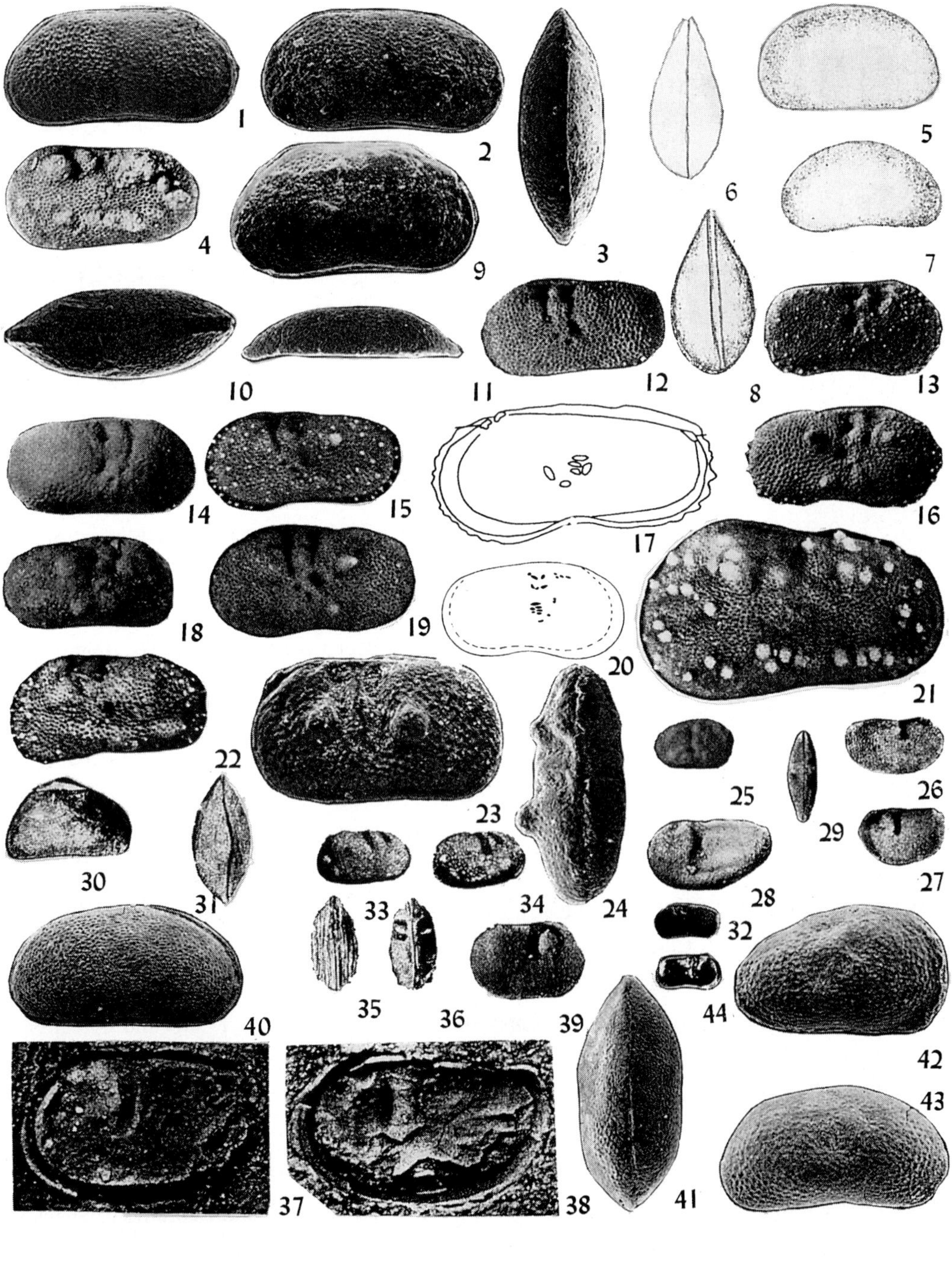

Plate 24

Figures 1-4. Kassinina mesoreticulata Swain and Wagoner, 1985. 1, R side, holotype male, L, 1.03; 2, RV interior, L, 0.58? X 63; 3, dorsal view, L, 0.73?; 4, L side, L, 0.75? Miocene, Esmeralda Formation, Blair Junction section, Esmeralda County, Nevada; 1, 2, 3, sample C1C; 4, sample C1P (after Swain and Wagoner, 1985). (2)

Figures 5-7. Kassinina microreticulata Swain and Wagoner, 1985. 5, R side, female, L, 1.07; 6, RV interior immature, L, 0.77?; 7, dorsal view, male?, L, 0.73. Miocene, Esmeralda Formation, Blair Junction series, Esmeralda County, Nevada; 5, sample C1C, 6, sample C1C3, 7, sample C1B (after Swain and Wagoner, 1985). (2)

Figures 8-10. Kassinina paucireticulata Swain and Wagoner, 1985. 8, L side, female holotype, L, 1.17; 9, RV interior, male, L, 0.76?; 10, ventral view, male, L, 0.80? Miocene, Esmeralda Formation, Blair Junction series, Esmeralda County, Nevada; 8, sample C1J, 9, sample C1M, 10, sample C1R (after Swain and Wagoner, 1985). (2()

Figures 11-13. Limnicytheropteron cephalotoidea Swain, 1986. 11, L side, holotype, male?, L, 0.50; 12, L side, female?; 13, R side, male? Pliocene, Glenns Ferry Formation, sec. 14, T. 4 S., R. 1 W., Owyhee County, Idaho (after Swain, 1986a). (2)

Figures 14-17. Limnicytheropteron typica Swain, 1986a. 14, R side, holotype, female?, L, 1.00; 15, L side, male? (note adductor muscle scar trace), L, 0.88; 16, dorsal view, male?, L, 0.96; 17, LV interior, juvenile, L, 0.85. Pliocene, Glenns Ferry Formation, sec. 14, T. 4 S., R. 1 W., Owyhee County, Idaho (after Swain, 1986a). (2)

Figures 18-22. Limnicytheropteron unicornis (Swain, 1947). 18, LV, holotype; 19, dorsal view; 20, R valve; 21, elongate RV, male?; 22, RV interior, L, 0.54. Pliocene, Salt Lake Formation, sec. 15, T. 13 N., R. 12 W., Cache and Box Elder Counties, Utah (after Swain, 1947). (1)

Figure 23. Limnicytheropteron cf. unicornis (Swain), Swain, 1987. L side, L, 0.53. Pliocene, Salt Lake Group, Cache Valley Formation, T. 13 N., R. 3 W., Cache Valley, Utah, SW of Newton (after Swain, 1987a). (2)

Figures 24-27. Limnicytheropteron winecupensis Swain, 1986. 24, 25, L side and dorsal view, holotype, L, 0.47; 26, 27, R side and dorsal view, poorly preserved shell, L, 0.95. Miocene Humboldt Formation, Gulf Oil Company, no.1 Thousand Springs well, Elko County, Nevada; 1140-1150 feet (348-351 m) (after Swain, 1986b). (2)

Figures 28, 29. Limnocythere ceriotuberosa Delorme, Lister, 1975. 28, RV, female, L, 0.65; 29, LV male, L, 0.78. Pleistocene, Lake Bonneville Group, Saltair core hole, SE1/4 sec. 25, T. 1 N., R. 3 W., Salt Lake County, Utah (after Lister, 1975). (3)

Figures 30-32. Limnocythere chippewaensis Staplin, 1983. 3, RV, male holotype, L, 0.47; 31, LV female; 32, LV male, L, 0.52. Pleistocene, pre-Nipissing lower stages of Lake Michigan, core QW 101-B (after Staplin, 1963). (1)

Figures 33-35. Limnicytheropteron dibolba (Swain, 1947). 33, LV, holotype, L, 0.55; 34, LV interior; 35, ventral view, L, 0.57. Pliocene, Salt Lake Group, sec. 15, T. 13 N., R. 2 W., Cache County, Utah (after Swain, 1947). (1)

Figures 36-39. Limnocythere friabilis Benson and Macdonald, Lister, 1975. 36, RV, female, L, 0.65; 37, RV interior, female, L, 0.51; 38, LV male, L, 0.55; 39, LV, male interior, L, 0.52. Pleistocene, Lake Bonneville Group, Saltair core hole, SE1/4 sec. 25, T. 1 N., R. 3 W., Salt Lake County, Utah (after Lister, 1975). (3)

Figures 40, 41. Limnocythere herricki Staplin, 1963. 40, RV, female, L, 0.71; 41, LV, holotype female, L, 0.72. Pleistocene, Sankoty Lake Stage of Lake Michigan, Nebraskan Stage, core 78A (after Staplin, 1963). (1)

Figures 42, 43. Limnocythere huntingdonensis Dickinson and Swain, 1967. 42, RV, male holotype, L, 0.67; 43, LV male interior, L, 0.70. Pliocene, Hay Ranch Formation (=Upper Humboldt), 2 miles (3.2 km) north of Twin Bridges, Nevada on north side of Huntington Creek, Elko County (after Dickinson and Swain, 1967). (1)

Plate 24

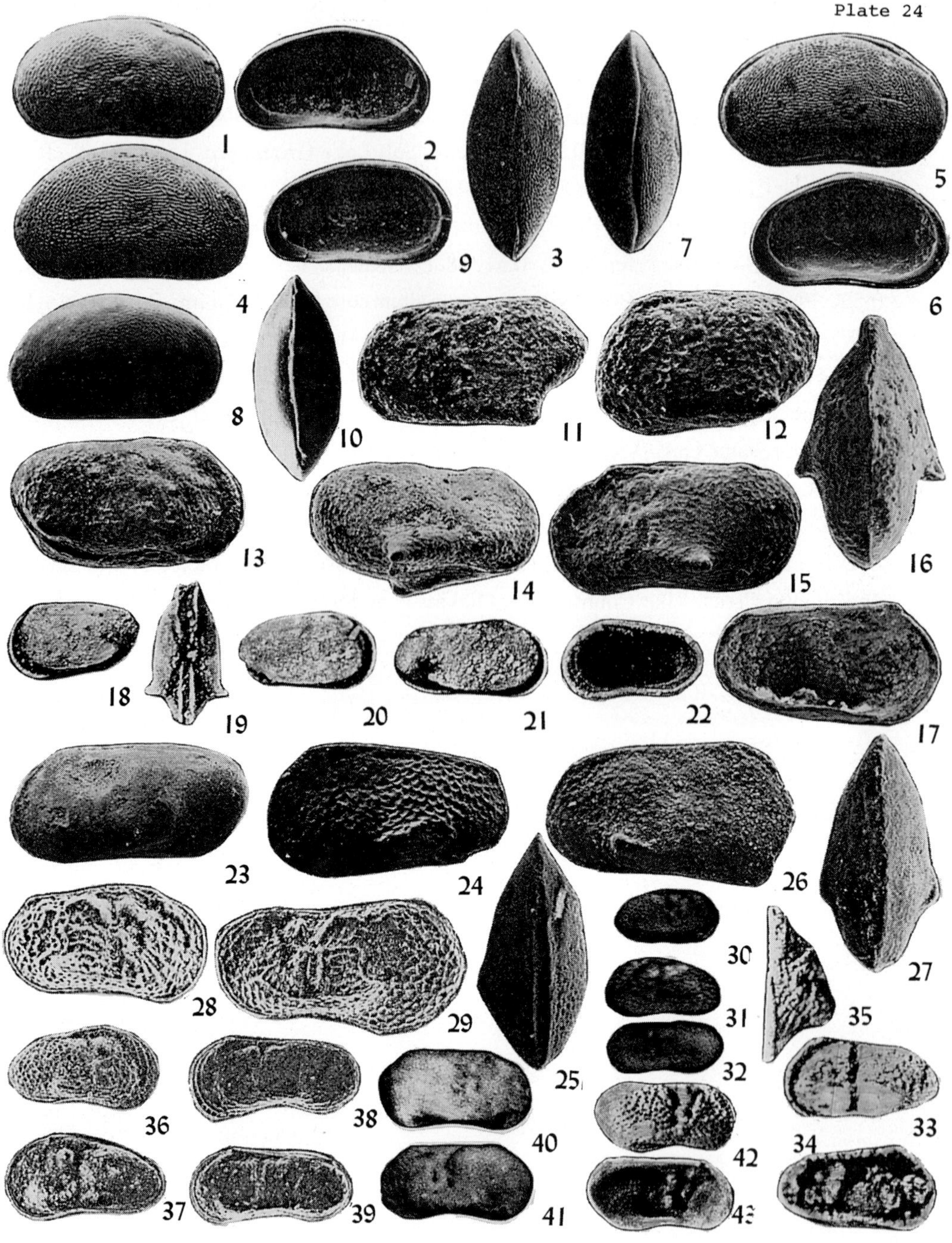

Plate 25

Figures 1-3. Limnocythere huntingdonensis Dickinson and Swain, 1967. 1, LV, female, L, 0.72; 2, LV interior, female, L, 0.60; 3, dorsal view, female. Pliocene, Hay Ranch Formation (=Upper Humboldt), 2 miles (3.2 km) north of Twin Bridges, Elko County, Nevada on west side of Huntingdon Creek, 65-67-5 feet (20-20.5 m) above base of section (after Dickinson and Swain, 1967). (1)

Figure 4. Limnocythere huntingdonensis Dickinson and Swain, Swain, 1987. R side, L, 0.65. Pliocene, Cache Valley Formation, Hoopes Creek section, NW1/4 SE1/4 sec. 10, T. 12 S., R. 40 E., Franklin County, Idaho (after Swain, 1987a). (12)

Figures 5-7. Limnocythere illinoisensis Sharpe, Lister, 1975. 5, LV, female; 6, LV, female, L, 0.76; 7, RV interior, female, L, 0.67. Pleistocene, Lake Bonneville Group, Saltair core hole, SE1/4 sec. 25, T. 1 N., R. 3 W., Salt Lake County, Utah, depth 2 feet (0.6 m) (after Lister, 1975). (3)

Figures 8-10. Limnocythere itasca Cole, Lister, 1975. 8, LV, female, L, 0.63; 9, LV, male, L, 0.61; 10, LV interior, male, L, 0.66. Pleistocene, Lake Bonneville Group, Saltair core hole, SE1/4 sec. 25, T. 1 N., R. 3 W., Salt Lake County, Utah; 8, 125 feet 7 inches (41.4 m), 9, 10, 35 feet 6 inches (10.8 m) (after Lister, 1975). (3)

Figures 11-13. Limnocythere ornata wabashensis Staplin, 1963. 11, RV, female, holotype, L, 0.78; 12, LV, male, L, 0.85; 13, juvenile, L, 0.62. Pleistocene, Mankato or late Cary Substage, Wisconsin Stage, 1.5-2 miles (2.4-3.6 km) east of Mt. Vernon, Indiana at steel bridge, Route 62 (after Staplin, 1963). (1)

Figures 14, 15. Limnocythere paraornata Delorme, Lister, 1975. 14, LV, male, L, 0.95; 15, RV, female, L, 0.85. Pleistocene, Lake Bonneville Group, Saltair core hole, SE1/4 sec. 25, T. 1 N., R. 3 W., Salt Lake County, Utah; 14, 63 feet 8 inches (19.4 m); 15, 483 feet 11 inches (147.5 m) (after Lister, 1975). (3)

Figures 16, 17. Limnocythere parascutariense Delorme, Lister, 1975. 16, L side, female, L, 0.65; 17, R side, male, L, 0.84. Pleistocene, Lake Bonneville Group, saltair core hole, SE1/4 sec. 25, T. 1 N., R. 3 W., Salt Lake County, Utah; depth 135 feet 7 inches (41.4 m) (after Lister, 1975). (3)

Figures 18, 19. Limnocythere posterolimba Delorme, Swain, 1990. 18, R side; 19, dorsal view, L, 0.85. Pliocene, below lava flow, 15 miles (24 km) west of Burns, Oregon (after Swain, 1990). (2)

Figures 20, 21. Limnocythere pseudoreticulata Staplin, 1963. 20, LV, female holotype, L, 0.57; 21, LV male, L, 0.56. Pleistocene, Cary or Mankato Substage, Wisconsin Stage,

Blue Island, Chicago, Illinois Brickyard, sec. 5, T. 37 N., R. 14 E. (after Staplin, 1963). (1)

Figures 22-24. Limnocythere pseudoreticulata Staplin, Lister, 1975. 22, LV, female, L, 0.55; 23, RV, female, interior, L, 0.51; 24, RV, male, L, 0.57. Pleistocene, Lake Bonneville Group, Saltair core hole, SE1/4 sec. 25, T. 1 N., R. 3 W., Salt Lake County, Utah; depth 35 feet 6 inches (10.8 m) (after Lister, 1975). (3)

Figures 25, 27. Limnocythere pterygoventrata Dickinson and Swain, 1967. 25, LV, female, holotype, L, 0.78; 26, LV, male, L, 0.62; 27, dorsal view, female, L, 0.61. Pliocene, Hay Ranch Formation (=Upper Humboldt), Gulf Oil Company, no.1 Wilkins Ranch well, sec. 21, T. 38 N., R. 61 E., Elko County, Nevada; 25, depth 957-987 feet (292-301 m); 26, depth 3250-3281 feet (991-1040 m); 27, depth 1801-1831 feet (549-558 m) (after Dickinson and Swain, 1967). (1)

Figures 28, 29. Limnocythere reticulata Sharpe, Staplin, 1963. 28, LV, female, L, 0.68; 29, RV, male, L, 0.76. Pleistocene, Mankato or late Cary Substage, Wisconsin Stage, T. 5 S., R. 13 W., just south of railroad track at Solitude, Posey County, Indiana (after Staplin, 1963). (1)

Figures 30, 31. Limnocythere cf. L. sanctipatricii Brady and Robertson, Swain, 1963. 30, RV, female, L, 0.73; 31, LV, immature?, L, 0.68. Pleistocene, Gubik Formation, Arctic coastal plain, Alaska; 30, Teshekpuk Lake shot holes, party 47, line 8, shot point 12, depth 50-60 feet (15-18 m); 31, Ikpikpuk shot holes, line 8 shot point 7, depth 60-70 feet (18-21 m) (after Swain, 1963). (1)

Figures 32, 33. Limnocythere sappaensis Staplin, 1963. 32, LV, female, holotype, L, 0.61; 33, RV, male, L, 0.65. Pleistocene, Sappa Silt, Kansan to early Yarmouthian Stage, SW1/4 NE1/4 sec. 11, T. 2 N., R. 20 W., Harlan County, Nebraska (after Staplin, 1963). (1)

Figures 34, 35. Limnocythere scotti Staplin, 1963. 34, RV, male; 35, LV, female, holotype, L, 0.70. Pleistocene, Mankato or late Cary Substage, Wisconsin Stage, sec. 24, T. 7 S., R. 14 W., near Mt. Vernon, Indiana; 100-150 yards SE of bridge on Indiana 69 (after Staplin, 1963). (1)

Figures 36-38. Limnocythere sharpei Staplin, 1963. 36, LV female, holotype; 37, RV, male, L, 0.82; 38, RV, female, L, 0.72. Pleistocene, Mankato or late Cary Substage, Wisconsin Stage, sec. 26, T. 9 S., R. 9 E., 2.2 miles (3.5 km) west of Shawneetown, Gallatin County, Illinois (after Staplin, 1963). (1)

Figures 39-42. Limnocythere staplini Gutentag and Benson, 1962. 39, LV, female, holotype, L, 0.63; 40, LV, female, L, 0.54?; 41, RV, interior, L, 0.70?; 42, LV, interior, L, 0.76?

Pleistocene, Illinoian-Wisconsin Odee Formation, Meade County, Kansas (after Gutentag and Benson, 1962). (4)

Figures 43, 44. Limnocythere staplini Gutentag and Benson, Lister, 1975. 43, LV, male, L, 0.65; 44, LV, female, L, 0.68. Pleistocene, Lake Bonneville Group, Saltair core hole, SE1/4 sec. 25, T. 1 S., R. 3 W., Salt Lake County, Utah; depth 32 feet 11 inches (14.7 m) (after Lister, 1975). (3)

Figures 45, 46. Limnocythere itasca Cole, Staplin, 1963. 45, RV, female, L, 0.76; 46, LV male, L, 0.70. Pleistocene, Cary or Mankato Substage, Wisconsin Stage, sec. 5, T. 37 N., R. 14 E., Blue Island, Chicago, Illinois (after Staplin, 1963). (1) (as L. trapeziformis Staplin).

Figures 47, 48. Limnocythere varia Staplin, 1963. 47, LV, female, holotype, L, 1.00; 48, LV, male, L, 1.01. Pleistocene, Cary or Mankato Substage, Wisconsin Stage, Blue Island, Chicago, Illinois, sec. 5, T. 37 N., R. 14 W., Glenwood or Calumet Stage of Lake Chicago (after Staplin, 1963). (1)

Figures 49, 50. Limnocythere verrucosa Hoff, Staplin, 1963. 49, RV, male, L, 0.66; 50, RV, female, L, 0.61. Pleistocene, pre-Illinoian? Stage, SE1/4 SW1/4 sec. 5, T. 24 N., R. 4 W., Tazewell County, Illinois, 2.5 miles (4 km) east of Pekin (after Staplin, 1963). (1)

Figure 51. Limnocythere sp., Swain, 1985. LV, broken, immature?, L, 0.72. Miocene, Esmeralda Formation, Simon Mine area, 3.3 miles (5.2 km) east of Finger Rock Wash, Nye County, Nevada (after Swain, 1985). (2)

Plate 25

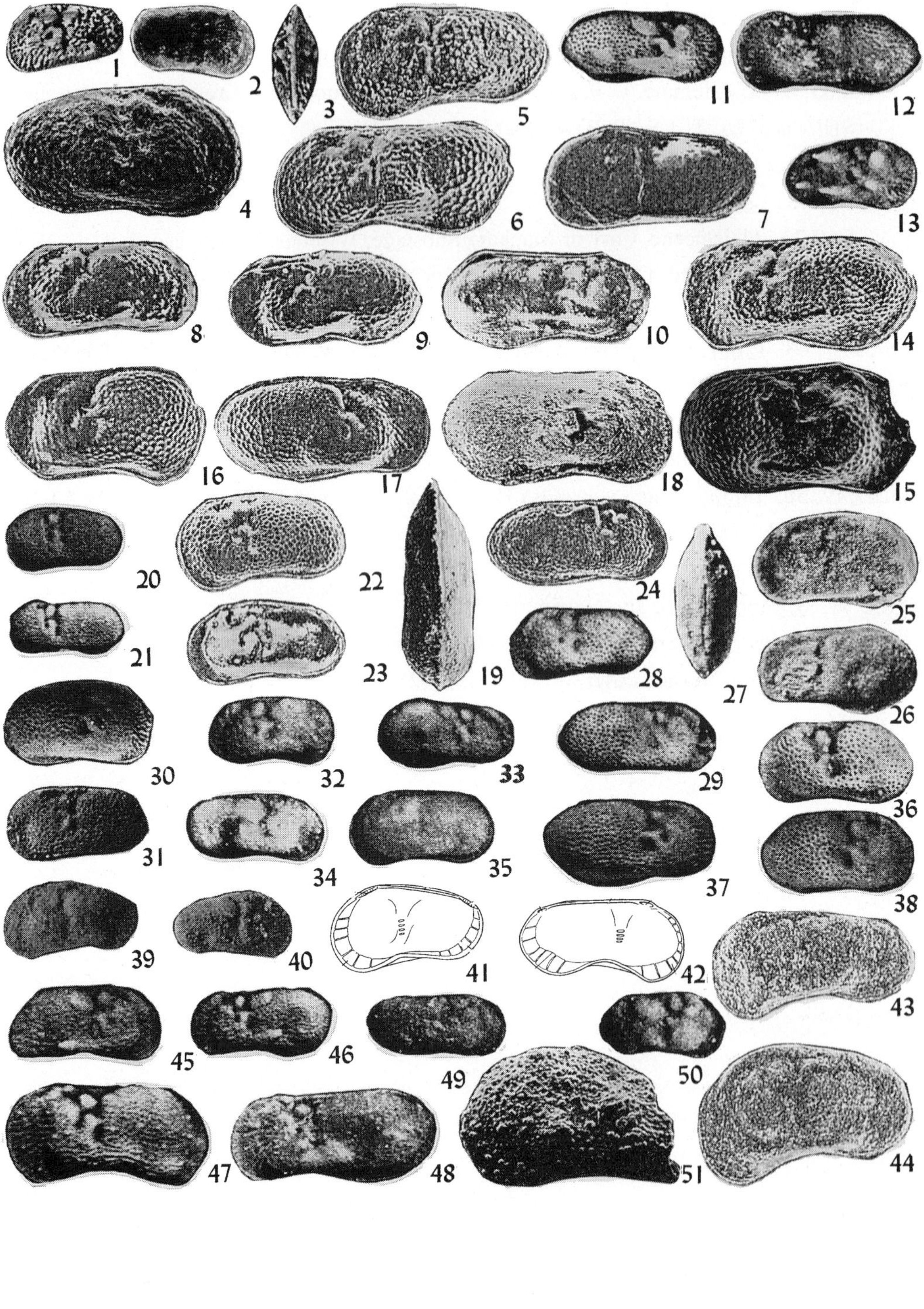

Plate 26

Figure 1. Limnocythere cf. varia Staplin, J.E. King specimen not previously figured. RV exterior, L, 0.67. Pleistocene, bone bed at Boney Spring, Benton County, Missouri.

Figures 2-4. Limnocythere? williamsi (Swain, 1947). 2, RV, holotype, L, 0.56; 3, LV; 4, dorsal view in which hinge surface is broken near anterior end, falsely suggesting presence of socket. Pliocene, Salt Lake Group, sec. 15, T. 13 N., R. 2 W., Cache County, Utah (after Swain, 1947, as Limnicythere williams1). (1)

Figure 5. Limnocythere? sp., Swain, 1964. RV, imperfect specimen, L, 0.65. Eocene, Green River Formation, Anvil Point section near Rifle, Colorado, 805 feet (245 m) above base of Green River Formation (after Swain, 1964). (1)

Figures 6, 7. Limnocythere sp., Benson, 1969. 6, RV, female, L, 0.61; 7, LV, male, L, 0.62. Early Pleistocene, Rita Blanca Lake beds, near Channing, Hartley County, Texas (after Benson, 1969). (3)

Figure 8. Limnocythere? sp. 1, Sohn, 1956. RV, L, 0.48. Pliocene, Teewinot Formation, Jackson Hole, Wyoming (after Sohn, 1956). (9)

Figure 9. Limnocythere? sp. 2, Sohn, 1956. LV, L, 0.60. Pliocene, Teewinot Formation, Jackson Hole, Wyoming (after Sohn, 1956). (9)

Figure 10. Limnocythere? sp. 3, Sohn, 1956. RV, L, 0.68. Pliocene, Teewinot Formation, Jackson Hole, Wyoming (after Sohn, 1956). (9)

Figures 11-13. Limnocythere sp. 1, Swain, 1986. 11, R side; 3, dorsal view, L, 0.57; 13, L side, L, 0.50. Pliocene, Glenns Ferry Formation, sec. 14, T. 4 S., R. 1 W., Owyhee County, Idaho (after Swain, 1986a). (2)

Figure 14. Limnocythere sp. 2, Swain, 1986. R side, L, 0.77. Pliocene, Glenns Ferry Formation, sec. 14, T. 4 S., R. 1 W., Owyhee County, Idaho (after Swain, 1986a). (2)

Figure 15. Limnocythere sp. indet., Staplin, 1963. RV, female, L, 0.77. Pleistocene, Mankato Substage, Wisconsin Stage, sec. 15, T. 9 S., R. 10 E., Gallatin County, Illinois (after Staplin, 1963). (1)

Figures 16-20. Limnocyprdea morrisonensis (Roth), Sohn, 1979. 16, R side, L, 2.9; 17, dorsal view, male?; 18, R side, male?, L, 1.60; 19, R side, L, 1.40; 20, LV interior, with nested valves, L, 1.36. Lower Cretaceous, Lakota Formation, 3 miles (4.8 km) south of Piedmont, Meade County, South Dakota (after Sohn, 1979). (8)

Figure 21. Lutkevichinella sp., Kietzke, 1989. R side, L, 0.60. Upper Triassic, late Carnian to early Norian, Bull Canyon, NW1/4 sec. 28, T. 9 N., R. 26 E., Guadalupe County, New Mexico (after Kietzke, 1987). (7)

Figures 22, 23. Lineocypris? Sp., Sohn, 1956. 22, LV; 23, RV, L, 0.65. Pliocene, Teewinot Formation, Jackson Hole, Wyoming (after Sohn, 1956). (9)

Figures 24-26. Longispinella asymmetrica Sohn, 1979. 24, L side; 25, dorsal view; 26, R side, L, 0.56. Lower Cretaceous, Lakota Formation, Chilson Member, Fall River County, South Dakota (after Sohn, 1979). (8)

Figures 27-29. Longispinella longispina (Peck), Sohn, 1979. 27, L side; 28, R side; 29, dorsal view, L, 0.90. Lower Cretaceous, Lakota Formation, U.S.G.S. coll. 31153, Fall River County, South Dakota (after Sohn, 1979). (8)

Figures 30, 31. Looneyella monticula (Jones), Peck, 1951. 30, R side; 31, L side, L, 0.95. Lower Cretaceous, Bear River Formation, on Sulphur Creek and in Shell Hollow, near Evanstown, Wyoming (after Peck, 1951). (1)

Figures 32, 33. Limnocythere robusta Delorme, Cameron and Lundin, 1977. RV, L, 0.81, male; RV, female, L, 0.65. Pleistocene, Lake Cochise deposits, Willcox Playa, Cochise County, Arizona (after Cameron and Lundin, 1977). (10)

Figure 34. Limnocythere? sp. 2, Swain, 1982. R side, L, 0.67. Lower Cretaceous, Washitan Stage, well in Palm Beach County, Florida; 7691-7696 feet (2344-2346 m) (after Swain, 1982). (14)

Figure 35. Limnocythere? sp. 1, Swain, 1982. LV, L, 0.70. Lower Cretaceous, Washitan Stage, well in Hendry County, Florida; 8260-8630 feet (2627-2630 m) (after Swain, 1982). (14)

Plate 26

1
2
3
4
5
6
7
8
9
10
11
12
13
14
15
16
17
18
19
20
21
22
23
24
25
26
27
28
29
30
31
32
33
34
35

Plate 27

Figures 1-4. Looneyella monticula (Jones), Peck, 1951. 1, LV interior, L, 0.85; 2, RV interior, L, 0.75; 3, ventral view, L, 0.91; 4, dorsal view, L, 0.94. Lower Cretaceous, Bear River Formation, on Sulphur Creek on Shell Hollow, near Evanstown, Wyoming (after Peck, 1951). (1)

Figures 5-7. Looneyella quadrispinosa Peck, 1951. 5, R side; 6, dorsal view; 7, ventral view, L, 1.01. Lower Cretaceous, Bear River Formation, 7 miles (11 km) north of Evanston, Wyoming (after Peck, 1951). (1)

Figures 8-12. Malheurocypris variabilonodosa (Swain), Swain, 1990. 8, LV, L, 0.62; 9, LV interior, L, 0.88; 10, RV interior, L, 0.80; 11, dorsal view, L, 0.94; 12, R side holotype. Pliocene, Glenns Ferry Formation, sec. 14, T. 4 S., R. 1 W., Owyhee County, Idaho (after Swain, 1986a). (2)

Figures 13-15. Malheurocypris sp., Swain, 1990. 13, LV; 14, dorsal view; 15, LV interior, L, 0.93. Pliocene, Idaho Group, NE1/4 SE1/4 sec. 9, T. 22 S., R. 46 S., Malheur County, Oregon (after Swain, 1990). (2)

Figures 16-21. Dryelba angularis (Peck, 1941). 16, L side, female, L, 0.98; 17, ventral view, female; 18, dorsal view, female; 19, R side, holotype, nodose female, L, 1.12; LV, male, L, 1.06; 21, ventral view, male, L, 0.92. Lower Cretaceous, Peterson Limestone, Freedom Quadrangle, Idaho-Wyoming; Kootenai Formation, near Great Falls, Montana (after Peck, 1941). (1)

Figures 22, 23. Metacypris? clarkstonensis Swain, 1987. 22, R side, holotype; 23, dorsal view, L, 0.73. Pliocene, Salt Lake Group, Cache Valley Formation, South Clarkston section, T. 13 N., R. 12 W., Cache County, Utah (after Swain, 1987a). (2)

Figures 24-27. Metacpyris paracordata Swain, 1964. 24, L side, holotype, L, 0.56; 25, R? side; 26, dorsal view; 27, ventral view. Eocene, Colton-Green River transition beds, section near east end of Raven Ridge, Rio Blanco County, Colorado; 345 feet (105 m) above base of transition beds (after Swain, 1964a). (1)

Figures 28, 29. "Metacypris" forbesi Jones, 1886. R side and dorsal view, female, L, 1.00. Upper Jurassic, Morrison Formation, "Atlantosaurus Beds", near Canyon City, Colorado (after Jones, 1886). (15)

Figures 30, 31. "Metacypris" bradyi Jones, 1886. R side and dorsal view, male, L, 1.19. Upper Jurassic, Morrison Formation, "Atlantosaurus Beds", near Canyon City, Colorado (after Jones, 1886). (15)

Figures 32, 33. "Metacypris" whitei Jones, 1886. L side and dorsal view, male?, L, 1.13. Upper Jurassic, Morrison Formation, "Atlantosaurus Beds", near Canyon City, Colorado (after Jones, 1886). (15)

Figures 34, 35. "Metacypris" consobrina Jones, 1893. L side and dorsal view, female, L, 0.45. Lower Cretaceous, Lower Bear Formation or Gannett Group, Cokeville, Wyoming (after Jones, 1893). (15)

Figures 36, 37. "Metacypris" subcordata Jones, 1893. L side and dorsal view, female, L, 0.60. Lower Cretaceous, Lower Bear Formation or Gannett Group, Cokeville, Wyoming (after Jones, 1893). (15)

Figures 38, 39. "Metacypris" simplex Jones, 1893. R side and dorsal view, L, 0.40. Lower Cretaceous, Lower Bear Formation or Gannett Group, Cokeville, Wyoming (after Jones, 1893). (15)

Figures 40, 41. "Metacypris" cunieformis Jones, 1893. R side and dorsal view, L, 0.65. Lower Cretaceous, Lower Bear Formation or Gannett Group, Cokeville, Wyoming (after Jones, 1893). (15)

Figures 42, 43. Looneyella monticula (Jones), Peck, 1941. L side, ventral view, L, 0.62. Lower Cretaceous, Lower Bear Formation or Gannett Group, Cokeville, Wyoming (after Jones, 1893). (15)

Figures 44, 45. Longispinella longispina (Peck, 1941). L side and ventral view, L, 0.92. Lower Cretaceous, Kootenai Formation, near Great Falls, Montana (after Peck, 1941). (1)

Plate 27

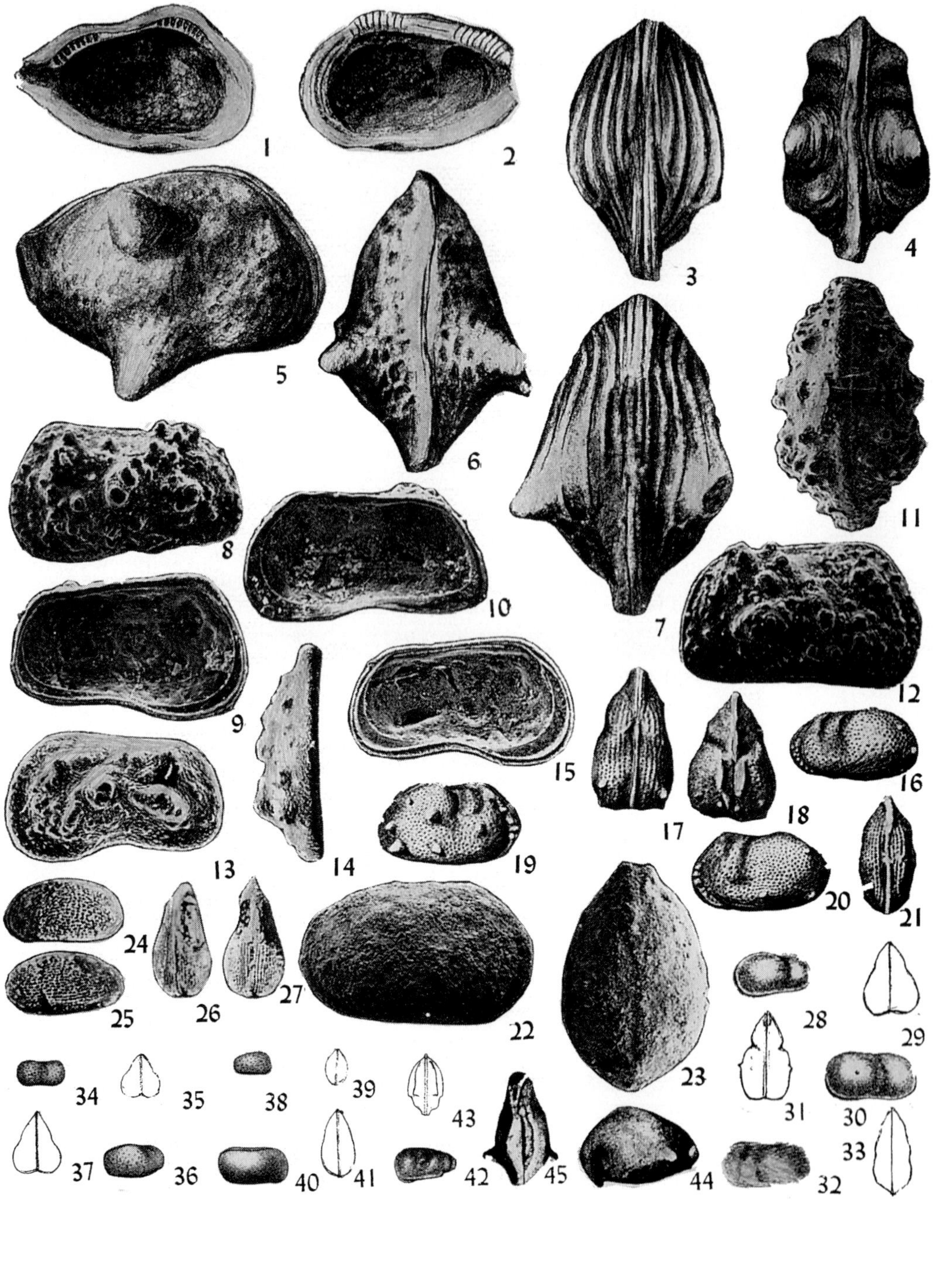

Plate 28

Figures 1, 2. "Metacypris?" sp. 1, L side; 2, ventral view, magnification unknown. Lower Cretaceous or Upper Jurassic, Morrison Formation?, Canyon City, Colorado? (after Nadeau, unpublished photographs, 1945.)

Figures 3-6. Neolimoncythere? prestonensis Swain, 1987. 3, L side, holotype, L, 0.59; 4, dorsal view; 5, R side; 6, RV interior, L, 0.51. Pliocene, Cache Valley Formation, Hoopes Creek loc., NW1/4 SE1/4 NE1/4 sec. 10, T. 12 S., R. 40 E., Franklin County, Idaho (after Swain, 1987a). (2)

Figures 7-11. Neolimnocythere? prestonensis Swain, Swain, 1990. 7, RV, L, 0.47; 8, RV interior; 9, LV, elongated, L, 0.47; 10, LV interior; 11, dorsal view, L, 0.52. Pliocene, Idaho Group, sec. 9, T. 22 S., R. 46 E., Malheur County, Oregon (after Swain, 1990). (2)

Figures 12, 13. Oreanocypris acutispinosus Swain, 1986. 12, R side, holotype; 13, dorsal view, L, 0.72. Pliocene, Glenns Ferry Formation, sec. 14, T. 4 S., R. 1 W., Owyhee County, Idaho (after Swain, 1986a). (2)

Figures 14, 15. Oreanocypris ascioformis Swain, 1986. 14, R side, holotype, L, 0.60; 15, LV interior, L, 0.55. Pliocene, Glenns Ferry Formation, sec. 14, T. 4 S., R. 1 W., Owyhee County, Idaho (after Swain, 1986a). (2)

Figures 16-18. Oreanocypris subtriangularis Swain, 1986. 16, RV, L, 0.56; 17, LV, immature, L, 0.36; 18, dorsal view. Pliocene, Glenns Ferry Formation, sec. 14, T. 4 S., R. 1 W., Owyhee County, Idaho (after Swain, 1986a). (2)

Figures 19-21. Pactolocypris biprojectus Swain, 1985. 19, LV, holotype, L, 0.81; 20, RV, L, 1.16; 21, RV ventral view. Miocene, Esmeralda Formation, 2.85 miles (4.5 km) west of Black Spring, Nye County, Nevada, Black Spring 1:24000 Quadrangle (after Swain, 1985). (2)

Figures 22, 23. Pactolocpyris biprojectus Swain, Swain, 1986. 22, R side, L, 0.66; 23, R side, L, 1.03. Miocene, Humboldt Formation, Cobre section, Elko County, Nevada; 22, 240-360 feet (104-110 m) above base of exposure; 23, 80-84 feet (24-26 m) above base (after Swain, 1986b). (2)

Figures 24, 25. Pactolocypris cancellatus pronodosus Swain, 1986. 24, R side; 25, dorsal view, L, 0.82, holotype. Miocene, Humboldt Formation, Gulf Oil Company, no.1 Thousand Springs well, center SE1/4 SW1/4 sec. 8, T. 40 N., R. 66 E., Elko County, Nevada; 2210-2220 feet (674-677 m) (after Swain, 1986b). (2)

Figures 26-30. Pactolocypris cancellatus (Swain and Wagoner, 1985). 26, R side, holotype, L, 0.98; 27, L side, L, 0.63; 28, R side, L, 0.66; 29, dorsal view small shell, L, 0.48; 30, ventral view, L, 0.46. Miocene, Esmeralda Formation, Blair Junction section, Esmeralda County, Nevada; 26, sample, C2J; 27, sample C1N; 28, sample C1M; 29, 30 sample C1D (after Swain and Wagoner, 1985). (2)

Figure 31. Pactolocypris centronotus (Swain, 1986). L side of imperfect shell, L, 0.90. Eocene or Oligocene, Elko Oil Shale, Catlin Works, Elko, Nevada (after Swain, 1986b). (2)

Figures 32, 33. Pactolocypris laevus Swain, 1986. 32, L side, holotype, partly exfoliated; 33, dorsal view, L, 0.88. Miocene, Humboldt Formation, Cobre surface section, Cobre, Elko County, Nevada; 72 feet (22 m) above base of exposure (after Swain, 1986b). (2)

Figures 34, 35. Pactoocypris laevus Swain, 1986. 34, R side; 35, dorsal view, L, 0.63. Miocene, Humboldt Formation, section on Medicine Creek, Winecup Ranch, T. 30 N., R. 65 E., Elko County, Nevada; 1883-1921 feet (574-586 m) below top of exposure (after Swain, 1986b). (2)

Plate 28

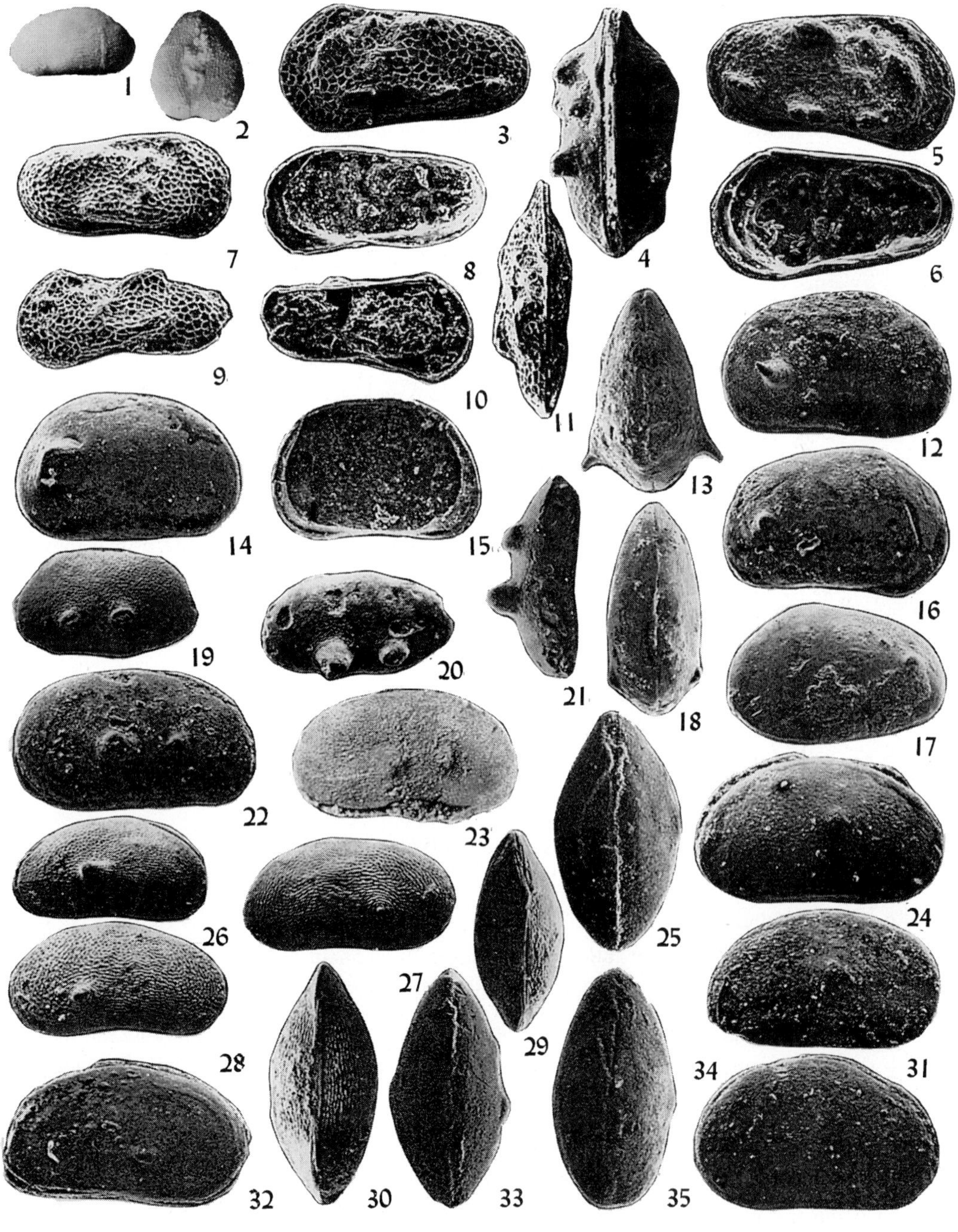

Plate 29

Figures 1-3. Pactolocypris pactolensis Swain, 1985. 1, R side, holotype, L, 0.94; 2, L side, L, 1.06; 3, R side, L, 0.92. Miocene, Esmeralda Formation, 2.85 miles (4.5 km) west of Black Spring, Black Spring 1:24000 Quadrangle, Nye County, Nevada (after Swain, 1985). (2)

Figure 4. Pactolocypris pactolensis Swain, Swain, 1990. LV internal mold, L, 0.63. Miocene, Truckee Formation, 2 miles (3.6 km) WSW of Hazen, Churchill County, Nevada (after Swain, 1990). (2)

Figures 5-7. Pactolocypris suborbicularis Swain, 1985. 5, R side, holotype, L, 1.07; 6, R side, L, 1.20; 7, dorsal view, L, 1.16. 5, Oligocene or Miocene, Humboldt Formation, just above Triassic, Winecup Ranch, Medicine Creek section, T. 39 N., R. 65 E., Elko County, Nevada; 6, 7, Miocene, Esmeralda Formation, 2.85 miles (4.5 km) west of Black Spring, Black Spring 1:24000 Quadrangle, Nye County, Nevada (after Swain, 1985). (2)

Figures 8-9. Pactolocypris suborbicularis Swain, Swain, 1986. 8, L side; 9, dorsal view, L, 1.21. Oligocene or Miocene, Humboldt Formation, near base of Medicine Creek section, T. 39 N., R. 65 E., just above Triassic, Winecup Ranch, Elko County, Nevada (after Swain, 1986b). (2)

Figures 10, 11. Pactolocypris suborbicularis attenuatus Swain, 1986. 10, R side, holotype; 11, dorsal view, L, 0.82. Miocene, Humboldt Formation, Gulf Oil Company, no.1 Thousand Springs well, center SE1/4 SW1/4 sec. 8, T. 40 N., R. 66 E., Elko County, Nevada; 1690-1700 feet (515-518 m) (after Swain, 1986b). (2)

Figure 12. Pactolocypris sp. aff. P. suborbicularis Swain. R side, L, 1.25. Lower Oligocene, Chadron Formation, Helmville, Nebraska; R. Konizeski coll., unpublished.

Figure 13. Paracandona euplectella (Brady and Norman), Staplin, 1963. LV, female, L, 0.61. Pleistocene, Mankato or late Cary Substage, Wisconsin Stage, 1.5 to 2 miles (2.4-3.6 km) east of Mt. Vernon, Indiana on Route 62 (after Staplin, 1963). (1)

Figures 14, 15. Paracypria declivoidea Swain, 1986. 14, R side, holotype; 15, dorsal view, L, 0.60. Pliocene, Glenns Ferry Formation, sec. 14, T. 4 S., R. 1 W., Owyhee County, Idaho (after Swain, 1986a). (2)

Figure 16. Paracypria?, sp., Swain and Wagoner, 1985. R side, L, 0.60. Miocene, Esmeralda Formation, Blair Junction section, Esmeralda County, Nevada, sample C1C2 (after Swain and Wagoner, 1985). (2)

Figure 17. "Paracypridea" sp., Sohn, 1969. RV, L, 2.12. Lower Cretaceous?, late Aptian?, Newark Canyon Formation, center NW1/2 NW1/4 sec. 21, T. 29 N., R. 53 E., Elko County, Nevada (after Sohn, 1969). (8)

Figures 18, 19. Paracypria? acuminata (Roth, 1933). 18, R side; 19, dorsal view, L, 1.08. Upper Jurassic or Lower Cretaceous, Lakota Formation by Sohn (1957), sec. 28, T. 4 N., R. 6 E., Lawrence County, South Dakota (after Roth, 19330. (1) (as Paracypris simplus Roth)

Figures 20-22. Paracypridopsis? curta Swain, 1986. 20, R side, holotype; 21, dorsal view, L, 0.52; 22, R side, L, 0.66. Miocene, Humboldt Formation, Gulf Oil Company, no.1 Thousand Springs well, center SW1/4 SW1/4 sec. 8, T. 40 N., R. 66 E., Elko County, Nevada; 2325-2350 feet (709-716 m) (figs. 20, 21); 2250-2260 feet (686-689 m) (fig. 22) (after Swain, 1986b). (2)

Figure 23. Paracypridopsis? fouchi Swain, 1987. L side, shell partly exfoliated, L, 0.88. Eocene, Twin Bridges Limestone, Twin Bridges, Elko County, Nevada (after Swain, 1987b). (2)

Figures 24-26. Paracypridopsis? fouchi Swain. 24, L side, L, 1.05; 25, R side, L, 1.51; 26, dorsal view, holotype, L, 1.15. Upper Cretaceous or Paleocene, Sheep Pass Formation, Johnson Canyon, Grant Range, Nevada (figs. 24, 26); Sheep Pass Canyon, Egan Range (fig. 25) (after Swain, 1987b). (2)

Figures 27, 28. Pontoniella? catlinensis Swain, 1987. 27, L side, imperfect shell only questionably in this species; 28, ventral view, L, 1.39. Upper Cretaceous or Paleocene, Sheep Pass Formation, Sheep Pass Canyon, Egan Range, Nevada (after Swain, 1987b). (2) (=Paracypridopsis fouchi [part] of Swain, 1987b).

Figure 29. Paracypridopsis? sp, Swain, 1987c. R side, L, 0.88. Miocene, Horse Camp Formation, sec. 12, T. 10 N., R. 58 E., Nye County, Nevada (after Swain, 1987c). (2)

Figures 30, 31. Pelocypris tuberculatum (Ferguson), Lister, 1975. 30, RV, 31, RV interior, L, 1.93. Pleistocene, Lake Bonneville Group, Saltair core hole, SE1/4 sec. 25, T. 1 N., R. 3 W., Salt Lake County, Utah, (after Lister, 1975). (3)

Figures 32, 33. Petrobrasia? sp., Sohn, 1969. 32, LV, broken, L, 1.31+; 33, RV adductor muscle scar area, X 120. Lower Cretaceous?, late Aptian?, Neward Canyon Formation, center W1/2 NW1/4 sec. 21, T. 29 N., R. 53 E., Elko County, Nevada (after Sohn, 1969). (8)

Figures 34, 35. Physocypria fadeewi Dubowsky, Staplin, 1963. 34, RV; 35, LV, L, 0.73 X 40. Pleistocene, very late Mankato Substage, Wisconsin Stage, Chicago region (after Staplin, 1963). (1)

Plate 29

1
2
3
4
5
6
7
8
9
10
11
12
13
14
15
16
17
18
19
20
21
22
23
24
25
26
27
28
29
30
31
32
33
34
35

Plate 30

Figures 1-3. Physocypria pustolosa (Sharpe), Staplin, 1963. 1, RV, L, 0.47; 2, LV, L, 0.52; 3, LV with dorsal bulge, L, 0.57. Pleistocene, Cary or Mankato Substage, Wisconsin Stage, Blue Island, Chicago, Illinois, sec. 25, T. 37 N., R. 4 W. (after Staplin, 1963). (1)

Figure 4. Physocypria pustolosa (Sharpe), Benson, 1967. RV interior, L, 0.55. Pleistocene, Yarmouthian Stage, Dickinson County, Kansas, U.S.N.M. 649610, loc. P (after Benson, 1967). (3)

Figures 5-7. Pontoniella? catlinensis Swain, 1967. 5, L side, holotype, L, 0.90; 6, R side, small shell broken anteriorly, L, 0.53; 7, dorsal view, L, 0.61. Eocene or Oligocene, Elko Oil Shale, Catlin Works, Elko, Nevada (after Swain, 1987b). (2)

Figure 8. Pontoniella? sp., Swain and Wagoner, 1985. L side, L, 0.40. Miocene, Esmeralda Formation, Blair Junction, Esmeralda County, Nevada, sample C10 (after Swain and Wagoner 1985). (2)

Figure 9. Pontoniella? sp., Swain, 1985. R side, L, 0.75. Miocene, Esmeralda Formation, 2.85 miles (4.5 km) west of Black Spring, Black Spring 1:24000 Quadrangle, Nye County, Nevada (after Swain, 1985). (2)

Figure 10. Pontoniella? sp., Swain, 1986. ? side, L, 0.58. Pliocene, Glenns Ferry Formation, sec. 14, T. 4 S., R. 1 W., Owyhee County, Idaho (after Swain, 1986a). (2)

Figure 11. Pontoniella? sp., Swain, 1986. L side, L, 0.75. Miocene, Humboldt Formation, Gulf Oil Company, no.2, sec. 21. T. 38 N., 61 E., Elko County, Nevada; 3130-3160 feet (after Swain, 1986b). (2)

Figures 12-15. Potamocypris williamsi Swain, 1964. 12, R side, holotype; 13, dorsal view; 14, R side; 15, RV interior, L, 1.06. Eocene, Colton-Green River transition beds, Dripping Rock Gulch section, east end of Raven Ridge, Rio Blanco County, Colorado; 211-315 feet (64-98 m) above base of transition beds (after Swain, 1964a). (1)

Figure 16. Potamocypris cf. comosa Furtos, Swain, 1987. LV, L, 0.61. Pliocene, Cache Valley Formation, Hoopes Creek loc., NW1/4 SE1/4 NE1/4 sec. 10, T. 12 S., R. 40 E., Franklin County, Idaho (after Swain, 1987a). (12)

Figure 17. Potamocypris granulosa Daday, Lister, 1975. R side, L, 0.54. Pleistocene, Lake Bonneville Group, Saltair core hole, SE1/4 sec. 25, T. 1 N., R. 3 W., Salt Lake County, Utah (after Lister, 1975). (3)

Figures 18, 19. Potomocypris minuta Staplin, 1963. 18, LV, holotype; 19, RV, L, 0.46. Pleistocene, Mankato or late Cary Substage, Wisconsin Stage, sec. 13, T. 9 S., R. 10 E., Gallatin County, Illinois (after Staplin, 1963). (1)

Figures 20-22. Potomocypris? moorei Swain, 1987. 20, L side, holotype, L, 0.88; 21, dorsal view, L, 0.45; 22, R side, L, 0.47. Miocene, Horse Camp Formation, approx. NW1/4 sec. 2, T. 10 N., R. 58 E., unsurveyed area, Nye County, Nevada (after Swain, 1987c). (2)

Figure 23. Potomocypris pallida Alm, Lister, 1975. RV, L, 0.60. Pleistocene, Lake Bonneville Group, Saltair core hole, SE1/4 sec. 25, T. 1 N., R. 3 W., Salt Lake County, Utah (after Lister, 1975). (3)

Figures 24, 25. Potomocypris? pellucidus Swain, 1963. 24, RV; 25, LV, holotype, L, 0.65. Pleistocene, Gubik Formation, Arctic coastal plain, Alaska, Ikpikpuk shot holes, line 12, shot point 5 at 30-40 feet (9.1-12.2 m) (after Swain, 1963). (1)

Figures 26, 27. Potomocypris smaragdina (Vavra), Daday, Staplin, 1963. 26, RV, L, 0.53; 27, LV, male?, L, 0.53. Pleistocene; 26, SE cor. Sec. 17, T. 37 N., R. 12 E., Chicago, Illinois; 27, sec. 13, T. 9 S., R. 10 E., Gallatin County, Illinois (after Staplin, 1963). (1)

Figures 28, 29. Potomocypris smaragdina (Vavra), Gutentag and Benson, 1962. 28, RV, L, 0.44; 29, LV interior, L, 0.46. Pleistocene, late Kansan to Wisconsin Stages, Meade County, Kansas and Beaver County, Oklahoma (after Gutentag and Benson, 1962). (4)

Figure 30. Potomocypris unicaudata Schafer, Lister, 1975. R side, L, 0.76. Pleistocene, Lake Bonneville Group, Saltair core hole, SE1/4 sec. 25, T. 1 N., R. 3 W., Salt Lake County, Utah (after Lister, 1975). (3)

Figures 31, 32. Potamocypris variegata (Brady and Robertson), Staplin, 1963. 31, LV, L, 0.56; 32, RV, L, 0.54. Pleistocene, Mankato Substage, Wisconsin Stage, sec. 13, T. 9 S., R. 10 E., Gallatin County, Illinois (after Staplin, 1963). (1)

Figure 33. Potamocypris? sp., Swain and Brown, 1972. R side, L, 0.65. Lower Cretaceous (probably) or Upper Jurassic, Unit H, Well NC-DA-OT-12, Dare County, North Carolina, 7836-7846 feet (2388-2391 m) (after Staplin , 1963). (1)

Figure 34. Potamocypris sp., Lister, 1975. L side, L, 0.55. Pleistocene, Lake Bonneville Group, Saltair core hole, SE1/4 sec. 25, T. 1 N., R. 3 W., Salt Lake County, Utah (after Lister, 1975). (3)

Figure 35. Prionocypris sp. aff. P. glacialis (Sars), Dickinson and Swain, 1967. LV, L, 1.73. Pliocene, Hay Ranch Formation (=Upper Humboldt), 1 mile (1.6 km) south of Elko, Nevada on Highway 46 (after Dickinson and Swain, 1967). (1)

Figure 36. Prionocypris pigra (Fischer), Swain, 1963. R side, L, 0.76. Pleistocene, Gubik Formation, Arctic coastal plain, Alaska, Ikpikpuk shot holes, line 10, shot point 31 at 40-50 feet (12.2-15.2 m) (after Swain, 1963). (1)

Figure 37. Procyprois? gracilis Swain, 1985. R side, holotype, L, 0.92. Miocene, Esmeralda Formation, 2.85 miles (4.5 km) west of Black Spring, Black Spring 1:24000 Quadrangle, Nye County, Nevada (after Swain, 1985). (2)

Figure 38. Procyprois ravenridgensis Swain, 1964. R side, holotype, L, 0.99. Eocene, Colton-Green River transition beds, 10 feet (3 m) above base, Dripping Rock Guch, east end of Raven Ridge, Rio Blanco County, Colorado (after Swain, 1964a). (1)

Plate 30

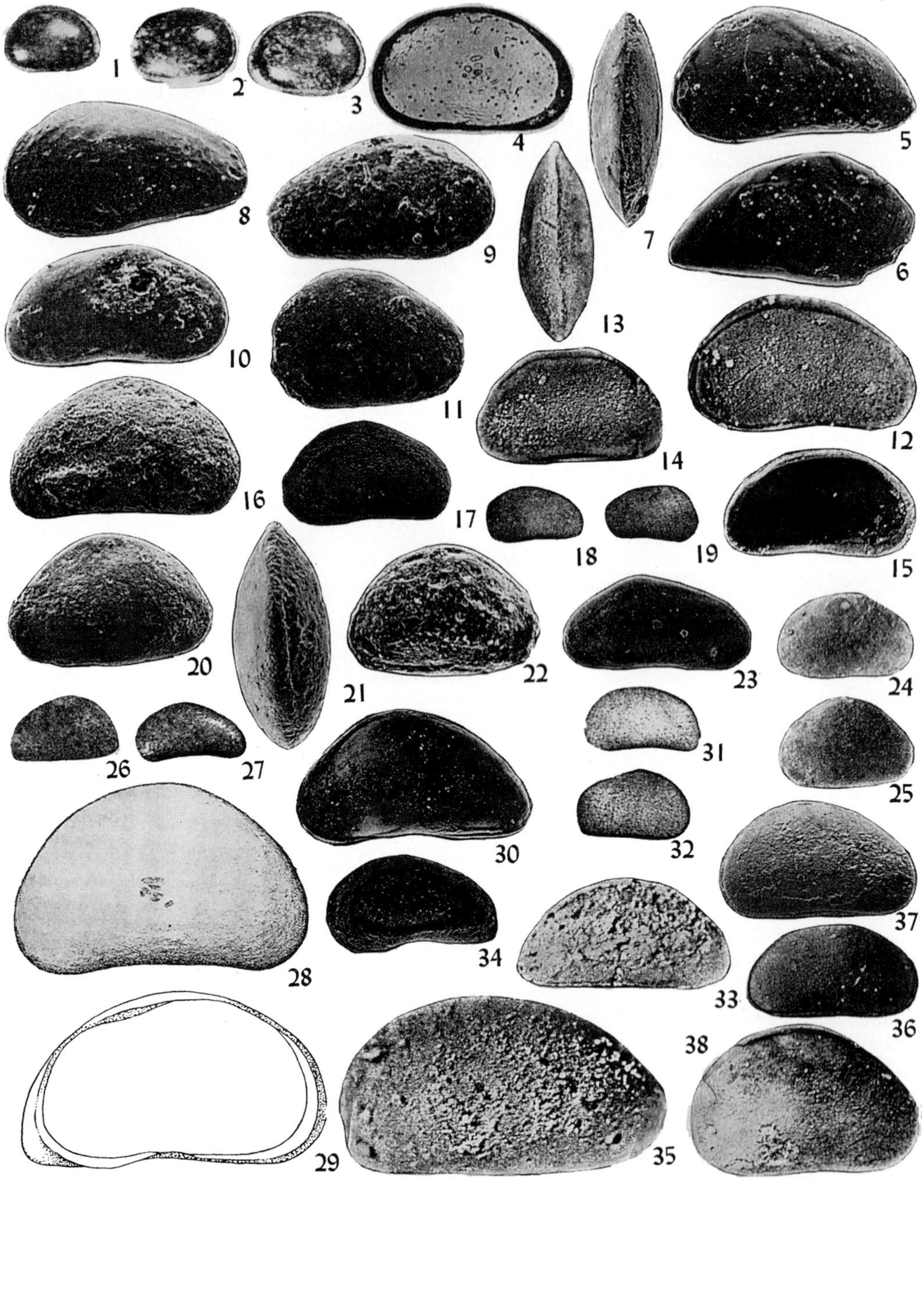

Plate 31

Figures 1-4. Procyprois twinbridgensis Swain, 1967. 1, R side, holotype, imperfect posteriorly, L, 1.03; 2, R side, exfoliated shell showing impression of midlayer of RV, polygonal structure and impression of pits on mid-layer surface, L, 1.02; 3, adductor muscle scar pattern, approx. X 185; LV internal mold showing polygonal shell structure, L, 0.65. Eocene, Twin Bridges Limestone, sec. 36, T. 32 N., R. 55 E., Elko County, Nevada, at Twin Bridges (after Swain, 1987b). (2)

Figures 5, 6. Procyprois? sp., Swain, 1985. 5, R side, L, 0.80; 6, dorsal view, L, 0.85. Miocene, Esmeralda Formation, 2.85 miles (4.5 km) west of Black Spring, Black Spring 1:24000 Quadrangle, Nye County, Nevada (after Swain, 1985). (2)

Figure 7. Procyprois? sp., Swain, 1986. R side, L, 0.96. Miocene, Humboldt Formation, Gulf Oil Company, no.1 Thousand Springs well, SE1/4 SW1/4 sec. 8, T. 40 N., T. 66 E., Elko County, Nevada; 1850-1860 feet (564-567 m) (after Swain, 1986b). (2)

Figures 8-12. Pruvostina wanlessi Scott and Summerson, 1943. 8, L side, pyritized, L, 0.93; 9, ventral view; 10, dorsal view; 11, R side, L, 0.92; 12, LV, L, 1.20. Lower Pennsylvanian, Scott Formation, Kanawha Series, near Newcomb, Campbell County, Tennessee (after Scott and Summerson, 1943). (11)

Figures 13, 14. Pruvostina? sp., Sohn, 1985. 13, L? side; 14, dorsal view, L, 0.85. Upper Mississippian, Bluestone Formation, Bramwell Member, Crumpler Quadrangle, Wyoming County, West Virginia (after Sohn, 1985). (1)

Figures 15-17. Pactolocypris centronotus (Swain). L side, ventral view, L, 1.00; R side, L, 0.92. Eocene, 75 feet (23 m) above base of Colton-Green River transition beds, near west end of Raven Ridge, Rio Blanco County, Colorado (after Swain, 1964a). (1)

Figures 18, 19. Phacocypris? cf. pisiformis Hou and Shan, Swain, 1985. R side and dorsal view, L, 0.95. Miocene, Esmeralda Formation, lower shale, Finger Rock Wash, 0.5 mile (0.8 km) south of Rawhide Ranch, Granny Goose 1:24000 Quadrangle, Mineral County, Nevada (after Swain, 1985). (2)

Figures 20, 21. Phacocypris? sp. aff. P. tongbinensis Hou and Shan, Swain, 1985. R side and dorsal view, L, 0.87. Miocene, Esmeralda Formation, Finger Rock Wash, 0.5 mile (0.8 km) south of Rawhide Ranch, Granny Goose 1:24000 Quadrangle, Mineral County, Nevada (after Swain, 1985). (2)

Figures 22, 23. "Potomocypris" affinis Jones, 1893. R side and ventral side views, L, 0.70. Lower Cretaceous, Lower Bear River Formation or Gannett Group, Cokeville, Wyoming (after Jones, 1893). (15)

Figures 24, 25. "Potomocypris" unisulcatus Jones, 1893. L side and dorsal view, L, 1.05. Upper Cretaceous, Montana Formation, Coaville, Utah (after Jones, 1893). (15)

Figures 26, 27. Trapezoidella trapezoidalis (Roth), Sohn, 1958. R side, holotype and dorsal view, L, 1.03. Lower Cretaceous, Lakota Formation, sec. 28, T. 4 N., R. 6 E., Lawrence County, South Dakota, by Sohn (1957, 1958) (after Roth, 1933). (1)

Figures 28-30. Theriosynoecum? minnekahtense (Roth), Sohn, 1979. RV, holotype, dorsal view, ventral view, L, 1.00. Lower Cretaceous, Lakota Formation, by Sohn (1979), sec. 28, T. 4 N., R. 6 E., Lawrence County, South Dakota (after Roth, 1933). (1)

Figures 31, 32. Whipplella sp., Sohn, 1977. R side, L, 1.25, adductor muscle scar area, approx. X 300. Permian, upper Rockport Limestone Member, Greene Formation, West Virginia (after Sohn, 1977). (8)

Figure 33. Whipplella? sp. 2, Sohn, 1985. L side, internal mold, L, 0.45. Upper Mississippian, Bluestone Formation, Bramwell Member, Crumpler Quadrangle, Wyoming County, West Virginia (after Sohn, 1985). (1)

Plate 31

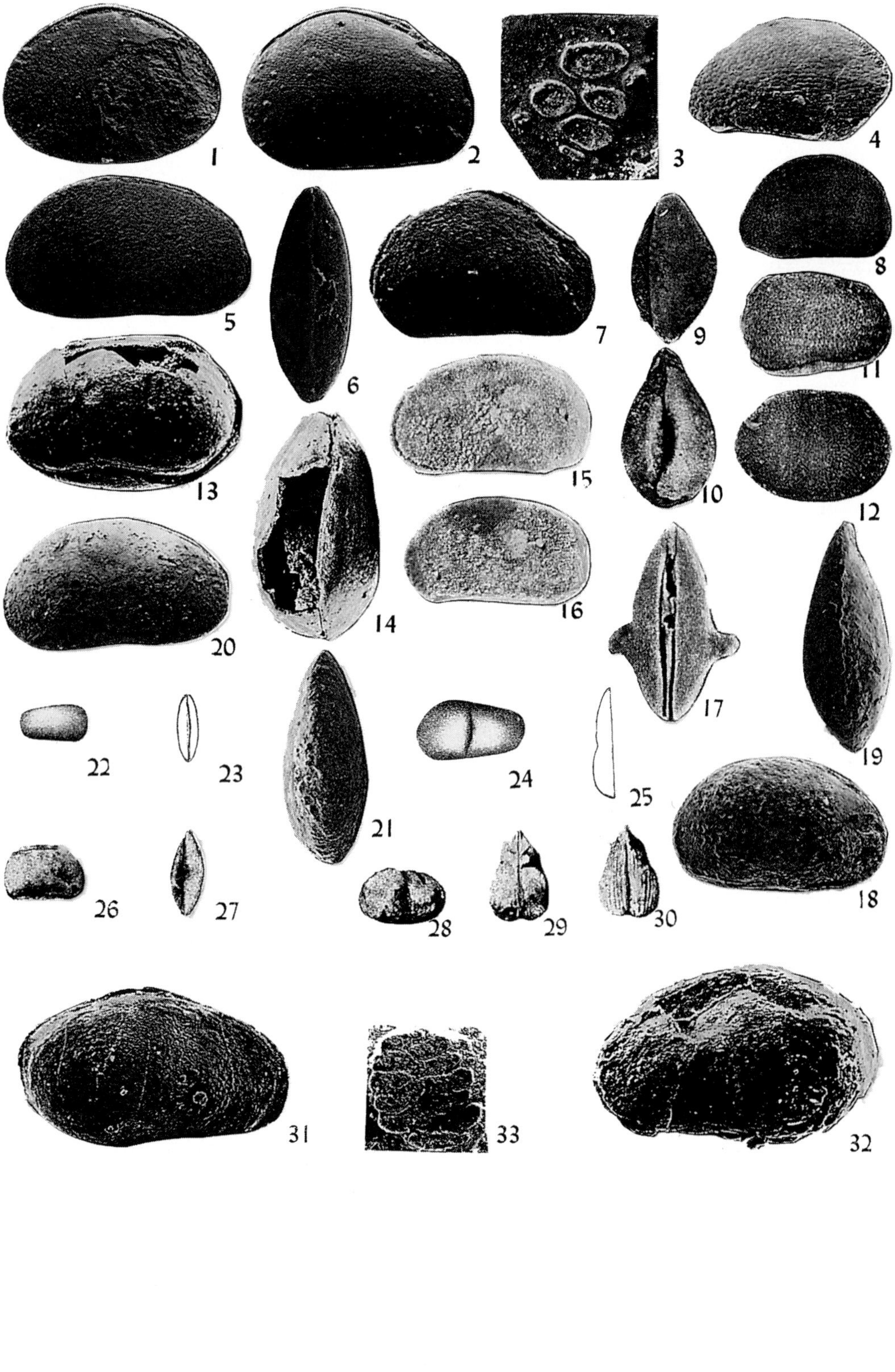
1
2
3
4
5
6
7
8
9
10
11
12
13
14
15
16
17
18
19
20
21
22
23
24
25
26
27
28
29
30
31
32
33

Plate 32

Figures 1-3. Whipplella depressa Holland, Scott, 1944. R side, dorsal view and ventral view, L, 0.76. Permian, Nineveh Limestone, Greene Group, Limestone Hill, Wood County, West Virginia (after Scott, 1944). (1)

Figures 4-6. Whipplella? sp., Sohn, 1985. L side, R side, dorsal view, L, 0.71. Upper Mississippian, Bluestone Formation, Bramwell Member, Crumpler Quadrangle, Wyoming County, West Virginia (after Sohn, 1985). (1)

Figure 7. Xiyingia? sp. aff. X. elliptica Hou and Shi, Swain, 1985. R side, L, 1.07. Miocene, Esmeralda Formation, Finger Rock Wash, 0.5 mile (0.8 km) south of Rawhide Ranch, Granny Goose 1:24000 Quadrangle, Mineral County, Nevada (after Swain, 1985). (1)

Figures 8, 9. Pseudoeucypris pagei (Swain, 1949). 8, L side, holotype; 9, ventral view, L, 1.36. Lower Eocene, sec. 3, 4, 9, 10, T. 14 S., R. 14 E., near Sunnyside, Carbon County, Utah (after Swain, 1949). (1)

Figure 10. Sollenella? sp., Friedman and Lundin, 1998. Damaged LV, L, 4.35. Middle Devonian, Gilboa Formation, East Windham, Greene County, New York (after Friedman and Lundin, 1998). (1)

Figures 11-13. Pseudoeucypris pagei (Swain), Swain, 1964. 11, R side, L, 1.30; 12, dorsal view, L, 1.12; 13, R side, immature molt, L, 0.76. Lower Eocene, 196 feet (60 m) above base of Colton-Green River transition beds, Dripping Rock Gulch section, east end of Raven Ridge, Rio Blanco County, Colorado (after Swain, 1964a). (1)

Figures 14, 15. Pseudoeucypris pagei (Swain), compressed variant, Swain, 1964. 14, LV; 15, dorsal view, L, 1.18. Eocene, 272-276 feet (83-84 m) above base of Colton-Green River transition beds, Dripping Rock Gulch, east end of Raven Ridge, Rio Blanco County, Colorado (after Swain, 1964a). (1)

Figure 16. Pseudoeucypris sp., Swain, 1986. L side, L, 0.63. Miocene, Humboldt Formation, Medicine Creek section, Winecup Ranch, T. 39 N., R. 65 E., Elko County, Nevada (after Swain, 1986b). (2)

Figures 17, 18. Reticulocypris reticulata Staplin, 1963. 17, LV, holotype, in transmitted light; 18, same showing marginal tubercles at different level of focus, L, 0.48. Pleistocene, Mankato or late Cary Substage, Wisconsin Stage, sec. 13, T. 9 S., R. 10 E., Gallatin County, Illinois (after Staplin, 1963). (1)

Figures 19-21. Scabriculocypris?, n. sp., Kietzke amd Lucas, 1995. 19, R side; 20, L side; 21, ventral view, L, 0.73. Lower Jurassic, Kayenta Formation, SW1/4 sec. 35, T. 27 N., R. 12 E., Coconino County, Arizona (after Kietzke and Lucas, 1995). (17)

Figure 22. Scottia sp. aff. S. eaglespringensis Swain, Swain, 1987. R side, L, 0.70. Eocene, Twin Bridges Limestone, Twin Bridges, Elko County, Nevada (after Swain, 1987b). (2)

Figure 23. Scottia subquadrata Swain, 1964. R side, L, 0.83. Eocene, Green River Formation, 11 miles (17.7 km) north of Castlegate, Utah, loc. 19 (after Swain, 1949). (1)

Figure 24. Scottia eaglespringensis Swain, 1964. L side, holotype, L, 0.43. Paleocene or late Cretaceous, Sheep Pass Formation, Shell Oil Company, no.1 Eagle Springs well, core 7450 feet (2270 m), Nye County, Nevada (after Swain, 1964a). (1)

Figures 25, 26. Scottia? sublanceolata Swain, 1964. 25, L side, holotype; 26, dorsal view, L, 0.35. Paleocene or late Cretaceous, Sheep Pass Formation, Shell Oil Company, no.1 Eagle Springs well, Nye County, Nevada; core 7534-7536 feet (2296-2297 m) (after Swain, 1964a). (1)

Figure 27. Scottia subquadrata Swain, 1964. R side, holotype, L, 0.83. Eocene, 66 feet (20 m) above base of Colton-Green River transition beds, Dripping Rock Gulch section east end of River Ridge, Rio Blanco County, Colorado (after Swain, 1964a). (1)

Figure 28. Scottia sp. aff. S. subquadrata Swain, Swain, 1985. R side, L, 0.81. Miocene, Esmeralda Formation, 2.85 miles (4.5 km) west of Black Spring, Black Spring 1:24000 Quadrangle, Nye County, Nevada (after Swain, 1985). (2)

Figure 29. Scottia trapezoidalis Swain, 1964. R side, holotype, L, 0.43. Paleocene or late Cretaceous, Sheep Pass Formation, Shell Oil Company, no.1 Eagle Springs well, 7450 feet (2270 m), Nye County, Nevada (after Swain, 1964a). (1)

Figures 30-34. Stenocypris? horsecampensis Swain, 1987. 30, R side, holotype, L, 0.99; 31, L side, L, 0.55; 32, dorsal view, L, 0.54; 33, dorsal view, L, 0.47; 34, RV interior, L, 0.56, showing broad terminal inner lamellae, valve partly filled with calcite crystals. Miocene, Horse Camp Formation, sec. 12, T. 10 N., R. 58 E., Nye County, Nevada (after Swain, 1987c). (2)

Figure 35. Stenocypris cf. sinuata Müller, Swain, 1964. R side, L, 0.90. Eocene, 475 feet (145 m) above base of Colton-Green River transition beds, Nine Mile Canyon section, Carbon County, Utah (after Swain, 1964a). (1)

Plate 32

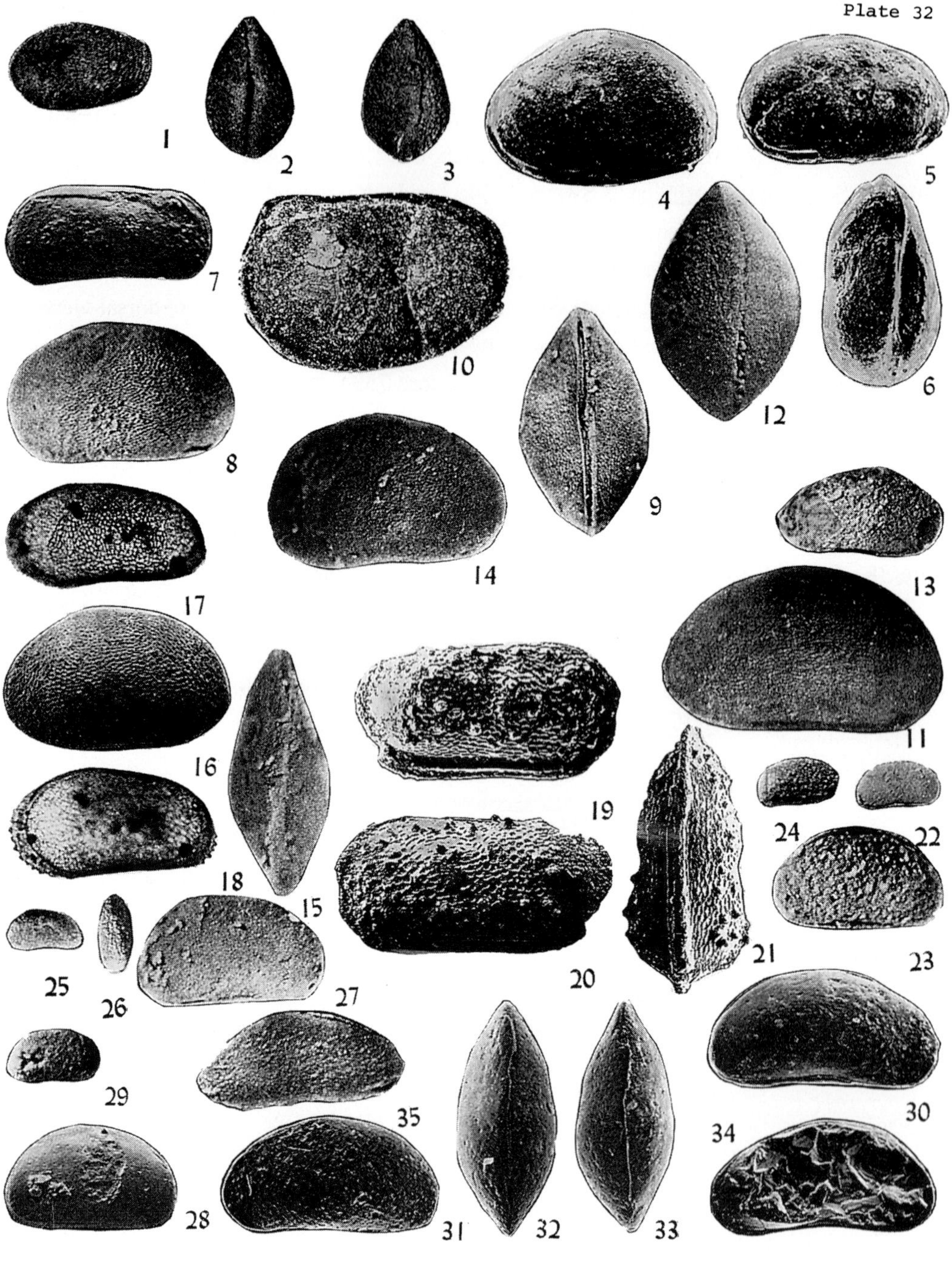
1
2
3
4
5
7
10
12
6
8
9
14
13
17
11
16
19
24
22
18
15
20
21
23
25
26
27
29
30
35
34
28
31
32
33

Plate 33

Figures 1-3. Theriosynoecum wyomingensis (Branson), Sohn, 1982. 1, RV; 2, LV; 3, dorsal view, female, L, 1.30. Upper Jurassic, Morrison Formation; 1, U.S.G.S. coll. 26406, Custer County, South Dakota; 2, U.S.G.S. coll. 26407, same loc.; 2, U.S.N.M. coll. 75502, Johnson County, Wyoming (after Sohn, 1982). (5)

Figures 4, 5. Theriosynoecum wyomingensis (Branson), Sohn, 1958. 4, LV; 5, LV ventral view, L, 1.41. Upper Jurassic, Morrison Formation, sec. 33, T. 56 N., R. 63 W., Crook County, Wyoming (after Sohn, 1958). (9)

Figures 6, 7. Theriosynoecum? sp., Swain, 1986. 6, L side; 7, dorsal view, L, 0.80. Pliocene, Glenns Ferry Formation, sec. 14, T. 4 S., R. 1 W., Owyhee County, Idaho (after Swain, 1986a). (2)

Figure 8. Theriosynoecum? sp. R side, magnification unknown. Upper Jurassic or Lower Cretaceous, Morrsion Formation, Canyon City, Colorado? (after Nadeau, 1945, unpublished).

Figures 9-12. Timiriasevia? heylmuni Swain, 1987. 9, L side, holotype, L, 0.55, imperfect posteriorly; 10, L side, imperfect posteriorly, L, 0.45; 11, R side, broken posteriorly, L, 0.54; 12, L side, internal mold, L, 0.53, showing adductor muscle scar. Eocene, Twin Bridges Limestone, Twin Bridges section, Elko County, Nevada (after Swain, 1987b). (2)

Figures 13-16. Timiriasevia? heylmuni Swain, 1987. 13, L side, L, 0.29; 14, LV interior, L, 0.24; 15, ventral view, L, 0.30; 16, adductor muscle scar, approx. X 705. Eocene, Twin Bridges Limestone, Elko County, Nevada (after Swain, 1987b). (2)

Figures 17-19. Trapezoidella rothi, Sohn, 1979. 17, R side, L, 1.37; 18, R side; 19, ventral view, L, 0.93. Lower Cretaceous, Lakota Formation, Chilson Member, Fall River County, South Dakota (after Sohn, 1979). (8)

Figures 20-22. Trapezoidella trapezoidalis (Roth), Sohn, 1979. 20, R side; 21, L side; 22, dorsal view, L, 1.03. Lower Cretaceous, Lakota Formation, 3 miles (4.8 km) south of Piedmont, Meade County, South Dakota (after Sohn, 1979). (8)

Figures 23, 24. Trapezoidella sp. aff. T. trapezoidalis (Roth). 23, RV; 24, LV, magnification unknown. Upper Jurassic or Lower Cretaceous, Morrison Formation?, Canyon City, Colorado? (B.K. Nadeau, unpublished).

Figures 25-28. Tuberocypris acuminatus Swain, 1947. 25, RV, holotype, L, 1.05; 26, LV, 1.08; 27, RV interior, L, 1.10; 28, ventral view, L, 1.08. Pliocene, Salt Lake Group, sec. 15, T. 13 N., R. 2 W., Cache County, Utah (after Swain, 1947). (1)

Figures 29-33. Tuberocypris brighti Swain, 1986. 29, R side, holotype, L, 0.70; 30, RV interior, L, 0.68; 31, adductor muscle scar area, X 295; 32, dorsal view, L, 0.70; 33, LV interior, L, 0.76. Pliocene, Glenns Ferry Formation, sec. 14, T. 14 S., R. 1 W., Owyhee County, Idaho (after Swain, 1986a). (2)

Plate 33

1
2
3
4
5
6
7
8
9
10
11
12
13
14
15
16
17
18
19
20
21
22
23
24
25
26
27
28
29
30
31
32
33

Plate 34

Figures 1-3. Wilkinsia gondoliformis Dickinson and Swain (Cladocera). 1, 2, L sides of juvenile ephippia, L, 0.70; 3a, c, L side and dorsal view, L, 0.75; 3b, L side, L, 0.65, of ephippia. Pliocene, Hay Ranch Formation, Gulf Oil Company no.1, Wilkins Ranch well, Elko County, Nevada, ranging in depth from 957 to 3584 feet (292-1052 m) (after Dickinson and Swain, 1967). (1)

Figures 4, 5. Tuberocypris ovalis Swain, 1947. 4, RV, holotype; 5, ventral view, L, 1.19. Pliocene, Salt Lake Group, sec. 15, T. 13 N., R. 2 W., Cache County, Utah (after Swain, 1947). (1)

Figures 6-10. Tuberocypris quadralis Swain, 1947. 6, LV, holotype; 7, ventral view, L, 0.82; 8, LV, L, 0.85; 9, RV interior, female, L, 0.88; 10, RV interior, male?, L, 0.88. Pliocene, Salt Lake Group, sec. 15, T. 13 N., R. 2 W., Cache County, Utah (after Swain, 1947). (1)

Figures 11-13. Tuberocypris quadralis Swain, Swain, 1987. 11, R side; 12, dorsal view, L, 0.61; 13, LV interior, L, 0.90. Pliocene, Cache Valley Formation, T. 13 N., R. 3 W., SW of Newton, Utah (after Swain, 1987a). (2)

Figures 14-16. Tuberocyproides dipleura Swain, 1947. 14, LV, holotype, L, 1.20; 15, LV interior, L, 1.35; 16, ventral view, L, 1.32. Pliocene, Salt Lake Group, sec. 15, T. 13 N., R. 2 W., Cache County, Utah (after Swain, 1947). (1)

Figures 17-24. Tuberocyproides oreanensis Swain, 1986. 17, LV, female, holotype, L, 1.26; 18, dorsal view, female, L, 0.59?; 19, RV, male, L, 0.61?; 20, LV, male, dorsal view, L, 0.60; 21, R side, female, L, 0.60?; 22, R side, female, truncated variant, L, 0.96; 23, dorsal view, L, 1.13; 24, RV interior, L, 1.11. Pliocene, Glenns Ferry Formation, sec. 14, T. 4 S., R. 1 W., Owyhee County, Idaho (after Swain, 1986a). (2)

Figures 25, 26. Turkmenella laeva Swain, 1987c. R side and dorsal view, holotype, L, 0.77. Miocene, Esmeralda Formation, Warrior Mine section, 7650 feet (2332 m) S 26° east of NW corner, Simon Quadrangle, Nye County, Nevada (after Swain, 1987c). (2)

Figures 27-29. Turkmenella warriorensis Swain, 1987. 27, R side; 28, dorsal view, holotype, L, 0.80; 29, R side, L, 0.51. Miocene, Esmeralda Formation, Warrior Mine section, 7650 feet (2332 m) S 26 degrees east of NW corner, Simon Quadrangle, Nye County, Nevada (after Swain, 1986c). (2)

Figure 30. Turkmenella? sp., Swain, 1987. R side, L, 0.88. Miocene Horse Camp Formation, sec. 2, T. 10 N., R. 59 E., Nye County, Nevada (after Swain, 1987c). (2)

Figures 31-33. Cypridea (Ulwellia) minuta Peck, 1951. L side, dorsal view and ventral view, holotype, L, 0.42. Lower Cretaceous, Minnewaste Limestone, near Hot Springs, South Dakota (after Peck, 1951). (1)

Figures 34, 35. Whipplella carbonaria Scott, 1944. 34, L side, L, 0.61; 35, dorsal view, holotype, L, 0.82. Pennsylvanian, McLeansboro Group, Cohn Cyclothem, Webbs Ferry, White County, Illinois (after Scott, 1944). (1)

Figures 36-39. Whipplella cunieformis Holland, Scott, 1944. 36, R side, holotype, L, 0.84; 37, R side, holotype; 38, LV, L, 0.83; 39, dorsal view. Permian, Greene Group, Nineveh Limestone, Limestone Hill, Wood County, West Virginia (after Scott, 1944). (1)

Plate 34

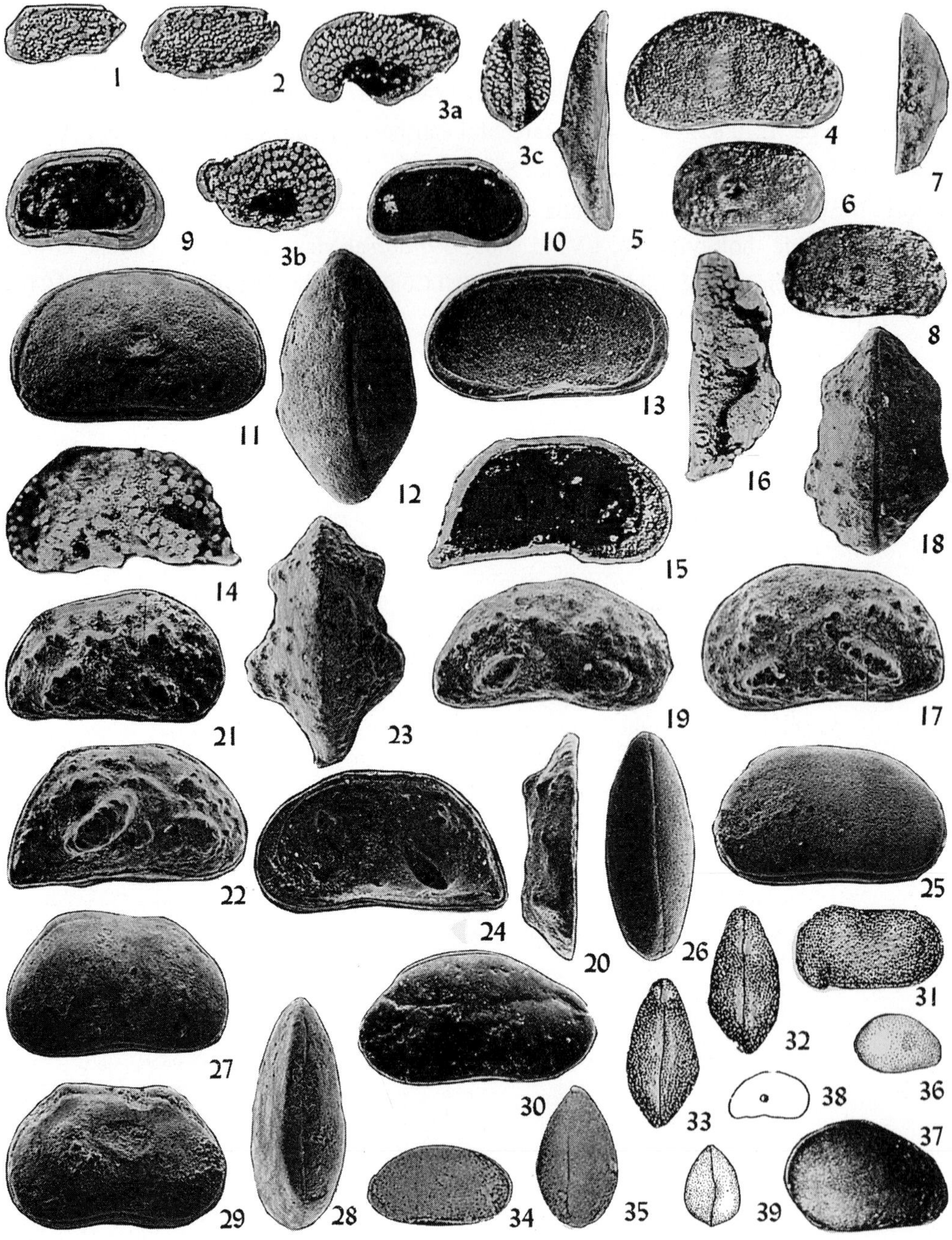
1
2
3a
3c
4
7
9
3b
10
5
6
8
11
12
13
16
18
14
15
21
23
19
17
22
24
20
26
25
31
27
28
30
32
36
33
38
37
29
34
35
39

Appendix A

Features of nonmarine Ostracoda animal and shell.

Some morphological features of three common modern nonmarine ostracodas are shown in the accompanying figures. Figure a, Limnocythera sanctipatricii (Brady and Robertson) represents the Superfamily Cytheracea. Figure b, Darwinula stevensoni (Brady and Robertson) represents the Darwinulacea. Figure c, Candona suburbana (Hoff) represents the Cypridacea. The chitinous appendages are seldom preserved, even in modern sediment accumulations, but a few such instances are known. All the figures are reproduced, courtesy of the Geological Society of America, from the Treatise in Invertebrate Paleontology, Part Q Anthropoda 3, Crustacea, Ostracoda, 1961, R. C. Moore, editor.

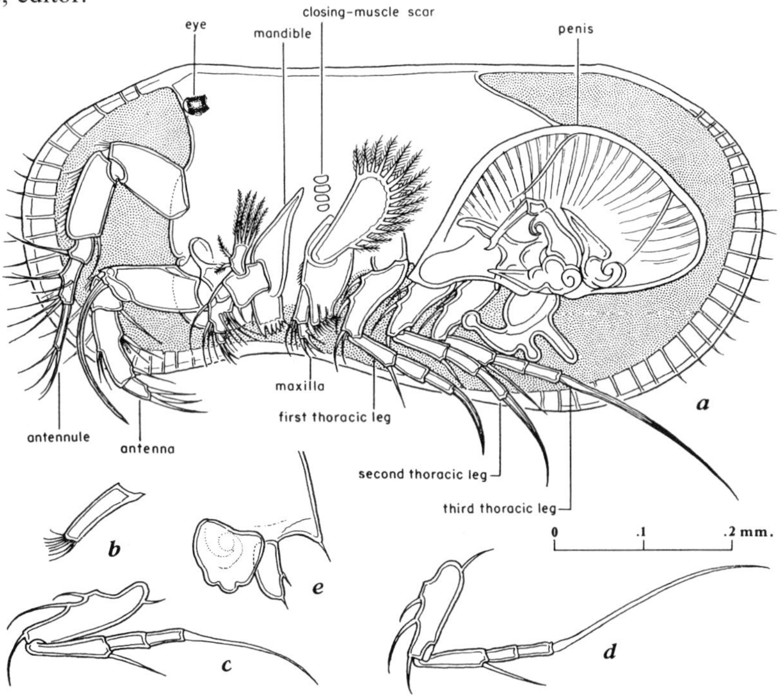

Fig. 1. Morphology of representative podocopid (podocopine) ostracode, Limnocythere sanctipatricii (Brady and Robertson), Recent (Limnocytheridae). a. Male with LV removed.--b. Male left brush-shaped organ, --c. Female left 3rd thoracic leg.--d. Female furca and genital lobe. (All figures to scale at right; from Kesling, after Sars, Moore et al., 1961)

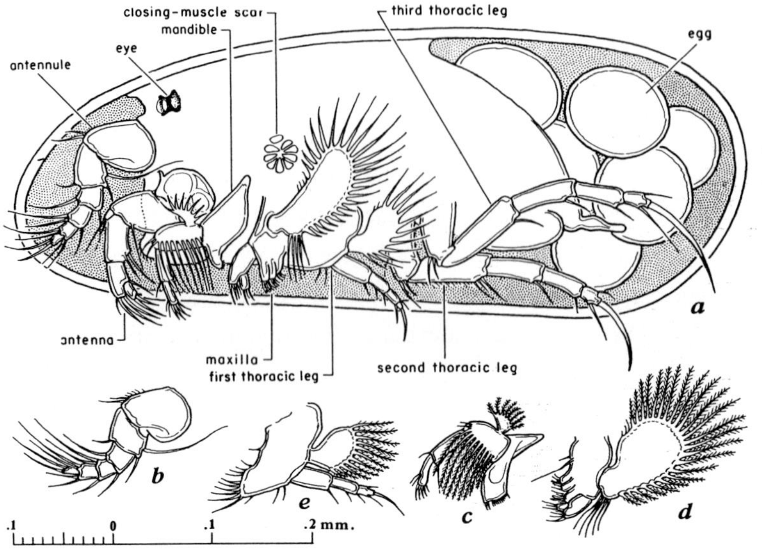

Fig. 2. Morphology of representative podocopid (podocopine) ostracode, Darwinula stevensonii (Brady and Robertson), Recent (Darwinulidae). a. Female with LV removed.--b. Left antennule.--c. Left mandible. --d. Left maxilla from front. --e. Left 1st thoracic leg. (All figures to scale in lower left corner; from Kesling, after Sars, Moore et al., 1961).

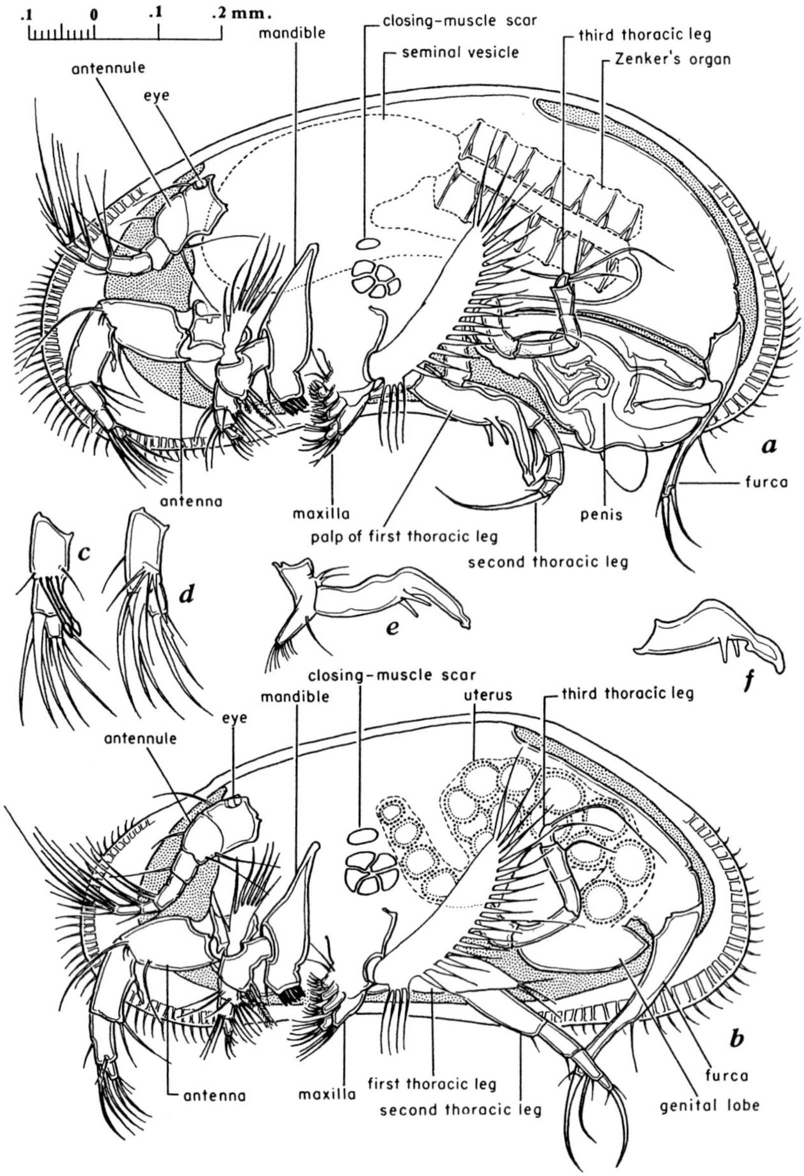

Fig. 3. Morphology of representative podocopid (podocopine) ostracode, Candona suburbana (Hoff), Recent (Cyprididae). a. Male with LV removed; palp of maxilla turned backward to show "masticatory processes" or endites. --b. Female with LV removed. --c. Inner face of distal part of male right antenna, showing "male" setae. --d. Inner face of distal part of female right antenna. --e. Outer face of male left lst thoracic leg. --f. Inner face of male right 1st thoracic leg. (Figs. a, b, e, and f to scale in upper left corner, from Kesling).

Appendix B

Distribution of Significant Fossil Non-Marine Ostracoda in the United States

Species	Dev.	Miss.	Penn.	Perm.	Trias.
a. Paleozoic					
Sollenella? Fr. and Lun.	x				
Carbonita?, n. sp., Sohn		xx			
Gutschickia spp, Sohn		xx			
Pruvostina? sp., Sohn		xx			
Whipplella? sp., Sohn		xx			
Geisinidae, indet., Sohn		xx			
Darwinula sp.		xx			
Cypridopsis fabulina J. & K.			xxx		
Pruvostina wanlessi S. & S.			xxx		
Candona kentuckyensis S. & S.			xxx		
Jonesina? arcuata (Bean)			xxx		
Cypridina radiata J., K. & B.			xxx		
Cypridina aciculata S. & S.			xxx		
Hilboldtina multipunctata S. & S.			xxx		
Whipplella carbonaria Scott			xx-------	xxx	
Gutschickia sp. , Sohn			xx		
Gutschickia sp., Sohn			xxx		
Darwinula sp., Sohn			xxx		
Whipplella cunieformis Hol.				xxx	
Whipplella depressa Hol.				xxx	
Cypridopsis ovata (Hol.)				xxx	
Cypridopsis parvula (Hol.)				xxx	
Candona magnitata (Hol.)				xxx	
Gutschickia deltoidea (Hol.)				xxx	
Darwinula hollandi Scott				xxx	
Whipplella sp., Sohn				xxx	
Gutschickia wellingtonensis T.				xx	
Gutschickia convexa Tasch				xx	
Cypridopsis dickinsonia Tasch				xx	
Candona paleostraca Tasch				xx	

Species	Dev.	Miss.	Penn.	Perm.	Trias.
b. Mesozoic					
Darwinula sp. A Kietzke					xx
Darwinula sp. B. Kietzke					xx
"Gerdalia" sp., Kietzke					xx
Darwinuloides sp., A. Kietzke					xx
Darwinuloides sp., B. Kietzke					xx
Darwinula rogersii (Jones)					xxx
Darwinula subquadrata S. & B.					xxx

Species	Miss.	Penn.	Perm.	Trias.	Jura.
Darwinula spp. Swain & Brown				xxx	
Darwinula aff. luilingchuanenesis				xxx	
"Gerdalia" cf. triassica Kietzke				xxx	
Darwinula? sp., Sohn				xxx	
Scabriculocypris, n. sp., K. & L.					?xxx?
Darwinuloides magna K. & L.					?xxx?
Darwinula sarytirmensis Sharapova					?xxx?
Cytheridella todiltensis (Swain)					xx
Cypridea aff. wyomingensis (J.)					xx
Theriosynoecum wyomingensis (B.)					xxxx
Pseudocypridina piedmonti Roth					xxx
Theriosynoecum bradyi (Jones)					xxx
Theriosynoecum forbesi (Jones)					xxx
Other species of Jones, 1896					xxx
Species of Betty K. Nadeau, 1945					xxx
Cypridea wyomingensis, Jones					xxx

Species	LCret.	UCret.	Paleo.	Eoc.	Oligo.
Cypridea anomala Peck	xxx				
Cypridea brevicornis Peck	xxx				
Cypridea compta Peck	xxx				
Cypridea nitidula Peck	xxx				
Dryelba persulcata (Peck)	xxx				
Longispinella longispina (P.)	xxx				
"Metacypris" angularis Peck	xxx				
Pseudocypridina inornata (P.)	xxx				
Trapezoidella trapezoidalis (R.)	xxx				
Limnocypridea morrisonensis (R.)	xxx				
C. (Ulwellia) minuta (Peck)	xxx				
Cypridea quadrata Peck	xxx				
Cypridea obesa Peck	xxx				
Cypridea pecki Sohn	xxx				
"Petrobrasia"? sp., Sohn	xxx				
"Paracypridea"? sp., Sohn	xxx				
Cypridea menevoides S. & B.	?xxxxx				
Cypridea dequeenensis S. & B.	?xxxxx				
Gomphocythere cf. silvana M.	?xxxxx				
Ilyocypris? tumida Swain	?xxxx				
Cypridea diminuta Vand.	?xxxx				
Cypridea wyomingensis Jones	xxxxxxx				

Species	LCret.	UCret.	Paleoc.	Eoc.	Oligo.
Looneyella quadrispina (Peck)		xxx			
Clinocypris? sp., Swain		?xxxxx?			
Paracypridopsis? sp., Swain		?xxxxx?			
Limnocythere? spp., Swain	xx				
C. (Bisulcocypridea) bicostata S.	?xxxxxxxxxxxxxxxx?				
C. (Bisulcocypridea) skeeteri (P.)		?xxx			
Looneyella monticula (Peck)		?xxx			

c. Paleogene

Species	LCret.	UCret.	Paleoc.	Eoc.	Oligo.
Herpetocypris winfreyi Swain			xxxxxxxxxxxxxxxx		
Paracypridopsis fouchi Swain			xxxxxxxxxxxxxxxx?		
Candoniella whitei (S. & B.)			xxxxxxxxxx		
C. (Bisulcocypridea) nyensis (S.)			xxxxxxx		
C. (Bisculococypridea) arvadensis (S.)			xxxxxxx		
Ilyocyprimorpha tuberculata (S.)			xxxxxxx		
Pseudoeucypris pagei (Swain)			xxxxxxxxxxx		
Candona sp., Swain			xxxxxxx		
Limnocythere sp., Swain			xxxxxxx		
Procyprois revenridgensis S.				xxx	
C. (Bisulcocypridea) bisulcata (S.)				xxx	
Cyprois ephraimensis Swain				xx	
Hemicyprinotus watsonensis S.				xxx	
Potamocypris williamsi Swain				xx	
Scottia? eaglespringensis S.				?xxxxxxx	
Cyprois whiteflatensis S.				?xxxxxxx?	
Procyprois twinbridgensis				?xxxxxxx?	
Pactolocypris centronotus (S.)				xxxxxxxxxxx?	
Chlamydotheca dixieflatensis S.				?xxxxxxxxx?	
Timiriasevia heylumi Swain				?xxxxxxxxx?	
Cyclocypris ketneri Swain				?xxxxxxxxx?	
Pontoniella catlinensis Swain				?xxxxxxxxx?	

Species	Oligo.	Mioc.	Plioc.	Pleistoc.
Cypricercus sacramentoensis S.	xxxxxxxxxxxxxxx			
Cypricercus toanensis S.	xxxxxxxxxxxxxxx			
Candonopsis aff. shahejiensis H.	xxx			
Cypridea? sp, Swain	xxx			
Pactolocypris suborbicularis S.	?xx?xxxxxxxxxxx			
Candona cf. ohioensis Furtos	?xxxxxx			
Candona cf. candida (M)	?xxxxxx			
Candona cf. distincta Furtos	?xxxxxx			
Candona cf. poseyensis Staplin	?xxxxxx			
Eucypris bobcatensis (Swain)	?xxxxxxxxxxxxxx			
Cypricercus palisadensis Swain	?xxxxxxxxxxxxxx			
Bradycypris? medicinecreekensis S.	?xxxxxxxxxxxxxx			
Limnicytheropteron winecup. S.	?xxxxxxxxxxxxxxx			
Cyprinotus aff. cingalensis B.		xxxxxx		
d. Neogene				
Heterocypris blairensis S. & W.		xxxxxxxxxxx		
Pactolocypris biprojectus (S.)		xxxxxxxxxxx		
Paracypridopsis? curta Swain		xxxxxxxxxxx		
Candoniella pinecreekensis S.		xxxxxxxxxxx		
Candona devexoidea D. & S.		xxxxxxxxxxxxxxxxxxx		
Cypricercus mineralensis S.		xxxxxxxxx		
Pactolocypris pactolensis S.		xxxxxxxxx		
Kassinina microreticulata S. & W.		xxxxxxxxx		
Hemicyprinotus ionensis Swain		xxxxxxxxx		
Ilyocypris thousandsprings, S.		xxxxxxxxx		
Procyprois gracilis Swain		xxxxxxxxx		

Species	Oligo.	Mioc.	Plioc.	Pleist.	Holo.
Eucypris fingerrockensis Swain		xxxxxxxxx			
Dongyingia lariversi Swain		xxxxxxxxx			
Stenocypris horsecampensis		xxxxxxxxx			
Potamocypris moorei Swain		xxxxxxxxx			
Kassinina gasspeakensis Swain		xxxxxxxxx			
Herpetocpypris franklinensis S.			xxxxxxxxx		
Cypria hoopescreekensis Swain			xxxxxxxxx		
Candona dipunta Swain			xxxxxxxxx		
Candona patzcuaro Tressler			xxxxxxxxxxxxxxxxxxxxxx		
Candona cachevalleyensis S.			xxxxxxxxx		
Fusocandona exotica Swain			xxxxxxxxx		
Neolimnocythere? prestonensis S.			xxxxxxxxx		
Elkocythereis postdorsoalata S.			xxxxxxxxx		
Candoniella elongatolactoidea D. &. S.			xxxxxxxxxxxxxxxxxxxxxx		
Limnocythere huntingdonensis D. & S.			xxxxxxxxx		
Elkocythereis bramletti D. & S.			xxxxxxxxx		
Candona chasei Dickinson & Swain			xxxxxxxx		
Oreanocypris acutispinosus S.			xxxxxxxx		
Tuberocyproides oreanensis S.			xxxxxxxx		
Candona catherinecreekensis S.			xxxxxxxx		
Limnicytheropteron cephaloto. S.			xxxxxxxx		
Tubercypris acuminatus Swain			xxxxxxxx		
Candona cachensis Swain			xxxxxxxx		
Species of Chapman, 1896			xxxxxxxx		
Limnocythere pterygoventrata D. & S.			xxxxxxxxxxxx		

* Partial listing of species

Species	Oligo.	Mioc.	Plioc.	Pleistoc.*
Candona leightoni Staplin				xxxxxxxxx
Candona wanlessi Staplin				xxxxxxxx
Limnocythere chippewaensis S.				xxxxxxxx
Eucypris trapeziformis S.				xxxxxxx
Limnocythere scotti Staplin				xxxxxxx
Limnocythere sappaensis S.				xxx
Candona sappaensis Staplin				xxx
Candona thomasi Staplin				xxxxxx
Candona illinoiensis S.				xx
Candona willmani Staplin				xx
Candona paraohioensis S.				xx
Candona bretzi Staplin				xx
Candona pearlensis Staplin				xx
Potamocypris minuta Staplin				xxxx
Limnocythere ornata wabash. S.				xxxx
Ilyocypris shafferi Staplin				xxxx
Candona havanaensis Staplin				xxxx
Candona poseyensis Staplin				xxxx
Reticulocypris tuberculata S.				xxxx
Candona shawneensis Staplin				xxxx
Chlamydotheca llanoensis Pierce				xxxx
Candona pronopa Lister				xxxx
Candona camuramarginis L.				xxxx
Candona adunca Lister				xxxx
Limncythere pseudoreticulata S.				xxxx
Cypericercus retic. wrenshall. S.				xxxx

* Partial listing of species; extant species not included

INDEX TO SPECIES ILLUSTRATED ON PLATES

Figures refer to plate numbers.

aciculata, Eocypridina, 15
acuminata, Candona lactea ,7
acuminata, Paracypria ,29
acuminatus, Tuberocypris, 33
acuta, Candona, 1, 2
acutispinosa, Oceanocypris, 28
acutula, Candona, 8
aequaetis, "Cytheridea", 8
adunca, Candona ,8
affinis, "Potomocypris", 31
agnes, Carbonita, 8
albicans, C. (Pseudocandona?), 1
alpinensis, Cytheridella, 16
alta, Candona elliptica, 2, 3
ampla, Cyclocypris, 8
andersoni, Herpetocypris, 22
angularis, Dryelba, 27
anomala, Cypridea, 12
anterosinuata, Cypria, 10
arcuata, Jonesina, 23
artesensis, Candona, 2
arvadensis, Cypridea (B.), 8, 12
ascioformis, Oreanocypris, 28
asymmetrica, Longispinella, 26
atlantosaurica, "Cytheridea", 8
attenuatus, Pactolocypris suborb., 17

---

bairdioides, Candona, 3
bergstromi, Candoniella, 9
berryi, Howellites
bicostata, C. (Bisculcocypridea), 12
biplicata, Ilyocypris, 23
biprojectus, Pactolocypris, 28
bisulcata, Cypridea (B.)., 12
blairensis, Heterocypris, 22
bobcatensis, Eucypris, 18
bradburyi, Limnocythere
bradyi, "Metacypris", 7
bradyi, Ilyocypris, 23
bramletti, Elkocythereis, 18
bretzi, Candona, 1
brighti, Tuberocypris, 33
brevicornis, Cypridea, 12

---

cachensis, Candona, 1
cachevalleyensis, Fusocandona, 19, 20
californica, Cyclocypris, 7
camuramarginis, Candona, 8
cancellatus, Pactolcypris, 28
candida, Candona, 1, 2
carbonaria, Whipplella, 34
catherinecreekensis, Candona, 1
catlinensis, Pontoniella, 29, 30
caudata, Candona, 1, 2, 8
centronotus, Pactolocypris, 28, 31
cephalotoidea, Limnicytheropteron, 24
ceriotuberosa, Limnocythere, 24
chippewaensis, Limnocythere, 24
chasei, Candona, 2
cingalensis, Cyprinotus, 15
clarkstonensis, Metacypris?, 27
clauseni, Candona, 2
coaldalensis, Dogelinella, 18

comosa, Potamocypris, 30
compressa, Candona, 2
compta, Cypridea, 12
concinnus, Advenocypris, 1
consobrina, "Metacypris", 27
convexa, Gutschickia, 20
crogmaniana, Candona, 2, 8
cunieformis, Whipplella, 34
cunieformis, "Metacypris", 27
curta, Paracypridopsis?, 29

---

declivoidea, Paracypria, 29
deltoidea, Gutschickia, 20
depressa, Candona candida, 7
depressa, Whipplella, 32
dequeenensis, Cypridea, 12, 13
devexoidea, Candona, 2
dibolba, Limnicytheropteron, 24
dickinsonia, Cypridopsis, 15
diffusoidea, Candona, 2
diminuta, Cypridea, 12
dipleura, Tuberocyproides, 34
dipunta, Candona, 2
distincta, Candona, 2
dixieflatensis, Chlamydotheca, 10

---

eaglespringensis, Scottia?, 32
elegantulus, Cypricercus, 10
elliptica, Candona, 2, 3
elliptica, Xiyingia, 32
elongata, Eucypris ornatoides, 19
elongatolactoides, Candoniella, 3
emaciata, Hemicyprinotus watsonensis, 21
ephraimensis, Cyprois, 16
eriensis, Candona, 3
esmeraldensis, Cypridopsella, 15
euplectella, Paracandona, 29
exotica, Fusocandona, 20

---

fabulina, Cypridopsis, 15
fadeewi, Physocypria, 29
fingerrockensis, Eucypris, 19
fluviatilis, Candona, 3
forbesi, Cyclocypris, 6, 10
forbesi, "Metacypris", 27
fouchi, Paracypridopsis, 29
fragilis, Cyclocypris, 9
franklinensis, Herpetocypris, 22
friabilis, Limnocythere, 24

---

gasspeakensis, Kassinina, 23
gibba, Ilyocypris, 23
glacialis, Prioncypris, 30
glaucus, Cyprinotus, 15
gondoliformis, Wilkinsia, 34
gracilis, Procyprois, 30
granulosa, Potamocypris, 30

---

havanaensis, Candona, 3
hendersoni, Disopontocypris, 18
henrybelli, C. (Pseudocypridina), 12
herricki, Limnocythere, 24
heylmuni, Timiriasevia, 33
hollandi, Darwinula, 17
hoopescreekensis, Cypria, 10
horsecampensis, Stenocypris, 32

houghi, Candona, 3
huidongensis, Huabeinia?, 23
humboldtensis, Daphnia
humboldtensis, Heterocypris, 22
hunterensis, Cypricercus, 11
huntingdonensis, Limnocythere, 24, 25
hyalina, Candona, 3

---

ikpikpukensis, Candona (P.) 7
illinoiensis, Candona, 3
illinoiensis, Limnocythere, 25
ilosvayi, Cytheridella, 16
incongruens, Heterocypris
inopinata, Candona
inornata, C. (Pseudocypridina), 7, 12
ionensis, Hemicyprinotus, 21
itasca, Limnocythere, 25

---

kingscreekensis, Candona, 3
kingsleyi, Candonopsis, 9
kentuckyensis, Candona, 3
ketneri, Cyclocypris, 10
kittlemani, Candona, 3
krafti, Hemicyprinotus, 21

---

lactea, Candona, 3
lacustris, Cytherissa, 16
laeli, C. (Pseudocypridina), 12
laeva, Turkmenella, 34
laevicula, Pseudocypridina, 7
laevis, Cyclocypris, 10
laevus, Pactolocypris, 28
lariversi, Dongyingia, 18
lata, Herpetocypris, 23
leguminella, Darwinula, 8
leightoni, Candona, 3
levis, Gutschickia, 19
linonodosa, Ilyocyprimorpha, 23
littoralis, Cypridea, 14
luilingchuanensis, Darwinula, 17
llanoensis, Chlamydotheca, 9
longispina, Longispinella, 13, 26, 27

---

maculata, Cypria, 10
magna, Darwinuloides, 16
magnitata, Candona, 3, 4
maladensis, Cyprideis, 14
marengoensis, C. (Pseudocandona?), 3, 7
marshii, "Cyprideis", 8
meadensis, Eucpyris, 19
mediana, Cypria, 10
medicinecreekensis, Bradycypris?, 1
menevoides, Cypridea (Ulwellia), 13
merriamiana, Herpetocypris, 23
mesoreticulata, Kassinina, 24
microreticulata, Eucypris, 19
microreticulata, Kassinina, 24
microstriata, Caspiollina, 9
minensis, Cypricercus, 10, 11
mineralensis, Cypicercus, 11
minnekahtensis, Theriosynoecum? 31
minuta, Potamocypris, 30
minuta, Ulwellia, 34
minuticandida, Candona, 4
minutidontis, Elkocythereis, 18
monticula, Looneyella, 26, 27
moorei, Potamocypris?, 30
morrisonensis, Limnocypridea, 23, 26

multipunctata, Hilboldtina, 22

---

nevadensis, Cypria, 10
ninevehensis, Gutschickia, 20
nitidula, Cypridea, 13
nyensis, C. (Bisulcocypridea), 13

---

obesa, Cypria, 10
obesa, Cypridea, 13
oblongatus, Cypricercus, 11
ohioensis, Candona, 4
owyheensis, Fusocandonopsis, 20
oneidensis, Hemicyprinotus, 21
orbiculata, Carbonita, 8
oreanensis, Tuberocypris, 34
ornata wabashensis, Limnocythere, 25
ornatoides, Eucypris, 19
osmondi, Cypris, 16
ovalis, Tuberocypris, 34
ovata, Gutschickia, 19
ovum, Cyclocypris, 10

---

pactolensis, Pactolocypris, 29
pagei, Pseudocypris, 32
pahasapensis, "Jonesina," 23
paleostraca, Candona, 4
palisadensis, Cypricercus, 11
pallida, Potamocypris, 30
parabretzi, Candona, 4
paracaudata, Candona, 4
parca, Candoniella, 9
paracordata, Metacypris, 27
parafuscatus, Cypricercus, 11
paraohioensis, Candona, 4
paraornata, Limnocythere, 25
parascutariense, Limnocythere, 35
parvula, Cypridopsis, 15
patzcuaro, Candona, 4
paucipunctata, Clithrocytheridea,
paucireticulata, Kassinina, 24
pearlensis, Candona, 4, 5
pecki, Cypridea, 13
pellucidus, Potamocypris?, 30
persulcata, Dryelba, 8
pervulgata, Hastifaba, 21
piedmonti, C. (Pseudocypridina), 7, 8, 13
pigra, Prionocypris, 30
pinecreekensis, Candoniella, 9
pisiformis, Phacocypris, 31
planidorsata, Candona, 3
pliocaenica, Cypridopsis, 7
poseyensis, Candona, 5
postdorsoalata, Elkocythereis, 18
posterolimba, Limnocythere, 25
potamides, Cypris, 16
prestonensis, Neolimnocythere, 28
procera, Cypris, 7
pronodosus, Pactolocypris cancellatus, 28
pronopa, Candona, 8
protzi, Candona, 5
pseudoreticulata, Limnocythere, 25
pterygoventrata, Limnocythere, 25
punctata, D. (Pseudocandona?), 5
punctatula, Eucypris, 19
purbeckensis, "Cypris", 7, 8, 16
pustulosa, Candona (P.?), 5
pustulosa, Physocypria, 18, 30

quadralis, Tuberocypris, 34
quadrata, Cypridea, 13
quadrispina, Looneyella, 27

radiata, Eocypridina, 15
rangliensis, Candona, 5
ravenridgensis, Procyprois, 30
rawsoni, Candona, 4, 5, 8
reniformis, Hemicyprinotus, 21
renoensis, Candona, 5
reticulata, Limnocythere, 25
reticulata, Reticulocypris, 32
reticulatus, Cypricercus, 10, 11
riflensis, Hemicyprinotus, 22
robusta, Hastifaba, 22
robusta, Limnocythere, 26
rothi, Trapezoidella, 33
rogersii, Darwinula, 17

sacramentoensis, Cypricercus, 11
salebrosa, Cypridea, 8, 14, 15
salterina, Candona, 5
saltlakensis, Cypridopsis, 15
sanctipatricii, Limnocythere, 25
sappaensis, Candona, 6
sappaensis, Limnocythere, 25
sarytirmensis, Darwinula, 16
scopulosa, Candona, 5
scotti, Limnocythere, 25
serena, Cyclocypris, 8, 10
serrata, Eucypris, 17
shafferi, Limnocythere, 23
shahejiensis, Candonopsis, 9
sharpei, Cyclocypris, 10
shawneensis, Candona, 5
shawneetownensis, Ilyocypris, 23
sigmoides, Candona, 5
simplissima, Cytherissa, 17
simplus, Darwinula, 17
simpsoni, Candona, 5
simplex, Hemicyprinotus, 21
sinuata, Stenocypris, 32
skeetori, C. (Bisulcocypridea), 13
smaragdina, Potamocypris, 30
spinosa, Hastifaba, 38
spinosa, Jonesina?, 23
stami, Cypricercus, 11
staplini, Limnocythere, 25
stenoformis, Herpetocypris, 22
stevensoni, Darwinula, 17
subangulata, Cypria, 7
subcordata, "Metacypris", 27
sublanceolata, Scottia?, 32
subovata, Candona, 1
subquadrata, Darwinula, 17
suborbicularis, Pactolocypris, 29
subreniformis, Candona, 7
subreniformis, Ussuricypris?, 21
subtriangularis, Oreanocypris, 28
suzini, Candoniella, 9
symmetrica, Candona, 6

tenuis, "Cytheridea", 8
tenuis, Carbonita, 8
thomasi, Candona, 5
thousandspringsensis, Ilyocypris, 23
toanoensis Cypricercus, 11

todiltensis, Cytheridella, 7
tongbinensis, Phacocypris, 31
trapeziformis, Eucypris, 19
trapezoidalis, Scottia, 32
trapezoidalis, Trapezoidella, 31, 33
triassica, Gerdalia, 16
tuberculatum, Pelocypris, 29
tuberculatus, Cypricercus, 11
turneri, Cypria, 10
truncata, Candona, 6
truncata, "Cytheridea", 8
twinbridgensis, Procyprois, 31
typica, Limnicytheropteron, 24

---

umbonata, Advenocypris, 1
unicaudata, Potamocypris, 30
unisulcata, "Potomocypris", 31
unicornis, Limnicytheropteron, 24

---

varia, Limnocythere, 25, 26
variabilonodosa, Malheurocypris, 26
variegata, Potamocypris, 30
verrucosa, Limnocythere, 25
vidua, Cypridopsis, 8, 15

---

wabashensis, Candona, 6
wabashensis, Limnocythere ornata, 25
wanlessi, Candona, 6
wanlessi, Pruvostina, 31
warriorensis, Turkmenella, 34
watsonensis, Hemicyprinotus, 21
wellingstonensis, Gutschickia, 20
whiteflatensis, Cyrpois, 16
whitei, Candoniella, 10, 22
whitei, "Metacypris", 7
whiteriverensis, Hemicyprinotus, 22
wilkinsensis, Heterocypris, 22
williamsi, Limnocythere?, 26
williamsi, Potamocypris, 30
willmani, Candona, 6
winecupensis, Limnicytheropteron, 24
winfreyi, Herpetocypris, 22
wrenshallensis, Cypricercus retic., 10
wyomingensis, Theriosynoecum, 33

---

yuchengensis, Cyprinotus, 16

STRATIGRAPHIC INDEX

Aftonian 32, 33
Albian 10
Algonquin 37
Allegheny 4, 6
Anisian 6
Annelly 5
Aptian 9
Arctic 6
Atlantosaurus Bed 12
Atokan 4

Blackford 37
Blackwater Draw 40
Bloomington 32, 34, 35
Bloomsburg 1
Bear River 8, 10
Bechler 13
Bluestone 3
Bogota 4
Bramwell 3
Breathitt 3
Briceville 3
Brule 16
Brunhes 37
Brussels 34
Buck Hill 15
Buckland 6

Cache Valley 24, 27, 29, 30
Calumet 32
Cannonball 11
Cary 32, 35-37
Catskill 4
Centralia 4
Chadron 16
Chesterian 3
Chinle 6
Chippewa 35
Clarendonian 24
Cloverly 9
Coahuila 10
Cohn 4
Collinston 27
Colton 12
Cotton Valley 7, 10
Culpeper 6
Cumnock 5

De Queen 10
Des Moinesian 4
Douglas Creek 12
Draney 8, 9, 10
Draper 40
Duff 15
Dunkard 4, 5

Edwards Creek 15
Elko 13
Ephraim 13
Esmeralda 15, 16, 19, 20, 23, 43
Evacuation Creek 12

Farmdale 32
Flagstaff 20
Flat Tire 29
Fort Union 11
Frederick 34
Freeport 4

Gannett 8, 10
Garden Gulch 12
Garrett Ranch 26
Gilboa 3
Gimlet 4
Glenns Ferry 24, 26, 29
Glen Rose 10
Glenwood 32, 35
Grand Island 33
Greene 4
Green River 1, 12, 13
Gubik 43

Hance 3
Hay Ranch 17, 24, 25, 31
Hells Bells Canyon 15
Hemphillian 24
Hiko 27
Horse Camp 23
Horse Spring 23
Humboldt 14, 16, 17, 19, 24, 25, 27, 28, 29

Idaho 31
Illinoian 32, 33, 38, 39
Illipah 14
Indian Well 14, 17
Iowan 32

Jamestown 4
Jellico 3
Jessup 37

Kanawha 3
Kansan 32, 33, 37
Kayenta 7
Kootenai 8, 9

Lake Bonneville 40
Lake Chicago 32, 39
Lake Illinois 53
Lake Tinley 32, 35
Lakota 7, 8, 9
Lance 29
Laverne 30
Little Cottonwood 40
Loveland 32, 34
Lower Bear River 9

Mahogany Ledge 23
Mankato 32, 34-37, 39
Matayuma 38
McLeansboro 6
Meade 33
Merced 30
Midway 20
Mingo 3
Mink Creek 24
Minnewaste 8, 9
Missourian 4
Moenkopi 6
Monongahela 4
Morrison 6, 7, 9

Namurian 3
Navajo 7
Nebraskan 31, 3, 37
Newark 5
Newark Canyon 9
New Pass 23
Newton 4
Nineveh 4
Nipissing 32
Norian 6
North Creek 16
Norwood 27

Ogalalla 30
Omega 4
Oreana 25

Painted Desert 6
Palisades Canyon 17
Parachute Creek 12
Payette 27, 31
Pearlette 33, 38
Pekin 5
Peterson 8
Petrified Forest 6
Pierre 16
Pruett 15

Rexroad 30
Rita Blanca 39
Rockport 5

Salt Lake 24, 26-28
Sanford 5
Sangamon 32
Sankoty 32
Scott 3
Seville 4
Sheep Pass 11, 14
Shongaloo 7
Shublik 6
Sinemurian 7
Sloan Canyon 6
Snake River 31
Summum 4
Sundance 7
Sythian 6

Tascotal 15
Tazewell 32, 34, 35, 39
Tecovas 6
Teewinot 30
Titus Canyon 15
Todilto 6, 7
Toleston 32, 35
Tongue River 11

Trinity 10
Truckee 23, 24
Twin Bridges 14, 15
Tygee 13

Uinta 13
Unkpapa 9

Virgilian 4

Wasatch 29, 21
Washita 10
Waterfall 6
Wellington 5
Westphalian 3
West Spring 24
White River 28
Windy Gap 5
Wingate 6
Wisconsin 31, 32, 34, 36
Wrenshall 39

Yarmouth 32, 33